# Biosafety and Biosecurity

There are many guidelines, protocols, and advisories that outline how biosafety and biosecurity can be adopted by institutions around the world. While helpful, many of these are tailored to affluent Western nations. This leaves developing nations far behind since their laboratories and institutions are resource-scarce and biosafety and biosecurity are not mainstreamed entirely among the different laboratory workers, healthcare professionals, researchers, and academics.

*Biosafety and Biosecurity: Practical Insights and Applications for Low and Middle-Income Countries* aims to bridge this gap by comprehensively summarizing the state and development of biosafety and biosecurity in developing and developed nations in a comparative analysis. This book includes basic concepts and principles of biosafety and biosecurity, including certification and legal frameworks, both international and local, and biosafety and biosecurity across disciplines including environmental, medical, and special topics that are relevant to countries with comparable conditions. This proposed book solves the problem of the lack of a prescribed professional title that comprehensively summarizes the state and development of biosafety and biosecurity throughout the world, allowing the reader a 360 view of the subject area.

This book will appeal to a global audience of biorisk officers, health and safety professionals and specialists in the life sciences, health and allied fields, environmental science, engineering, and plant and animal agriculture.

# Biosafety and Biosecurity

## Practical Insights and Applications for Low and Middle-Income Countries

Edited by
Jonathan Jaime G. Guerrero, Rohani Cena-Navarro,
Raul V. Destura, Marian P. De Leon,
Kin Israel R. Notarte, and
Mark Angelo O. Balendres

CRC Press is an imprint of the
Taylor & Francis Group, an **informa** business

Designed cover image: © Karl Joseph Lagmay Hufalar

First edition published 2025

by CRC Press
2385 NW Executive Center Drive, Suite 320, Boca Raton FL 33431
and by CRC Press
4 Park Square, Milton Park, Abingdon, Oxon, OX14 4RN

*CRC Press is an imprint of Taylor & Francis Group, LLC*



ISBN: 978-1-032-54405-2 (hbk)
ISBN: 978-1-032-54669-8 (pbk)
ISBN: 978-1-003-42621-9 (ebk)

DOI: 10.1201/9781003426219

Typeset in Times New Roman
by Deanta Global Publishing Services, Chennai, India

# About the Editors

**Jonathan Jaime G. Guerrero** is a certified biorisk officer (CBO), trained under the third cohort of the Philippine Advanced Biorisk Officer Training Program (PhABOT) of the National Training Center for Biosafety and Biosecurity of the University of the Philippines Manila in 2015. He has a master's degree in Plant Pathology from the University of the Philippines Los Baños and is currently pursuing a dual degree in Doctor of Medicine–Master in Public Health (MD-MPH), a joint program of the University of the Philippines Manila College of Medicine and College of Public Health.

**Rohani Cena-Navarro** is a veterinarian and Assistant Professor at the National Institute of Molecular Biology and Biotechnology, National Institutes of Health-University of the Philippines Manila, and Assistant Director at the National Training Center for Biosafety and Biosecurity (NTCBB) NIH-UP Manila. She is the UP Manila Chair of the Institutional Animal Care and Use Committee and also served as the Chair of the Institutional Biosafety and Biosecurity Committee (IBBC) from 2019 to 2022. Dr. Navarro received her certification as a Biosafety Officer under the Philippine Advanced Biorisk Officers Training in 2015 and is also a Certified Biorisk Professional by the International Federation of Biosafety Associations. As an advocate of biosafety, she led several capacity-building projects in promoting biosafety and biosecurity awareness in the Philippines through various local and international partners.

**Raul V. Destura** is a molecular microbiologist, medical doctor and infectious disease specialist, and industry leader. He underwent a fellowship in Emerging Infectious Diseases at the University of Virginia, USA. Joining the National Institutes of Health Philippines, he served as the Director of the National Institute of Molecular Biology and Biotechnology in 2006 and later became the Director of the National Training Center for Biosafety and Biosecurity in 2018. A course developer and coordinator since 2008, he continues to contribute significantly to biosafety training in the Philippines through the Advanced Biorisk Officer's Training and the Basic Biosafety Online Training.

**Marian P. De Leon** is Director of the Museum of Natural History and concurrent Assistant to the Vice Chancellor for Research and Extension of the University of the Philippines Los Baños. She also serves as the University Researcher and Curator of the UPLB MNH Microbial Culture Collection. She pioneered research on exploration of Philippine Caves for pathogenic, novel, and potential microorganisms. In 2019, she was conferred as Certified Biorisk Officer (CBO) given by the National Training Center for Biosafety and Biosecurity and the US Department of State Biosecurity Engagement Program. As CBO, she was given an appointment to serve as Biosafety Officer and Scientist Member of the UPLB Institutional Biosafety Committee. Her contributions in Microbiology have been recognized by the Philippine Academy of Microbiology, Inc. and certified her as a Registered Microbiologist. Currently, she is the Corporate Secretary and Chair of Division 5—Biological Sciences of the National Research Council of the Philippines and holds several positions and active membership in professional organizations.

**Kin Israel Notarte** holds a Doctor of Medicine and a Master of Science in Microbiology, both from the University of Santo Tomas (Philippines). Currently, he is a PhD candidate in Pathobiology (Microbiology and Infectious Disease) at the Johns Hopkins University School of Medicine (USA). With over sixty publications, his current research focuses on post-COVID-19 syndrome, HIV genotyping, and humoral immune responses associated with HIV viremic control, utilizing Next-Generation Sequencing, Phage Immunoprecipitation Sequencing, and Molecular Indexing of Proteins by Self-Assembly. Additionally, he serves as a reviewer for over thirty journals, including *The Lancet* and the *New England Journal of Medicine*, and as a review editor at *Frontiers in Medicine* and a guest editor at *Frontiers in Immunology*. An accomplished scientist, he received training in chemical biology from Waseda University (Japan) and molecular mycology from Seoul National University (Korea) and served as a Bayer Fellow for the 72nd Lindau Nobel Laureate Meetings in Medicine and Physiology (Germany).

**Mark Angelo O. Balendres** is Full Professor at the De La Salle University Department of Biology. He obtained his PhD in Agricultural Science (Plant Pathology) from the University of Tasmania, Australia. He is also a member of the Global Young Academy. His work focuses on the etiology, epidemiology, impact, and management of plant diseases. He is on the editorial board of several scientific journals and has written papers on plant pathology, crop protection, plant science, and agricultural mycology.

# Contributors

**Rhea G. Abisado-Duque**
Ateneo de Manila University
Department of Biology
Quezon City, Philippines

**Bianca Mae Adalem**
Independent Researcher, Metro Manila, Philippines

**Angelo R. Agduma**
University of Southern Mindanao Kabacan, College of Science and Mathematics
Department of Biological Sciences, Ecology and Conservation Research Laboratory
Kabacan, Cotabato, Philippines
Guangxi University, College of Forestry
Nanning City, Guangxi, P.R. China

**Michael O. Baclig**
Trinity University of Asia, St. Luke's Medical Center College of Medicine-William H. Quasha Memorial
Quezon City, Philippines

**Mark Angelo O. Balendres**
Department of Biology, College of Science, De La Salle University
Manila, Philippines
Plant and Soil Health Research Unit
Center for Natural Sciences and Environmental Research, De La Salle University
Manila, Philippines

**Reynand Jay Canoy**
Scilore LLC, Muntinlupa City, Philippines
Institute of Biology, College of Science
University of the Philippines Diliman
Quezon City, Philippines

**Timothy Hudson David Culasino Carandang**
Pamantasan ng Lungsod ng Maynila College of Medicine
Manila, Philippines

**Jesus Alfonso Catahay**
Department of Medicine
Saint Peter's University Hospital
New Brunswick, New Jersey, USA

**Fresthel Monica Climacosa**
Department of Medical Microbiology
College of Public Health
University of the Philippines
Manila, Philippines

**Christian Joseph Cumagun**
Parma Research and Extension Center
University of Idaho
Idaho, USA

**Jan Irving A. Bibay**
Biological Resource Centre (BRC)
Agency for Science, Technology and Research (A*STAR)
Singapore

**Patrick Ryan Bello**
National Institutes of Health
University of the Philippines,
Manila, Philippines

**Jann Eldy L. Daquioag**
Institute of Biology, National Science Complex, College of Science
University of the Philippines
Diliman, Quezon City, Philippines

**Jessica Joyce De Guia**
Department of Biology
De La Salle University
Manila, Philippines

**Marian P. De Leon**
Museum of Natural History
University of the Philippines Los Baños College
Los Baños Laguna, Philippines

**Sheriah Laine de Paz-Silava**
Department of Medical Microbiology
College of Public Health
University of the Philippines
Manila, Philippines

**Angelo dela Tonga**
National Training Center for Biosafety and Biosecurity
Institute of Molecular Biology and Biotechnology
NIH UP Manila

**Raul V. Destura**
National Training Center for Biosafety and Biosecurity
National Institutes of Health
University of the Philippines, Manila

**Fedelyn P. Estrella**
Lyceum of the Philippines University Cavite and
National Training Center for Biosafety and Biosecurity
National Institutes of Health
University of the Philippines
Manila, Philippines

**Allan L. Fellizar**
Mariano Marcos Memorial Hospital and Medical Center Molecular Biology Laboratory
City of Batac, Ilocos Norte

**Mary Jane Flores**
Department of Biology
De La Salle University
Manila, Philippines

**Janiza Lianne Foronda**
Department of Virology
Research Institute for Tropical Medicine

**Mheljor A. General**
Department of Biology
College of Science, Bicol University
Legazpi City, Philippines

**Connie Gibas**
Department of Pathology and Laboratory Medicine
University of Texas Health
San Antonio, Texas, USA

**Anna Gibson**
Scilore LLC.
Muntinlupa City, Philippines

**Jonathan Jaime G. Guerrero**
University of the Philippines, Manila College of Medicine and
College of Public Health
Manila, Philippines

**Daisy Ilagan-Tagarda**
Faculty of Medicine and Surgery
University of Santo Tomas
Manila, Philippines
Department of Internal Medicine,
University of Santo Tomas Hospital
Manila, Philippines

**Alleni T. Junsay**
Industrial Technology Development Institute
Metrology in Chemistry, National Metrology Laboratory of the Philippines, Department of Science and Technology
Bicutan, Taguig City, Philippines

**Toni Rose Lamata-Porras**
Office for Health Laboratories
Department of Health
Manila, Philippines
Department of Pathology and Laboratory, Quirino Memorial Medical Center
Quezon City, Philippines
College of Allied Health Professions
University of the East Ramon Magsaysay Memorial Medical Center, Inc.
Quezon City, Philippines

**Don Eliseo Lucero-Prisno III**
Department of Global Health and Development
London School of Hygiene and Tropical Medicine
London, UK

**Charmaine A. Malonzo**
Department of Biology
College of Science, Bicol University
Legazpi City, Philippines

**Llewelyn Moron-Espiritu**
Department of Biology
De La Salle University
Manila, Philippines

**Frederick John B. Navarro**
Scilore, LLC
Ayala Alabang, Muntinlupa City

**Rohani Cena-Navarro**
National Training Center for Biosafety and Biosecurity
National Institutes of Health
University of the Philippines
Manila, Philippines

**Kin Israel R. Notarte**
Department of Pathology
Johns Hopkins University School of Medicine
Maryland, USA

**Adriel Pastrana**
Faculty of Medicine and Surgery
University of Santo Tomas
Manila, Philippines

**Gerard Lorenz M. Penecilla**
Department of Pharmacology
College of Medicine
West Visayas State University
Iloilo City, Philippines

**Gil M. Penuliar**
Institute of Biology, National Science Complex, College of Science
University of the Philippines
Diliman, Quezon City, Philippines

**Ric Ryan H. Regalado**
National Institute of Molecular Biology and Biotechnology
College of Science
University of the Philippines
Diliman, Quezon City, Philippines

**Anelyn Reyes**
Department of Pediatrics
University of Santo Tomas Hospital
Faculty of Medicine and Surgery
University of Santo Tomas
Manila, Philippines

**Melissa Marie R. Rondina**
College of Veterinary Medicine
University of the Philippines Los Baños
Laguna, Philippines

**Alexander Sadiasa**
Department of Microbiology
Research Institute for Tropical Medicine

**Edsel Allan G. Salonga**
Cytogenetics Laboratory
Institute of Human Genetics
University of the Philippines
Manila, Philippines

**Ma. Socorro Edden P. Subejano**
Department of Microbiology, Anatomy, Physiology and Pharmacology
School of Agriculture, Biomedicine and Environment
La Trobe University
Melbourne, Victoria, Australia

**Krizler C. Tanalgo**
University of Southern Mindanao Kabacan, College of Science and Mathematics
Department of Biological Sciences, Ecology and Conservation Research Laboratory,
Kabacan, Cotabato, Philippines

**Ourlad Alzeus Tantengco**
Department of Biology
De La Salle University
Department of Physiology
College of Medicine,
University of the Philippines
Manila, Philippines

**Paul Taylor**
Faculty of Science
University of Melbourne
Victoria, Australia

**Gianne Eduard L. Ulanday**
National Training Center for Biosafety and Biosecurity
National Institutes of Health,
University of the Philippines
Manila, Philippines
Department of Medical Microbiology
College of Public Health
University of the Philippines
Manila, Philippines

**Mayan Uy-Lumandas**
Department of Virology
Research Institute for Tropical Medicine
Muntinlupa City, Philippines

**John Mark Velasco**
Institute of Molecular Biology
National Institutes of Health and Department of Clinical Epidemiology
College of Medicine
University of the Philippines
Manila, Philippines

**Rodel Jonathan S. Vitor II**
National Training Center for Biosafety and Biosecurity Training, National Institutes of Health,
University of the Philippines
Manila, Philippines
Department of Biology
College of Science
De La Salle University
Manila, Philippines

# Preface

This first edition of the book covers the concepts and principles of biosafety and biosecurity as applied in low- and middle-income countries, using the Philippines as the example. It documents and reviews the advances in the understanding and application of these concepts and principles across different disciplines: academe, health, and agriculture.

The book also looks into the development of biosafety and biosecurity in the Philippines, a relatively new but an essential aspect of laboratory management. Finally, it takes into account the recent COVID-19 pandemic which necessitated a review of current practices and guidelines to ensure a safe and secure environment for laboratory workers and the public in general.

# Acknowledgments

The editors and authors acknowledge the valuable contributions of Mackinley Graham Go Ngo and Karl Joseph Lagmay Hufalar.

# 1 Introduction to Biosafety and Biosecurity in the Philippines

*Raul V. Destura*
National Training Center for Biosafety and Biosecurity, National Institutes of Health
University of the Philippines
Manila, Philippines

*Fedelyn P. Estrella*
Lyceum of the Philippines University Cavite and National Training Center for Biosafety and Biosecurity, National Institutes of Health
University of the Philippines
Manila, Philippines

*Patrick Ryan Bello*
National Institutes of Health
University of the Philippines
Manila, Philippines

## 1.1 HISTORY IN BRIEF

In 1908, Charles-Edward Amory Winslow made a groundbreaking contribution to the field of microbiological safety by introducing the 'Microbial impingement air sampler', a pioneering method for quantifying airborne bacteria (Bayot & Limaiem, 2023). This innovation laid the foundation for what we now recognize as the principle of 'Biosafety'. Fast forward to 1947, post-World War II, the US National Institutes of Health (NIH) established the first research laboratory dedicated to studying microbiological safety, marking a significant milestone in biosafety development (Wedum & Kruse, 2017). The inaugural Biological Safety Conference in 1955 at Camp Detrick, Maryland, USA, addressed emerging biocontainment challenges and discussed precursor systems to the modern biosafety cabinet, including bacteriological cabinet systems. Notably, this conference gave rise to the American Biological Safety Association (ABSA) in 1984. ABSA plays a pivotal role in promoting and advancing biosafety and biosecurity, exemplified by its publication and distribution of the quarterly journal *Applied Biosafety*. This historical trajectory underscores the continuous evolution and global significance of biosafety principles.

DOI: 10.1201/9781003426219-1

In the Philippines, concerns related to genetically modified organisms (GMO) in agricultural produce led to the creation of the Joint Ad-hoc Committee on Biosafety, composed of the University of the Philippines (UP) Los Banos, the International Rice Research Institute, and the Department of Agriculture (DA), releasing a report in 1987 that served as the precursor to the Philippine Biosafety Guidelines (PBG) (Lantican, 1991). The National Committee on Biosafety of the Philippines (NCBP), attached to the Department of Science and Technology (DOST), was constituted on October 15, 1990, through Executive Order (EO) No. 430. The attached agency formulated and adopted the Philippine Biosafety Guidelines (PBG) in 1991, the first in Southeast Asia and one of the most stringent guidelines in the world. PBG covered work on *'genetic engineering, and activities requiring the importation, introduction, field release and breeding of nonindigenous or exotic organisms even though these are not genetically Modified'*. The PBG required Institutional Biosafety Committees (IBCs) in *'all institutions engaged in genetic engineering and/or potentially hazardous biological and/or genetic engineering work'*.

The Cartagena Protocol on Biosafety was discussed during the Convention on Biological Diversity Conference of Parties in 2000. It considered the impacts of modern biotechnology on living-modified organisms to ensure the 'safe handling, transport and use of living modified organisms'. The Philippines became a party to the Cartagena Protocol on May 24, 2000, which entered into force on September 11, 2003 (Office of the President of the Philippines, 2006).

In line with the Philippines' commitment and compliance with the Cartagena Protocol, the country developed the National Biosafety Framework (NBF) in 2004 (Office of the President of the Philippines, 2006). EO 514 formally established the NBF in 2006, strengthened NCBP, and upheld the country's policy to *'promote the safe and responsible use of modern biotechnology and its products as one of the several means to achieve and sustain food security, equitable access to health services, sustainable and safe environment and industry development'*. Our commitments were enhanced in 2016 through the DOST–DA–Department of Environment and Natural Resources (DENR)–Department of Health (DOH)–Department of the Interior and Local Governance (DILG) Joint Department Circular No. 1, which instructs government agencies and departments to hold more public consultations and consider environmental and health impacts in making decisions on biosafety (GRP, 2016).

### 1.1.1 The UP National Institutes of Health (NIH) National Training Center for Biosafety and Biosecurity (NTCBB): From a Program to a Center

The center was born out of a need to translate another unmet need in the medical field: Infectious Disease Diagnostics. The then director of the National Institute of Molecular Biology and Biotechnology (NIMBB) of the UP NIH, Dr. Raul Destura, created the Molecular Diagnosis for Infectious Disease Program of the institute. Working with infectious diseases requires a safe environment for the scientist and the environment, hence the need to establish a biosafety program of the NIMBB. While this initiative is at play, an opportunity opens. Four faculty members of the University of the Philippines (Dr. Irma Makalinao, Dr. Vicente Belizario, Dr. Regina Berba, and Dr. Raul Destura) were sent to attend the training workshop on biosafety and biosecurity of the American Biological Safety Association (ABSA) through the support from Biosecurity Engagement Program of the US Department of State (DOS-BEP). Returning from the country, the four training recipients conceptualized the creation of the 'train the trainer program' in Biosafety to ensure the availability of a critical amount of expertise in the field. Dr. Raul Destura took the lead as a project proponent for the 'train the trainer Program' called the Philippines Advanced Biorisk Officers Training (PhABOT) and also the reconstitution of the Institutional Biosafety Committee (IBC) of UP Manila.

Nineteen graduates comprise the Philippines' first roster of Certified Biorisk Officers (CBOs) on September 12, 2011, through PhABOT. US DOS-BEP funded and partnered with the UPM-IBC to organize PhABOT. Experts from several national and international health organizations facilitated the 6- to 8-week PhABOT course. Graduates underwent training on Biosafety and Biosecurity, and Biorisk Assessment Management. PhABOT graduates served as the trainers of the succeeding batches. After a successful run, PhABOT's second to fourth iterations were organized by UPM-NIH NIMBB. During the fifth PhABOT iteration, the faculty composed of PhABOT 1 and 2 graduates successfully created the first edition of the instructional module covering ten core competencies. Since the demand for the program has grown to a level, program proponents initiated the establishment of the UPM-NIH NTCBB, with Dr. Raul V. Destura as its first center director. The program became a success story in the Asian region, sparking interest in the program and gathering its graduates' reputation, not only passing 100% of its graduates in the international certification program but also the top passers provided by the International Federation of Biosafety Association.

As the program's demands increased, the development of a dedicated center was conceptualized, studied, and rolled out. On February 22, 2018, through the 1333rd meeting of the UP Board of Regents, the National Training Center for Biosafety and Biosecurity was approved (UP, 2018). The PhABOT and Certification Program is one of NTCBB's flagship programs. This has produced over a hundred certified biorisk officers from key institutions under the Department of Health; regional government hospitals and private hospitals; and UP Diliman, UP Manila, and other universities in Luzon, Visayas, and Mindanao.

Today, the NTCBB is the principal source of information and expertise related to the practice of biosafety and biosecurity in the Philippines, providing the capacity for training for biosafety and biosecurity practitioners in the Philippines, linking the country with counterpart organizations globally. Graduates of the program became leaders in the field and paved the way for establishing two private organizations, the Philippine Biosafety and Biosecurity Association (PhBBA) and the Biorisk Association of the Philippines (BRAP).

The center's take on advocacy and leadership, training and awareness, development of applicable biosafety standards, and the generation of best practices in laboratory safety and security played a significant role in the COVID-19 pandemic. NTCBB offered the Biosafety Education and Awareness Training (BEAT) COVID-19 Online Training Program in collaboration with the Department of Health, specifically for institutions planning to establish laboratories for the molecular detection of SARS-CoV-2. Several PhABOT graduates helped in establishing and developing the curriculum used in BEAT COVID-19. By 2020, a total of 3371 trainees from over 100 DOH-accredited COVID-19 testing laboratories obtained certificates from the free online training (Navarro et al., 2022). This increased to a total of 4,308 graduates as of June 16, 2023.

Moving forward, the center seeks to advance the creation of mechanisms for non-degree programs and degree-granting programs that will enhance the professional practice of biosafety and biosecurity in the Philippines. The Center continues to lead in integrating biosafety principles among research, academic, and healthcare institutions.

Furthermore, the NTCBB actively engaged in international collaborations, hosting the International Fellowship Program on Biosafety and Biosecurity in cooperation with the Medical City and the University of the Philippines Manila—Institute of Molecular Biology and Biotechnology. From December 2018 to March 2019, Dr. Nawras Mahir Farhan, a visiting fellow, familiarized himself with the standard operating procedures of different laboratories in the country, imparting knowledge on biorisk management.

The center's commitment extends beyond education to infrastructure development, with involvement in the construction and renovations of Regional Animal Disease Diagnostic Laboratories (RADDL) in Central Luzon, Central Visayas, and Mindanao's northern, southern, and central regions. The research faculty of the NTCBB has showcased its expertise in conferences and workshops globally, contributing to disaster vulnerability, risk reduction management, and biosecurity diplomacy.

Anticipating the future needs of biosafety education, the NTCBB has initiated the drafting of the curriculum for the proposed Master of Science in Public Health (Biorisk Management) program, aligning with its mission to advance knowledge in the field. Despite challenges posed by the pandemic, the center pivoted to online platforms, conducting various training workshops and symposiums, including the Personnel Reliability Program (PRP) and Insider Threat Management Course in Ortigas Center, Mandaluyong—the first PRP Training in the Philippines.

Establishing secure connections, the NTCBB has fostered partnerships with institutions like the Philippine Public Safety College (PPSC) and the Central Visayas Consortium for Health Research and Development, providing annual biosafety training programs since 2018. The center also contributed its expertise to the Department of Science and Technology—Industrial Technology Development Institute (DOST-ITDI) for the establishment of the Virology and Vaccine Science and Technology Institute of the Philippines.

### 1.1.2 The Need for a National Biorisk Framework for the Philippines

Among the first deliverables of NTCBB was a Biosafety and Biosecurity draft legislation, the *Biorisk Management Act of 2019*, a proposed National Framework for the Philippines (UP, 2018). During a meeting with the DOST-National Committee on Biosafety of the Philippines, NTCBB shared the results of their policy analysis on EO 514, which showed a lack of mandate given to government agencies on biosafety curricula, association, professional competence and credentialing, and individual mentoring and organizational twinning (National Committee on Biosafety of the Philippines, n.d.). Both organizations concurred that the existing NBF covered only LMOs as defined by the Cartagena Protocol and that a framework capturing biosecurity and biosurveillance should be established.

With the center tasked by UP to take the lead in putting together all stakeholders for the drafting of the National Strategic Framework for Biosafety and Biosecurity, NTCBB initiated a consultative conference to assess the biosafety and biosecurity landscape in the Philippines. Engaging relevant and principal institutions, the center provided expertise in drafting the framework, conducting charter group meetings, and tabletop exercises to delineate essential tasks for its development. NTCBB worked with UPM-NIH Institute of Health Policy and Development Studies (IHPDS) to assess the country's biosafety and biosecurity landscape and help develop a national biorisk management framework through a joint research project. The project included policy reviews, multi-stakeholder meetings with government agencies and private entities, tabletop exercises, and the drafting of the framework (Destura et al., 2021). Table 1.1 shows gaps identified by the parties during the tabletop exercises based on given case scenarios. As of this writing, the framework is still in the process of being presented to the legislative branch of the Philippine government for possible enactment into a law.

## 1.2 BIOSAFETY

*'**Biosafety** aims to protect people from bad bugs while **Biosecurity** aims to protect bugs from bad people'.*—Common adage

Biosafety and Biosecurity are critical elements that demand careful consideration in any research involving infectious organisms, whether conducted in the field or within laboratory settings. These components are integral to the overarching principle of biorisk management.

Biosafety involves adopting secure practices for biological materials, specifically infectious agents. It encompasses the principles, technologies, and practices of containment designed to prevent inadvertent exposure to pathogens and toxins and their accidental release. Responsible laboratory practices are crucial, including measures for the protection, control, and accountability of valuable biological materials. This prevents unauthorized access, loss, theft, misuse, diversion, or intentional release.

Important definitions applied in the practice of biosafety and biosecurity come from or are based on the International Organization for Standardization (ISO). This international non-government organization sets global industry standards, even for laboratories specifically handling biological materials (ISO, n.d.-a). The following ISO 35001:2019 (Biorisk management for laboratories and other related organizations) definitions are crucial to understand the book:

## TABLE 1.1
## Gap Analysis Results of the Tabletop Exercises adapted from Assessment of the Biosafety and Biosecurity Landscape in the Philippines and the National Biorisk Management Framework Development with Hypothetical Human Transmission (Destura et al., 2021)

| Gap Analysis | | |
|---|---|---|
| **Case Scenarios** | | **Gaps Identified** |
| Response Phase | | |
| Event 1: Dead piglet brought to the RADDL for testing | Subevent 1.1: Initial laboratory testing for suspecting ASF<br>Subevent 1.2: Samples negative for ASF and stored in refrigerator for next day<br>Subevent 1.3: 36 pigs not reported to RADDL during weekend<br>Subevent 1.4: Dead pigs of two or more farmers brought in for testing<br>Subevent 1.5: 6 farms affected with high pig mortality | • Weak implementation of existing SOPs and policies<br>• Lack of quality assurance of skills and training of veterinarians<br>• The need to regulate laboratory practices among government and private facilities through strict monitoring and evaluation processes<br>• The need to ensure availability of controls in the laboratory, in relation to the validity of testing<br>• The need to strengthen HR capacity for 24/7 emergency staff response<br>• The need for guidance and reliable reporting mechanisms to RADDL through the adoption of possible strategies such as electronic geographic information systems-enabled reporting mechanism<br>• Lack of protocols for waste disposal |
| Activation Phase | | |
| Event 2: Samples brought to another laboratory for testing 250km away | Subevent 2.1: Meat of dead pigs sold in the market<br>Subevent 2.2: Children getting sick<br>Subevent 2.3: Analyst getting sick and not reporting in; sample missing from laboratory<br>Subevent 2.4: 17 affected farms, pig mortality >50% | • Noncompliance to SOPs for the handling, transport, and referral of samples in larger distances to reference laboratories<br>• The need for coordinated action among different government agencies<br>• The need for clear coordinating mechanisms between animal and human health responders<br>• Shortage of personal protective equipment for responders<br>• Incorporating proper documentation as part of laboratory SOPs; recording and tracking system should be in place<br>• There is a need for restraining and re-educating laboratory personnel on safety practices |

(*Continued*)

**TABLE 1.1 (CONTINUED)**
**Gap Analysis Results of the Tabletop Exercises adapted from Assessment of the Biosafety and Biosecurity Landscape in the Philippines and the National Biorisk Management Framework Development with Hypothetical Human Transmission (Destura et al., 2021)**

| Gap Analysis Case Scenarios | | Gaps Identified |
|---|---|---|
| Coordination Phase | | |
| Event 3: Confirmed diagnosis | Subevent 3.1: Local media asking for update<br>Subevent 3.2: Samples from veterinary reference laboratory positive for ASF<br>Subevent 3.3: Pig carcasses buried on-site<br>Subevent 3.4: Visit to analyst's home reveal samples used in 'DIY home experiment' | • There is a need to ensure compliance to laboratory SOPs on divulging information to media<br>• There is an absence of facility for burying of carcass and waste decontamination and disposal<br>• Implementation of policy to monitor laboratory personnel, as part of safety and security precaution |

1. **Biological agents** are 'any microbiological entity, cellular or non-cellular, naturally occurring or engineered, capable of replication or of transferring genetic material that may be able to provoke infection, allergy, toxicity or other adverse effects in humans, animals, or plants'.
2. **Biological materials** are 'any material comprised of, containing, or that may contain biological agents and/or their harmful products, such as toxins and allergens'. These include the entire fauna or flora or their segments or parts.
3. **Toxins** are 'substance, produced by plants, animals, protists, fungi, bacteria, or viruses, which in small or moderate amounts produces an adverse effect in humans, animals, or plants'.
4. **Biohazards** are 'potential sources of harm caused by biological materials'.

According to the 2020 World Health Organization (WHO) Laboratory Biosafety Manual (4th Ed.), biosafety constitutes a comprehensive set of technologies, protocols, or practices implemented to prevent unintentional exposure to or release of biological agents or materials. Biosafety includes the safe handling, containment, and management of biological materials, particularly those that are infectious or could pose potential hazards to human health and the environment.

Biosafety practices are designed to protect researchers, laboratory workers, the public, and the environment from the potential risks associated with manipulating biological agents, including bacteria, viruses, toxins, and genetically modified organisms (Coelho & Diez, 2015). This involves using containment measures, protective equipment, and adherence to established guidelines to minimize the likelihood of laboratory-acquired infections or unintended release of biohazards (Coelho & Diez, 2015). A practical illustration of the importance of biosafety measures is evident in scenarios where zoonotic viruses, such as Nipah viruses or Coronaviruses, might inadvertently spill over to humans (Ellwanger & Chies, 2021).

### 1.2.1 Biocontainment

Biocontainment is an important principle of biosafety. It is the confinement of infectious agents or toxins to reduce exposure to them or their accidental release into the surroundings (Public Health Emergency [PHE], n.d.-a). This includes safety equipment, laboratory operations, procedures, and physical design of laboratories and facilities. Containment procedures, in general, can be classified as primary or secondary.

1. Primary containment protects workers or researchers. Containment barriers and biosafety cabinets shield researchers. Facilities housing animals use enclosures to protect those in the laboratory (PHE, n.d.-a).
2. Biological safety cabinets (BSCs) are vital barriers containing infectious aerosols or splashes generated by manipulating biological agents or materials. These shield the researchers and their equipment and other biological agents. BSCs have different classifications (refer to Table 1.2) with varying modifications, which are factored into laboratories' Biosafety Level (BSL) accreditation.
3. Secondary containment protects society, communities and the environment from exposure to biological risks. The building's walls protect the community (PHE, n.d.-a).

Biocontainment laboratories play a pivotal role in safeguarding public health, the environment, and researchers themselves by establishing standardized protocols and protective measures when handling biological agents. These laboratories are designated to handle materials that pose varying levels of risk to human health and the environment. The importance of biocontainment laboratories lies in their ability to mitigate these risks through a tiered system that dictates the level of containment required for specific agents.

One key aspect is preventing accidental exposure to infectious agents. Biocontainment laboratories are equipped with specialized containment measures, including ventilation systems, personal protective equipment, and controlled access, minimizing the likelihood of accidental release or exposure to pathogens. This is especially critical when dealing with highly infectious agents or those with unknown risks.

**TABLE 1.2**
**Differences Between BSC Classifications Adapted from the Biosafety Cabinet classification from the WHO Laboratory Biosafety Manual (3rd Ed.) (WHO, 2004)**

| BSC | Face Velocity (m/s) | Airflow (%) | | Exhaust System |
|---|---|---|---|---|
| | | Recirculated | Exhausted | |
| Class I* | 0.36 | 0 | 100 | Hard duct |
| Class IIA1 | 0.38–0.51 | 70 | 30 | Exhaust to room or thimble connection |
| Class IIA2 vented to the outside | 0.51 | 70 | 30 | Exhaust to room or thimble connection |
| Class IIB1[a] | 0.51 | 30 | 70 | Hard duct |
| Class IIB2[a] | 0.51 | 0 | 100 | Hard duct |
| Class III | NA | 0 | 100 | Hard duct |

NA, not applicable
[a] All biologically contaminated ducts are under negative pressure or are surrounded by negative pressure ducts and plenums.

Additionally, biocontainment laboratories contribute significantly to research and development in fields such as medicine, microbiology, and biotechnology. They provide a secure environment for studying and manipulating potentially hazardous biological materials, facilitating advancements in diagnostics, vaccine development, and the understanding of infectious diseases. The controlled conditions within these laboratories ensure that experiments are conducted safely, preventing the unintentional spread of harmful agents.

Moreover, biocontainment laboratories play a crucial role in biosecurity, protecting against intentional misuse of biological materials. By strictly regulating access and monitoring activities, these laboratories contribute to global efforts to prevent the development of biological weapons and ensure the responsible use of potentially dangerous biological agents.

Biocontainment Laboratories are accredited in four Biosafety Levels (BSLs), requiring more stringent controls than the preceding level (Meechan & Potts, 2020).

1. **BSL-1** laboratories have the basic level of protection and undergo research on biological agents or toxins that are not consistently pathogenic. BSL-1 laboratories do not require advanced or specialized equipment.
2. **BSL-2** laboratories are equipped to study moderate-risk biological agents or toxins that may cause harm through inhalation, ingestion, or skin contact or absorption. Procedures generating aerosols are conducted in primary contaminants such as biosafety cabinets (BSCs). Hand-washing sinks, eye-washing stations, and automatic doors are key features, and equipment for decontamination, such as autoclaves or incinerators, should also be present (PHE, n.d.-B).

3. **BSL-3** laboratories have a higher emphasis on primary, especially BSCs, and secondary containments. These are equipped to study high-risk biological agents or toxins that can be transmitted as aerosols. PPE and other primary containment strategies should be implemented if work within BSCs is not feasible. A key difference in the equipment of these laboratories is enhanced ventilation with controlled or directional airflow.
4. **BSL-4** laboratories can undergo research or work on microbes with the highest biorisks that pose aerosol-transmitted laboratory infections and life-threatening diseases that have no available treatment. Research is undergone in Class III Biosafety Cabinets.

Biocontainment laboratories are indispensable in maintaining public health, preserving the environment, and advancing scientific knowledge. They establish a structured framework that enables researchers to work safely with biological materials, fostering innovation in various fields while minimizing the risks of handling hazardous agents.

## 1.3 BIOSECURITY

Biosecurity, on the other hand, refers to measures and practices implemented to safeguard against the unauthorized access, theft, or accidental or intentional release of biological materials that could threaten human, animal, plant, or environmental health and safety (PHE, n.d.-C). The primary objective of biosecurity is to prevent the misuse of biological agents for malicious purposes, such as bioterrorism or biowarfare (Xue et al., 2021). This involves implementing protocols and security measures to control and monitor access to laboratories, facilities, and research environments where biological materials are handled.

According to Grainger (2012), biosecurity has five essential pillars in biorisk management.

1. **Physical security** is the 'assurance of safety from physical intrusion'. This includes a relevant and important concept of 'Graded protection' wherein areas within the facility will have varying levels of security. Usually, the innermost areas have the highest risk assets (biological materials or information) and thus would require the highest level of physical security and can be accessed only by individuals with the highest clearances or access controls. There are four concepts associated with physical security.

   a. **Detection:** determining if an incident or action that is occurring or has transpired is unauthorized. Important physical features include sensors, such as motion or infrared, that can activate alarms to alert personnel. This may also include training personnel to identify suspicious or unauthorized people.
   b. **Response:** can include alerting personnel, safeguarding biological assets and staff, and neutralizing the intruder or threat.
   c. **Delay:** impeding or slowing down unauthorized people to help in detection and response

2. **Personnel management** is a set of protocols ensuring that only authorized personnel have access to biological materials or information to reduce the risk of theft and fraud, scientific misconduct, or unwanted incidents. This includes the management of internal and external factors affecting personnel, which may influence them to commit unauthorized purposes.
3. **Material Control and accountability** includes the documentation, cataloging, or inventory of materials and their corresponding accountable or responsible personnel. Biological materials left unchecked are harder to detect and may be used for unauthorized purposes.
4. **Transport Security** relies on chain of custody principles. It includes rigorous protocols from the internal movement of biological materials from the workstations to the facility's entrance and their safeguarding during external transport until they reach another local or international facility. Commercial carriers may be contracted for the transport of biological materials.
5. **Information Security** is the safeguarding of sensitive and valuable physical or digital information from theft or diversion which may lead to severe consequences in terms of securing pathogens or toxins. Sensitive information should be labeled as restricted, have limited distribution, and have access controls.
   a. **Identification:** designation of sensitivity level informed by a review and approval process which is vital prior to public circulation
   b. **Control:** identifying the personnel responsible for control of sensitive information
   c. **Marking:** inscription of the sensitivity level designation on the top and bottom of each page and on the cover sheet
   d. **Communication Security:** can be digital (mail, email or fax) or physical (limited discussions in the open). Reproduction of information should be as needed with the original and its copies being controlled.
   e. **Network Security:** personal computer security, remote and wireless access controls, user authentication, virus protection, firewalls, layered network access
      i. Administrators should have full control over all information.
      ii. Systems should have procedures, such as password controls or two-person controls, to increase protection.
      iii. Operator privileges should have restrictions.
      iv. Equipment should have physical protection.
      v. Provide backup equipment and procedures to maintain security.
      vi. Computers should have emergency power and uninterrupted power supplies

Notably, Biosecurity is pivotal in 'dual-use research', a concept defined by the US NIH—Office of Intramural Research (OIR) (n.d.) as it refers to research activities that have the potential to furnish knowledge, information, products, or technologies that could be deliberately misapplied, posing a substantial threat to public health and safety, crops, animals, the environment, resources, or national security. In essence, both biosafety and biosecurity standards and regulations are indispensable safeguards that ensure the safe conduct of research activities involving infectious agents by ensuring responsible and ethical handling of biological materials (United States Department of Agriculture - Animal & Plant Health Inspection Service [USDA-APHIS], 2023).

## 1.4 BIORISK MANAGEMENT

*Biorisk management aims to provide the highest practical protection and the lowest practical exposure.*

Biorisk management is a comprehensive and systematic approach to identifying, assessing, and mitigating the risks associated with the handling, storage, transport, and disposal of biological materials. Biorisk management includes fundamental principles covering biosafety and biosecurity measures, emergency response planning, waste management, and personnel training programs. The goal is to prevent accidental exposure to infectious agents, laboratory-acquired infections, and the unintentional release of biological hazards.

ISO adopted the Plan, Do, Check, Act (PDCA) cycle, a recurring process approach utilized in management systems, including biorisk management. Below are the processes as described in ISO 9001 (ISO, n.d.-B):

- **Plan**: establish the objectives of the system and its processes, and the resources needed to deliver results following customers' requirements and the organization's policies, and identify and address risks and opportunities;
- **Do**: implement what was planned;
- **Check**: monitor and (where applicable) measure processes and the resulting products and services against policies, objectives, requirements and planned activities, and report the results;
- **Act**: take actions to improve performance, as necessary.

A typical PDCA cycle in laboratory biorisk management can be:

- **Plan**: Identify biorisks and formulate standard operating procedures (SOPs) and protocols;
- **Do**: Dry-run established SOPs and protocols;
- **Check**: Assess compliance and consistency of laboratory personnel to protocols and procedures, quantitatively explore biorisks mitigated with the protocol;
- **Act**: Revise SOPs as needed, strengthen engineering or administrative controls

Biorisk management aims to create a culture of safety, responsibility, and security in laboratories, research facilities, and other environments where biological materials are handled. This proactive approach helps prevent accidents, enhance laboratory safety, and mitigate the potential impact of biological hazards on public health and the environment.

Biorisk management involves three components: Biorisk Assessment, Biorisk Mitigation, and Performance (Menard, 2014).

1. **Biorisk Assessment:** Biorisk assessment is a systematic and structured process of evaluating the potential risks associated with handling, storing, and working with biological agents, toxins, and other hazardous materials. The primary objective of biorisk assessment is to identify and analyze the hazards and vulnerabilities inherent in a specific biological process or activity. This includes assessing the likelihood of exposure to biological agents, the severity of potential harm, and the effectiveness of existing safeguards. Several risk assessment tools, both qualitative and quantitative risk assessment matrices, may be utilized. Qualitative risk analysis tends to be more subjective; using a more objective approach to risk assessment, such as a quantitative risk assessment model (Figure 1.1) is generally the preferred route. A numerical value may be assigned to guide the mitigation planning based on the degree of risks identified per each scenario or condition. Biorisk assessment includes:

a. **Identification of Biological Agents:** Listing and characterizing the biological agents involved in a particular process or activity, including their potential hazards.

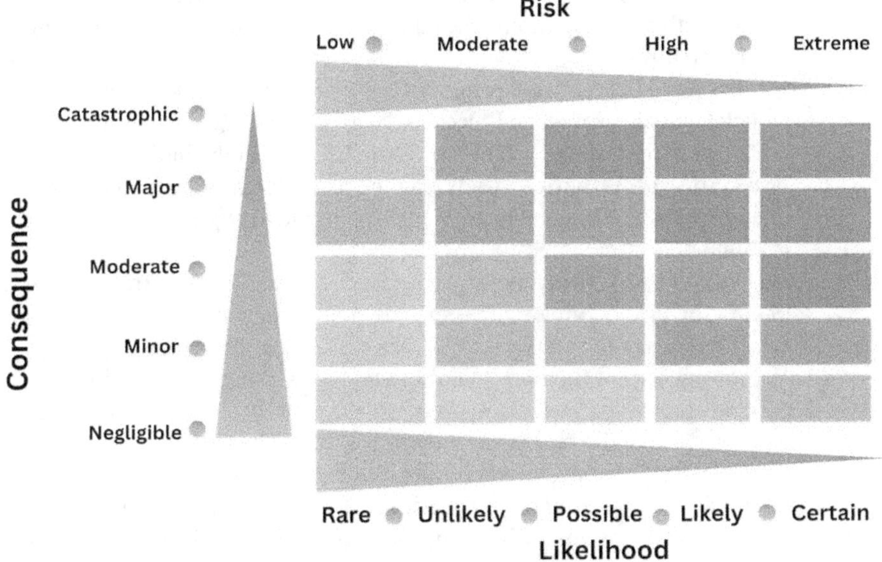


**FIGURE 1.1** Risk assessment quantitative matrix. 'Risks' and 'Likelihood' scores increase from left to right while the 'Consequences' score rises from bottom to top. The highest priority is given to extreme risks that are certain to happen and would lead to catastrophic outcomes if left unaddressed.

b. **Identification of Controls:** Any organization can employ different controls to minimize or manage hazards on personnel or biorisks. A vital model is the five-tiered reverse triangle Hierarchy of Controls (CDC, n.d.-B). This hierarchy is a guiding principle in selecting and implementing controls to reduce risks, emphasizing that the order should be considered when designing risk mitigation strategies. Although elimination and substitution may be the most straightforward and cheapest options, they are also the most difficult to integrate into existing systems or processes.
   i. **Elimination**, the highest and most effective tier, removes the hazard at the origin. This could include the revision or termination of processes involving the use of specific chemicals and equipment. This form of control is most preferred since workers are not exposed to hazards.
   ii. **Substitution** is the preference of using alternatives to certain processes while weighing the risks of the previous option with the proposed alternative.
   iii. **Engineering controls** minimize or mitigate hazards by modifying the workspace or equipment, such as stabilizing work desks. Engineering controls also come into play in biosecurity by installing biometric doors to allow only certain personnel to enter areas inside the laboratory.
   iv. **Administrative controls** minimize or mitigate hazards through policies, procedures or personnel training. In biosafety, professionals with adequate training establish and exercise standard operating procedures.
   v. **Personal Protective Equipment (PPE)** includes safety glasses or goggles, nitrile gloves, and face masks. These are the easiest to implement in the five tiers, but hazards are lessened at the individual level.

c. **Evaluation of Activities:** Analyzing the procedures and activities associated with handling biological materials to identify potential points of risk and exposure.

d. **Assessment of Containment Measures:** Evaluating the effectiveness of existing biosafety measures, such as containment equipment, laboratory facilities, and safety protocols, in preventing accidental exposure.

e. **Personnel Training and Competency**: Assessing the training and competency of personnel involved in the activity, ensuring they are equipped to handle biological materials safely.

f. **Biosecurity Measures:** Evaluating the security measures to prevent unauthorized access and intentional misuse of biological agents.

g. **Emergency Response Planning:** Assessing the readiness and effectiveness of emergency response plans in case of accidental spills, exposures, or other incidents.

h. **Environmental Impact:** Considering the potential impact of biological activity on the environment, including releasing genetically modified organisms or hazardous waste.

Biorisk assessments are dynamic and should be regularly reviewed and updated based on changes in processes, technologies, regulations, and lessons learned from incidents. This proactive approach is integral to maintaining high safety and security in laboratories and other settings where biological materials are handled.

2. **Biorisk Mitigation:** Biorisk mitigation refers to the strategic and systematic efforts to reduce, control, or eliminate the risks associated with handling, manipulating, and storing biological materials. Biorisk mitigation aims to implement measures and practices that minimize the likelihood and impact of accidental exposure, release, or intentional misuse of biological agents. As underscored by Salerno (2021), it's essential to recognize that when considered in isolation, none of the mitigation controls or measures can comprehensively control or reduce all risks. The effectiveness of risk mitigation hinges on the synergy achieved by combining these diverse measures in a coordinated manner. Furthermore, exercising prudence and avoiding overzealous implementation is crucial, as undoing measures can prove cumbersome and costly. Key aspects of biorisk mitigation include:

a. **Implementation of Biosafety Measures:** Integrating and enforcing effective biosafety practices, such as using appropriate containment equipment, safety protocols, and facility design to prevent accidental exposure to biological agents.

b. **Enhancement of Biosecurity Measures:** Strengthening security protocols to control access to laboratories, facilities, and biological materials, thereby reducing the risk of unauthorized access or intentional misuse.

c. **Training and Education:** Personnel handling biological materials are provided with comprehensive training programs, which ensure they are well informed about potential risks and equipped with the necessary skills to follow safety protocols.

d. **Regular Risk Assessments:** Conduct routine biorisk assessments to identify new hazards, evaluate the effectiveness of existing mitigation measures, and adjust protocols accordingly.

e. **Implementation of Emergency Response Plans:** Establishing and regularly reviewing emergency response plans ensures a swift and effective response to accidents, spills, or other incidents involving biological materials.

f. **Waste Management:** Proper disposal and treatment of biological waste to prevent environmental contamination and minimize the risk of spreading infectious agents.

g. **Continuous Improvement:** Promoting a culture of continuous improvement by actively seeking and implementing advancements in biosafety and biosecurity practices based on emerging knowledge and technological developments.

By integrating these measures into laboratory procedures and research activities, biorisk mitigation aims to create a safer working environment, protect the health of personnel, prevent environmental contamination, and contribute to overall public safety and security.

3. **Biorisk Performance:** Evaluating performance in biorisk management involves assessing how well an organization or laboratory implements its biorisk management program to prevent and control biological hazards. Here are key elements to consider when evaluating biorisk management performance:

a. **Compliance with Standards and Regulations:**
   i. Assess adherence to local, national, and international regulations and standards related to biorisk management.
   ii. Ensure that the organization is meeting the requirements outlined in relevant biosafety guidelines.

b. **Documentation and Record-Keeping:**
   i. Review the completeness and accuracy of documentation related to biorisk management practices.
   ii. Ensure that standard operating procedures (SOPs), risk assessments, incident reports, and training records are well-maintained.

c. **Risk Assessment and Mitigation:**
   i. Evaluate the effectiveness of risk assessments in identifying and characterizing biological hazards.
   ii. Assess the implementation and effectiveness of control measures to mitigate identified risks.

d. **Training and Competency:**
   i. Evaluate the training programs in place for personnel handling biological materials.
   ii. Ensure that personnel are adequately trained and demonstrate competency in following safety protocols.

e. **Incident Reporting and Investigation:**
   i. Assess the reporting system for incidents, accidents, and near misses.
   ii. Review the thoroughness and effectiveness of incident investigations and the implementation of corrective actions.

f. **Emergency Response Preparedness:**
   i. Evaluate the organization's readiness to respond to biological incidents and emergencies.
   ii. Assess the effectiveness of emergency response plans through drills and exercises.

g. **Monitoring and Surveillance:**
   i. Implement systems to monitor laboratory practices, equipment, and facilities regularly.
   ii. Conduct regular inspections and audits to identify areas of improvement and ensure ongoing compliance.

h. **Biosecurity Measures:**
   i. Evaluate the implementation of measures to control access to laboratories and facilities handling high-risk biological agents.

   ii. Assess the effectiveness of security protocols to prevent unauthorized access.

i. **Waste Management:**
   i. Review procedures for the safe collection, treatment, and disposal of biological waste.
   ii. Ensure compliance with waste management regulations and assess the organization's efforts to minimize biohazardous waste generation.

j. **Continuous Improvement:**
   i. Assess the organization's commitment to continuous improvement in biorisk management.
   ii. Review the results of internal and external audits and the organization's responsiveness to identified areas for improvement.

k. **Communication and Reporting:**
   i. Evaluate communication channels for disseminating information about potential risks and safety protocols.
   ii. Assess the effectiveness of communication in promoting a culture of safety and awareness among personnel.

Performance evaluation in biorisk management is a process that requires regular reviews, assessments, and updates to adapt to changes in procedures, technologies, and regulations. It helps ensure the organization continually improves its ability to manage biological risks effectively.

Biorisk management aims to create a safe and secure working environment where biological research and applications can be conducted responsibly, minimizing the potential risks of handling biological agents. It involves a continuous risk assessment, mitigation, and ongoing improvement process to ensure the highest safety and security standards in biological research and related activities.

## 1.5 BIORISK MANAGEMENT AND THE INTERLINKAGES ACROSS DISCIPLINES AND SECTORS

Biosafety and biosecurity encompass several disciplines, such as public health, veterinary health and medicine, interior design, and engineering, to name a few. Table 1.3 shows how biosafety and biosecurity problems concern other disciplines. Both concepts are also heavily intertwined with 'One Health' wherein the health of humans, animals and the environment and ecosystems are closely interdependent. Global issues and anthropogenic activities, including human travel, global trade, and climate change, pose biosafety hazards or biosecurity threats.

### 1.5.1 One Health

The One Health Initiative emphasizes the interconnectedness of human, animal, and environmental health, recognizing the complex relationships between these

**TABLE 1.3**
**Points of Entry for Various Disciplines or Sectors on Issues or Concerns Associated with Biosafety and Biosecurity Concepts.**

| Problems in Biosafety and/or Biosecurity | Associated Issues or Concerns | Involved Disciplines or Sectors |
|---|---|---|
| Unauthorized access to biological agents or toxins | • Theft<br>• Terrorism<br>• Geopolitical conflicts<br>• Cybersecurity | • Interior design<br>• Electrical<br>• Civil Engineering<br>• Architecture<br>• Information, Communication, Technology<br>• Global and national security<br>• Transportation |
| Cultivation of biological agents for violence or biowarfare (Anthrax, fungal poison) | • Geopolitical conflicts<br>• Terrorism | • Global and national security<br>• One health<br>• Biodiversity<br>• Agriculture<br>• Biotechnology |
| Introduction of invasive alien species | • Climate change<br>• Environmental degradation, deforestation, land conversion<br>• Global trade<br>• International tourism<br>• Creation of destructive living-modified organisms (LMOs) | • Transportation and travel<br>• Trade<br>• Agriculture<br>• Forestry<br>• Biodiversity<br>• One health<br>• Biotechnology |
| Spread of microbes resistant to antibiotics | • Runoff pollution<br>• Drug overprescription<br>• Irresponsible antibiotics intake<br>• Creation of drug-resistant LMOs | • One health<br>• Agriculture<br>• Biodiversity<br>• Pharmaceuticals<br>• Biotechnology |

domains. In this context, Biosecurity and Biosafety play crucial roles in ensuring the success and sustainability of the One Health approach. Biosecurity involves measures designed to prevent the introduction and spread of harmful biological agents, addressing potential threats to humans, animals, and ecosystems. This includes the implementation of strict protocols in laboratories, monitoring and regulating the movement of animals and animal products and safeguarding against intentional or accidental release of pathogens. Biosafety, on the other hand, focuses on the safe handling, containment, and disposal of biological materials to protect laboratory workers, researchers, and the broader community from potential risks associated with biological agents. Both biosecurity and biosafety contribute to preventing and mitigating the impact of infectious diseases that can affect humans, animals, and the environment. Adopting a holistic One Health perspective aims to enhance global health security, reduce the risk of zoonotic diseases, and promote sustainable coexistence between humans, animals, plants, and ecosystems. Integrating biosecurity and biosafety principles into the One Health framework ensures a comprehensive and collaborative approach to addressing emerging infectious diseases and other health challenges.

Another key relevant concept to this interlinkage is **'Select Agents or Toxins'** which are 'determined to have the potential to pose a severe threat to public health and safety, to animal and plant health, or to animal or plant products' (Centers for Disease Control and Prevention [CDC], n.d.-a). The United States, through the Public Health Security and Bioterrorism Preparedness and Response Act (Bioterrorism Act), established and maintained a list of select agents or toxins that undergo biennial review. The same act requires the consideration of the following criteria:

1. The effect on human health of exposure to the agent or toxin;
2. The degree of contagiousness of the agent or toxin and the methods by which the agent or toxin is transferred to humans;
3. The availability and effectiveness of pharmacotherapies and immunizations to treat and prevent any illness resulting from infection by the agent or toxin; and
4. Any other criteria, including the needs of children and other vulnerable populations, that are considered appropriate.

Specific agents may have varying risk rankings in certain scenarios based on the organisms' epidemiological prevalence in a community or country. Therefore, efforts must be made to have regional or country-specific detection agent and toxin listings. It may or may not be similar to other regions, but more importantly, each agent or toxin underwent the process of agent identification, classification, risk assessment, and categorization.

### 1.5.2 Human Health

Biosafety and biosecurity are integral components in safeguarding human health, with a crucial impact on the broader field of public health. Biosafety focuses on the safe handling, containment, and management of biological materials to prevent accidental exposure and the release of hazardous agents, particularly in laboratory and healthcare settings. On the other hand, biosecurity involves measures to prevent intentional misuse or theft of biological materials that could threaten public health. These principles protect communities and individuals from infectious diseases and potential bioterrorism threats.

In human health, biosafety is paramount in laboratories and healthcare facilities where biological agents are handled, tested, and researched. Strict protocols ensure the safe conduct of experiments, diagnostic procedures, and the development of vaccines or treatments. This protects laboratory personnel and prevents the accidental release of pathogens that could lead to outbreaks or public health crises.

From a public health perspective, biosafety measures contribute to preventing and controlling infectious diseases. Properly handling specimens, samples, and infectious agents in healthcare settings reduces the risk of healthcare-associated infections, protecting patients and healthcare workers. Additionally, biosafety practices are crucial in containing emerging infectious diseases and preventing their spread within communities.

Biosecurity, on the other hand, is critical for averting intentional acts that could compromise public health. This includes safeguarding against the theft or unauthorized access to dangerous biological materials that could be used in bioterrorism attacks. Strict control and monitoring of access to laboratories, research facilities, and biological repositories are essential to biosecurity.

Biosafety and biosecurity measures become even more crucial in the event of a public health emergency. Rapid and secure handling of potentially harmful biological agents, combined with effective communication and coordination, is vital for minimizing the impact of outbreaks and ensuring a prompt response.

Overall, the role of biosafety and biosecurity in human health, especially within the public health framework, is central to preventing the unintentional spread of infectious diseases and protecting communities from intentional threats. These measures form a cornerstone in building resilient and responsive public health systems.

### 1.5.3 Emerging Infectious Diseases, Re-emerging Infectious Diseases and Deliberately Emerging Infectious Diseases

From the early days, attempts to use microorganisms and toxins as weapons have been evident, evolving from crude methods such as contaminating water supplies with infected cadavers catapulted to enemy lines to the development of specialized munitions for battlefield and covert use.

Assessing the history of biological warfare is complex due to various confounding factors. Verification challenges arise from propaganda-driven claims of biological attacks, a lack of pertinent microbiological or epidemiological data, and the overlap with naturally occurring diseases during hostilities.

Distinguishing between a natural infectious disease event and an intentional act is a significant challenge in outbreak investigation. In highly developed countries like the United States, where extensive investigations are well-supported, the question arises: given our economic constraints, how capable are we of recognizing deviations in the natural patterns of infectious diseases? Can we promptly identify causative agents to enable real-time interventions, especially when dealing with previously unrecognized agents?

The Centers for Disease Control and Prevention in Atlanta, Georgia, has issued several statements and guidelines on bioterrorism's medical and public management tailored to their country's capacities. The challenge lies in adapting these recommendations to our setting, focusing on prevention and treatment.

While pursuing cutting-edge technology for rapid detection is commendable, a well-defined referral and communication system remains paramount. Given our current economic condition, relying on the inventiveness and resourcefulness that Filipinos are known for becomes crucial.

Unlike conventional warfare, the local health service and infectious disease specialists are on the frontlines of bioterrorism. Awareness of potential threats and a high index of suspicion for diseases caused by likely bioterrorism agents, such as anthrax or plague, are essential. Even the best-laid plans may falter if the local health service needs to be adequately prepared and vigilant.

### 1.5.4 Agriculture

In the Philippine agricultural context, Biosafety and Biosecurity play pivotal roles in safeguarding public health and the integrity of the country's agricultural ecosystems. Biosafety in agriculture involves the responsible management of genetically modified organisms (GMOs) and other potentially hazardous biological materials to prevent unintended environmental consequences and ensure the safety of farmers and consumers. Implementing stringent protocols for developing, testing, and releasing genetically modified crops is essential to prevent unintended cross-breeding and uphold the ecosystem's health and well-being.

Another vital agricultural problem associated with biosafety is agricultural runoffs, where pathogens on farmlands, especially those that may cause waterborne zoonotic diseases, may enter bodies of water and multiply (Atwill et al., 2012). Antimicrobial resistance can also develop from the feces of animals treated with antibiotics that deposit into the ground or runoff into bodies of water.

Simultaneously, Biosecurity measures are crucial to prevent the introduction and spread of harmful pests, diseases, or pathogens that can devastate crops and threaten food security. Strict quarantine procedures, surveillance, and control mechanisms are essential for agricultural biosecurity. This is particularly relevant given the Philippines' dependence on agriculture, making protecting crops and livestock from introducing exotic pests and diseases imperative.

In the face of emerging challenges such as climate change and globalization, the role of biosafety and biosecurity has become even more critical. Adapting and applying international best practices, the Philippines must ensure that agricultural research and practices align with safety standards to protect biodiversity, preserve the environment, and sustain farmers' livelihoods. Integrating Biosafety and Biosecurity measures into agricultural practices is essential for promoting a resilient and sustainable agricultural sector in the Philippines, contributing to both food security and environmental preservation.

### 1.5.5 Transportation, Travel, and Trade

Biological agents and toxins are at risk of being seized or stolen in transit, necessitating that transportation procedures and routes be properly established before the delivery of reagents. Biocontainment design and packaging are also important factors in transporting goods since there are risks of their unintentional release into the surrounding environment and in communities. Toxins can also be delivered or transported, such as in the case of ricin. Letters laced with ricin, a toxin from the castor bean plant, were delivered to US officials Barack Obama, Roger Wicker, Sadie Holland, and Michael Bloomberg (Hayoun et al., 2023).

An unintended consequence of global trade is the introduction of alien species. The Intergovernmental Science-Policy Platform on Biodiversity and Ecosystem Services (n.d.) estimates that 37% of 37,000 identified alien species have been reported since 1970, largely caused by increasing global trade and human travel. They also estimate that the increase in total numbers is projected under business-as-usual conditions.

### 1.5.6 Engineering and Architecture

Engineering and architecture are critical in performing laboratory tasks, ensuring safety and security. In the context of biosafety, ensure that the facility and laboratory designs are up to standards with local and national guidelines and follow standards concerning the construction of biological laboratories. In the recent update of the WHO Laboratory Biosafety Manual 4th edition, risk-based laboratory design features (core requirements, heightened control measures and maximum containment measures) have superseded the traditional laboratory biosafety levels (BSLs 1 to 4) to reflect the changing paradigm that risk assessment should precede the lists of required equipment and features within a laboratory.

On the other hand, in biosecurity, the laboratory should be considered to be placed within a graded security such that the laboratory will not be infiltrated immediately by threats and actors that can steal valuable biological materials and data housed within the institution. However, this may be a problem with safety, especially within an incident and emergency response, when responders may need help reaching the area. Though biosafety and biosecurity concepts aligned with engineering and architecture may sometimes contradict each other, there should be a balance between them to ensure that both goals are reached and satisfied.

## 1.6 FUTURE DIRECTION FOR THE PHILIPPINES

The future of Biosafety and Biosecurity advocacy and training in the Philippines is contingent on developing continued training mechanisms responsive to resource-limited settings while upholding stringent principles and practices. Integrating policies for interagency coordination is crucial to maximize expertise in the field. Simultaneously, the importance of standardized, quality, and robust training initiatives cannot be overstated, ensuring proficiency in both the implementation and professional practice of Biosafety and Biosecurity. As technology advances, a proactive approach is necessary; revisiting policies, including the 2006-established NBF, is vital to adapt to shifting paradigms. Strengthening the NBF to encompass biosecurity and biosurveillance lenses is imperative, given our nation's status as the fifth most biodiverse place globally. Furthermore, developing a comprehensive list of select agents and toxins is pivotal to fortifying our preparedness against evolving biosecurity risks in the ever-changing landscape.

## REFERENCES

Atwill, E. R., Partyka, M. L., Bond, R. F., Li, X., Xiao, C., Karle, B. M., & Kiger, L. E. (2012). *Introduction to Waterborne Pathogens in Agricultural Watersheds.* United States Department of Agriculture. https://directives.sc.egov.usda.gov/OpenNonWebContent.aspx?content=32935.wba

Bayot, M., & Limaiem, F. (2023). *Biosafety Guidelines.* Stat Pearls Publishing. https://www.ncbi.nlm.nih.gov/books/NBK537210/

Centers for Disease Control and Prevention. (n.d.-a). *Division of Regulatory Science and Compliance: What Is a Select Agent?* https://www.cdc.gov/orr/dsat/what-is-select-agents.htm

Centers for Disease Control and Prevention. (n.d.-b). *Hierarchy of Controls.* CDC. https://www.cdc.gov/niosh/topics/hierarchy/default.html

Coelho, A. C., & Díez, J. G. (2015). Biological risks and laboratory-acquired infections: A reality that cannot be ignored in health biotechnology. *Frontiers in Bioengineering and Biotechnology, 3*(56). https://doi.org/10.3389/fbioe.2015.00056

Destura, R. V., Lam, H. Y., Navarro, R. C., Lopez, J. C. F., Sales, R. K. P., Gomez, M. I. F. A., dela Tonga, A. D., & Ulanday, G. E. L. (2021). Assessment of the biosafety and biosecurity landscape in the Philippines and the development of the national Biorisk management framework. *Applied Biosafety, 26*(4), 232–244. https://doi.org/10.1089/apb.20.0070

Ellwanger, J. H., & Chies, J. A. B. (2021, April 5). Zoonotic spillover: Understanding basic aspects for better prevention. *Genetics and Molecular Biology, 44*(1). https://doi.org/10.1590/1678-4685-GMB-2020-0355

National Committee on Biosafety of the Philippines. (n.d.). *National Committee on Biosafety of the Philippines 2nd Quarter Report.* https://ncbp.dost.gov.ph/download/annual-report/category/59-2018-quarterly-report

Office of the President of the Philippines. (2006). *Executive Order No. 514, s. 2006.* Official Gazette. https://www.officialgazette.gov.ph/2006/03/17/executive-order-no-514-s-2006/

Government of the Republic of the Philippines. (2016). *DOST-DA-DENR-DOH-DILG Joint Department Circular No. 1, Series of 2016.* World Trade Organization. https://members.wto.org/crnattachments/2017/SPS/PHL/17_2782_00_e.pdf

Grainger, L. A. (2012). *Laboratory Biosecurity.* OSTI.GOV. https://www.osti.gov/servlets/purl/1117406

Hayoun, M. A., Kong, E. L., Smith, M. E., & King, K. C. (2023). *Ricin Toxicity.* Stat Pearls Publishing LLC. https://www.ncbi.nlm.nih.gov/books/NBK441948

Intergovernmental Science-Policy Platform on Biodiversity and Ecosystem Services. (n.d.). *Media Release: IPBES Invasive Alien Species Assessment.* https://www.ipbes.net/IASmediarelease

International Organization for Standardization. (n.d.-a). *ISO: Global Standards for Trusted Goods and Services.* ISO - International Organization for Standards. https://www.iso.org/home.html

International Organization for Standardization. (n.d.-B). *The Process Approach in ISO 9001.* International Organization for Standardization. ISO. https://www.iso.org/files/live/sites/isoorg/files/archive/pdf/en/iso9001-2015-process-appr.pdf

Lantican, R. M. (1991). *Philippine Biosafety Guidelines (PBG).* https://dost-bc.dost.gov.ph/download/category/13-philippine-biosafety-guidelines-series-of-1991

Meechan, P. J., & Potts, J. (2020). *Biosafety in Microbiological and Biomedical Laboratories.* https://www.cdc.gov/labs/pdf/SF__19_308133-A_BMBL6_00-BOOK-WEB-final-3.pdf

Menard, K. L. (2014). *Orientation to Biorisk Management.* Office of Scientific and Technical Information. https://www.osti.gov/servlets/purl/1502307

National Institutes of Health. (n.d.). *Dual-Use Research.* U.S. Department of Health and Human Services, National Institutes of Health, Office of Intramural Research. https://oir.nih.gov/sourcebook/ethical-conduct/special-research-considerations/dual-use-research

Navarro, R. C., Vitor, R. J. S., II, Canoy, R. J. C., dela Tonga, A. D., Ulanday, G. E. L., Silva, M. R. C., & Destura, R. V. (2022). Biosafety Capacity Building During the COVID-19 Pandemic: Results, Insights, and Lessons Learned from an Online Approach in the Philippines. *Applied Biosafety, 27*(1), 42–50. https://doi.org/10.1089%2Fapb.2021.0021

Public Health Emergency. (n.d.-a). *Biocontainment - Biorisk Management.* Science Safety Security. https://www.phe.gov/s3/BioriskManagement/biocontainment/Pages/default.aspx

Public Health Emergency. (n.d.-b). *Biosafety Levels.* Science Safety Security. https://www.phe.gov/s3/BioriskManagement/biosafety/Pages/Biosafety-Levels.aspx

Public Health Emergency. (n.d.-c). *Biosecurity FAQ.* Science Safety Security. https://www.phe.gov/s3/BioriskManagement/biosecurity/Pages/Biosecurity-FAQ.aspx

Salerno, R. M., & Gaudioso, J. (2021). *Laboratory Biorisk Management: Biosafety and Biosecurity.* Taylor & Francis Group.

United States Department of Agriculture - Animal & Plant Health Inspection Service. (2023). *APHIS Biorisk Management Manual.* https://www.aphis.usda.gov/library/manuals/pdf/aphis-biorisk-management-manual.pdf

University of the Philippines. (2018). *Decisions of the Board of Regents 1333rd Meeting, 22 February 2018.* University of the Philippines Gazette. https://osu.up.edu.ph/wp-content/uploads/2019/01/1333_UP-Gazette.pdf

Wedum, A., & Kruse, R. (2017, March 17). *A History of the American Biological Safety Association Part II: Safety Conferences 1966–1977.* ABSA International. https://absa.org/about/hist02/

World Health Organization. (2004). *Laboratory Biosafety Manual*, 3rd ed. https://iris.who.int/bitstream/handle/10665/42981/9241546506_eng.pdf

World Health Organization. (2020). *Laboratory Biosafety Manual*, 4th ed. https://www.who.int/publications/i/item/9789240011311

Xue, Y., Shang, L., & Zhang, W. (2021). Building and implementing a multi-level system of ethical code for biologists under the Biological and Toxin Weapons Convention (BTWC) of the United Nations. *Journal of Biosafety and Biosecurity, 3*(2), 108–119. https://doi.org/10.1016/j.jobb.2021.09.001

# 2 Philippine Biosafety and Biosecurity Regulations
## *From Global Response to Domestic Realities*

*Edsel Allan G. Salonga*
Cytogenetics Laboratory
Institute of Human Genetics
University of the Philippines
Manila, Philippines

## 2.1 INTRODUCTION

The establishment of comprehensive legal frameworks is imperative for safeguarding lives, the environment, and national security from biosafety and biosecurity threats. Biosafety regulations aim to prevent unintentional exposure to dangerous biological agents and toxins, as well as unplanned releases into the environment. Simultaneously, biosecurity regulations focus on reducing the risk of such materials being intentionally misused as weapons.

This chapter examines the origins of regulations instituted in response to pivotal biosafety and biosecurity incidents, both globally and within the Philippines specifically. We will explore how international guidance shapes national and local policies in the Philippines, delving into the web of implications created when regulations designed for biosafety and biosecurity intersect, underscoring the need for harmonized risk management.

## 2.2 KEY HISTORICAL EVENTS PROMPTING REGULATIONS

Throughout history, catastrophic events have catalyzed the development of laws aimed at preventing, mitigating, and responding to biosafety and biosecurity risks. The 1984 Bhopal disaster, the 1995 case of Larry Wayne Harris, and the 2001 anthrax attacks demonstrate how such incidents can expose gaps, vulnerabilities, and weaknesses, prompting corrective action.

### 2.2.1 1984 Bhopal Disaster

On December 3, 1984, over 40 tons of toxic methyl isocyanate gas leaked from a pesticide plant in Bhopal, India, owned by the American company Union Carbide

DOI: 10.1201/9781003426219-2

Corporation (UCC). The immediate effects were devastating—over 3,800 deaths and over 550,000 injuries. Estimates suggest that overall, the Bhopal Gas Tragedy caused over 15,000 premature deaths. The incident pointed to a lack of safety standards and preventative maintenance protocols among multinational corporations operating in developing countries (Broughton, 2005). Outrage over the disaster spurred India to pass legislation like the 1986 Environment Protection Act, empowering the Ministry of Environment and Forests to monitor industrial activities. However, some note that communities near the abandoned UCC plant still suffer adverse health effects and that environmental rehabilitation remains incomplete, suggesting more work is needed. The tragedy brought unprecedented international awareness to industrial safety precautions and emphasized the need for enforceable global standards and preparedness for such disasters.

### 2.2.2 1995 Larry Wayne Harris Case

In 1995, an alarming biosecurity loophole was exposed when Larry Wayne Harris, a private citizen trained in microbiology, ordered three vials of *Yersinia pestis* (plague) bacteria from the American Type Culture Collection (ATCC) using fraudulent credentials. The non-profit ATCC distributes biological materials to scientific institutions but initially filled Harris's order believing it legitimate. Their decision to later notify the Centers for Disease Control and Prevention led to recovery of the vials before harm was done. However, the ease with which Harris obtained deadly pathogens raised concerns that human pathogens were not as strictly regulated as plant and animal agents (Marks, 2011). This case directly led to new requirements for registration, accountability, justification of need and use, and other oversight protocols specifically for access to hazardous biological agents that could impact human health if misused.

### 2.2.3 2001 Anthrax Attacks

The 2001 anthrax letter attacks further revealed vulnerabilities when envelopes containing *Bacillus anthracis* spores were mailed to media outlets and Congressional offices, killing five people. The crisis exposed problems with interagency coordination, laboratory testing capacities, stockpiles of medicines, and more (Roos & Schnirring, 2011). This prompted major investments and regulatory changes like the Public Health Security and Bioterrorism Preparedness and Response Act to close preparedness gaps. Relationships between law enforcement, public health, and national security agencies were also strengthened. The attacks made clear that bioterrorism was a formidable threat necessitating an integrated defense strategy combining prevention, detection, and response capabilities.

Each of these seminal moments exposed critical weaknesses in biosafety or biosecurity, catalyzing impactful regulations and systemic changes. However, history has shown that as technology advances, new risks continue to emerge. Remaining vigilant and proactive is essential to reduce preventable harm. Just as these catastrophes shaped 20th-century policies, new events will likely inform future improvements in oversight, coordination, and preparedness.

These examples demonstrate that biosafety and biosecurity regulations require continual adaptation to emerging threats. This chapter will examine both the historical roots and hierarchy of legal requirements in these areas, exploring how international guidance shapes national and local policies in the Philippines. The subsequent discussion will delve into the interconnected implications of regulations designed for biosafety and biosecurity intersect, underscoring the need for harmonized risk management.

## 2.3 INTERNATIONAL FRAMEWORKS INFLUENCING PHILIPPINE POLICIES

Biosafety and biosecurity regulations do not exist in isolation, but are part of an interconnected web of guidance with some hierarchy in place. International frameworks and standards established by treaties, conventions, and multilateral partnerships sit at the top of this hierarchy, shaping national laws and local policies globally.

This top-down relationship can be visualized as a pyramid (Figure 2.1). At the peak sit legally binding international agreements, which set expectations that influence national regulations below. In the middle layers are voluntary international standards and recommended best practices, which provide guidance that nations can customize based on their risk environment. At the base sit institutional policies, standard operating procedures, and training programs designed to comply with higher regulations while implementing them locally.

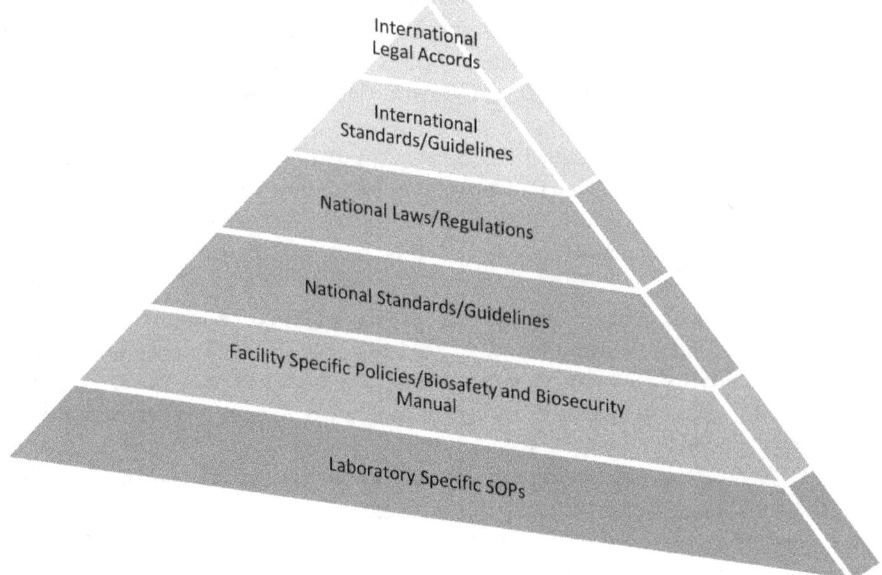


**FIGURE 2.1** Pyramidal hierarchy of biosafety and biosecurity laws and regulations showing the hierarchical relationship between international regulations, national laws, voluntary standards, and institutional policies related to biosecurity.

### 2.3.1 Legally Binding International Agreements

International law represents the pinnacle of the biosafety and biosecurity framework, providing a sturdy foundation upon which national policies and regulations are built. When enshrined in treaties and accords, nations' commitments become non-negotiable obligations backed by international authority. Five major legally binding agreements stand out for profoundly shaping today's landscape of biological oversight: the 1972 Biological Weapons Convention prohibiting biological weapons; the 1992 Convention on Biological Diversity and its 2000 Cartagena Protocol governing genetically modified organisms; the 2004 UN Security Council Resolution 1540 criminalizing weapons of mass destruction (WMD) assistance to non-state actors; the 2005 World Health Organization's Resolution WHA 58.29 strengthening biosafety after severe acute respiratory syndrome (SARS); and the 2005 revised International Health Regulations on global outbreak response. Together, these pioneering pacts comprise a robust global legal framework designed to reinforce biosafety and prevent biothreats. As rapid scientific advances create novel risks, updating and expanding international agreements through progressive negotiations will remain vital. Doing so erects guardrails protecting humanity while permitting science's safe and ethical advance.

#### 2.3.1.1 Biological Weapons Convention (BWC)

The groundbreaking 1972 Biological Weapons Convention (BWC) prohibits the development, production, stockpiling, retention, and use of biological weapons. This multilateral treaty criminalizes biological weapons activities and establishes means of multilateral cooperation (Tuzmukhamedov, 2021). By requiring national implementation and enforcement, the pioneering BWC significantly enhanced worldwide biosafety and biosecurity. It substantially curtailed the risk of biological warfare and malicious use.

Prior to the BWC, despite the 1925 Geneva Protocol's ban on biological weapons use, major state biological warfare programs had continued unchecked. With the Cold War superpowers rapidly advancing their biological capabilities, the threat of a devastating biological arms race loomed. It became imperative to put a stop to this dangerous trajectory.

After three years of negotiations at the Conference of the Committee on Disarmament in Geneva, the historic accord opened for signature. Upon entering into force in 1975, BWC member states made a solemn pledge to never, under any circumstances, develop, acquire, or retain biological agents, toxins, weapons, or equipment intended for hostile purposes.

The BWC was the first international treaty to ban an entire class of weapons of mass destruction. Its comprehensive prohibitions brought the era of unconstrained state biological weapons programs to a close. This new global norm against biological weapons was further strengthened through Review Conferences held every 5 years.

Nearly 50 years since its inception, the BWC remains a cornerstone of international law prohibiting biological weapons and reinforcing the biosafety and biosecurity framework globally as well as in the Philippines.

#### 2.3.1.2 UN Convention on Biological Diversity and the Cartagena Protocol

The 1992 Convention on Biological Diversity (CBD) emerged as countries recognized that our fates are bound to the planet's biodiversity riches. With species extinction accelerating, the CBD called for wiser stewardship of nature's life-sustaining treasures (Secretariat of the Convention on Biological Diversity, 2011).

The Convention requires parties to develop national strategies for conservation and sustainable use of ecosystems, species, and genes. It promotes measures from protected areas to research and public education to achieve its trifold objectives. The CBD has been ratified by 196 nations, reflecting unified concern for our biological inheritance.

But conserving biodiversity also requires caution in tampering with life's delicate designs. Rapid advances in biotechnology's ability to manipulate genetics warranted rules to avoid jeopardizing biodiversity or human health.

Thus, in 2000, the Cartagena Protocol on Biosafety was adopted under the CBD framework (Secretariat of the Convention on Biological Diversity, 2000). This landmark agreement governs transboundary movements of living modified organisms (LMOs) produced by modern biotech.

The Protocol introduced vital checks against jumping blindly into the gene-splicing age. It mandates risk assessments and notification procedures for LMOs that may have adverse impacts on biodiversity or health. Measures and infrastructure to enable safe transport, handling, and use of LMOs are required.

Together, the CBD and Cartagena Protocol propel profoundly consequential aims—preserving the web of life while safely harnessing biotech's bounties. By providing comprehensive guidance and constraints, these accords uphold biosafety and biosecurity in the perilous arena of genetic manipulation. Nearly 30 years on, their integrated framework remains essential for navigating biological perils and possibilities.

#### 2.3.1.3 UN Security Council Resolution 1540

In 2004, the United Nations (UN) Security Council took a pivotal step to reinforce global biosafety and biosecurity by unanimously adopting Resolution 1540 (United Nations Security Council, 2004). This landmark resolution addressed the growing danger of non-state actors obtaining or using nuclear, chemical, and biological weapons.

With the rise of sophisticated terrorist groups and advances in technology, the threat that a non-state actor could acquire or develop catastrophic weapons became real. If al-Qaeda, Aum Shinrikyo, or other violent extremists obtained access to weapons of mass destruction (WMD) materials, the consequences could be severe.

To confront this threat, Resolution 1540 established legally binding obligations on all UN member states. It prohibits support to non-state actors seeking WMD capabilities and requires domestic measures to prevent proliferation of these weapons. This includes securing sensitive materials, controlling exports, and adopting laws against WMD financing and transport.

Through these comprehensive requirements, Resolution 1540 aims to create a web of countermeasures across borders. By calling on states to strengthen regulatory and enforcement efforts at national, regional, and global levels, the resolution reduces vulnerabilities that could be exploited by resourceful terrorist groups or individuals.

A 1540 Committee was established to monitor implementation and maintain international attention on this vital issue. The UN Office for Disarmament Affairs (UNODA) also provides ongoing assistance to states and coordination among relevant organizations.

By recognizing the biosecurity threat of WMD proliferation and establishing concrete international expectations to address it, Security Council Resolution 1540 reinforced the biosafety and biosecurity framework both globally and in specific countries, like the Philippines. Reducing the risk of biological weapons acquisition by non-state actors remains an urgent concern 1540 seeks to combat.

#### 2.3.1.4 WHO International Health Regulations

The legally binding International Health Regulations (IHR) serve as the global framework governing how 196 countries prepare for and respond to dangerous epidemics and other public health emergencies with the potential for international spread (World Health Organization, 2016).

First adopted in 1969, the IHR were radically revised in 2005 after the SARS outbreak exposed glaring weaknesses in the global architecture for outbreak alert and response. The extensive revisions transformed the IHR into an instrument applicable to all public health risks rather than specific diseases.

However, COVID-19 has starkly revealed additional gaps. In July 2023, countries held intensive discussions in Geneva on over 300 proposed amendments to the IHR based on lessons from the pandemic. The aim is to further bolster the world's collective defenses against emerging biological threats.

Proposed revisions span surveillance, notification, verification, declaring emergencies, temporary recommendations, the Emergency Committee, and more. From mandating pathogen sharing to strengthening accountability, countries are scrutinizing every aspect of the IHR to identify changes that would facilitate earlier action on outbreaks.

Further rounds of complex negotiations lie ahead. But countries appear aligned on the need to remedy deficiencies COVID-19 exposed in international coordination, data sharing, resource mobilization, and political will to protect lives. Although revising any legal instrument presents challenges, most governments agree reforming the IHR will be essential preparation for the next pandemic.

By providing legally binding rules governing detection and assessment of outbreaks with international impacts, the IHR remains the foremost global treaty safeguarding against infectious disease crises. The ongoing process to amend the IHR demonstrates countries' commitment to proactive, collective action on the gaps laid bare by the coronavirus tragedy. With willingness to honestly confront weaknesses, we can build a safer world for all people.

#### 2.3.1.5 World Health Organization's Resolution WHA 58.29

In 2005, SARS exposed glaring holes in the world's biosafety defenses. As the mysterious respiratory illness spread globally, infecting thousands, it was traced to lab accidents in China (World Health Organization, 2005).

With biosafety breaches imperiling us all, the World Health Assembly unanimously adopted Resolution 58.29 calling for urgent actions to fortify laboratories worldwide.

Recognizing uneven biosafety standards, the resolution urged countries to implement programs using WHO guidance for safe handling of microbial agents. Resources were pledged to help nations lacking expertise or equipment to enhance biosafety and security.

The resolution mandated an active leadership role for WHO, tasking it with regularly updating biosafety guidance and supporting regional collaboration. Demonstrating resolve to prevent future laboratory-enabled outbreaks, the resolution required WHO to report progress on these matters to the Executive Board.

The commitments encapsulated a new global consciousness on biosafety after SARS. But words required follow-through. In 2009, renewed warnings came as H1N1 influenza leaked from labs, foreshadowing the pandemic.

Fifteen years after SARS, implementation of WHA 58.29 remains a work in progress, including in the Philippines. But the resolve it sparked continues motivating nations to lock down pathogenic perils. Because nature keeps no secrets, as SARS and COVID-19 proved. Our shared safety depends on the weakest link.

## 2.3.2 Regional Agreements

In addition to global accords, regional agreements play an important role in advancing biosafety and biosecurity aims within particular geographical areas. While universal in relevance, the legally binding treaties discussed earlier depend on regional collaboration for translation into action. Neighboring states with common interests can join together to implement international obligations, pool resources, and address context-specific issues. Regional accords are an example of such cooperation.

One such regional agreement is the Wassenaar Arrangement. This voluntary framework of participating states promotes transparency and responsibility in arms transfers to avoid destabilizing accumulations of weapons or dual-use items. By adapting export controls to prevent dangerous technologies from falling into the wrong hands, the Wassenaar Arrangement provides a valuable model for regional coordination supporting global biosecurity aims.

### 2.3.2.1 The Wassenaar Arrangement

The Wassenaar Arrangement on Export Controls for Conventional Arms and Dual-Use Goods and Technologies was established in 1996 by representatives from 33 nations gathered in Wassenaar, Netherlands (The Wassenaar Arrangement & The Australia Group, 1996). Born in the post-Cold War era, it filled a void left by outdated control regimes.

Under this voluntary framework, participating states promote transparency and responsibility in arms transfers to avoid destabilizing accumulations of weapons or dual-use items. The aim is preventing acquisition by terrorists or regional arms race actors.

Initially focused on stemming post-Cold War weapons proliferation, the arrangement became vital after 9/11 for impeding biological weapons development through export controls. Today, it has 42 participating states across every inhabited continent.

While not legally binding, the Wassenaar Arrangement promotes international security through harmonizing export policies on conventional arms and dual-use technologies. As technology advances, so too must multilateral export controls evolve. The arrangement now looks ahead to new horizons like surveillance and intrusion software controls.

By adapting to prevent dangerous technologies from falling into the wrong hands, the Wassenaar Arrangement provides a valuable model for regional coordination supporting global biosecurity aims, even for non-participants like the Philippines.

### 2.3.3 International Guidance Documents

While legally binding accords create obligations for state parties, international guidance serves a complementary role in translating high-level principles into practical action. Developed through multilateral processes, these technical documents codify consensus best practices on reducing biorisks. They provide detailed recommendations to assist countries and institutions in implementing robust oversight frameworks attuned to the latest scientific advances.

Key examples include the World Health Organization's Laboratory Biosafety Manual and the US CDC's Biosafety in Microbiological and Biomedical Laboratories (BMBL). First published in 1983 and 1984 respectively, these seminal documents outline risk-based approaches, containment levels, and engineering controls for safely handling hazardous pathogens in laboratories. Their extensive technical guidance has shaped regulations, facility design, training, and daily practices for manipulating infectious agents globally. By promoting a culture of safety and security, these manuals have been pivotal to preventing laboratory accidents worldwide.

As non-binding instruments, these guides lack punitive enforcement mechanisms. However, their evidence-based nature and broad expert input lend authority, driving voluntary adoption into national policies. Regular updating allows incorporation of new knowledge and emerging challenges. Ultimately, their influence flows from the rigor of science underpinning them and collective endorsement by the international community.

#### 2.3.3.1 WHO Laboratory Biosafety Manual

The World Health Organization (WHO) Laboratory Biosafety Manual is an international guidance document that outlines risk-based approaches and best practices for safely handling and containing hazardous biological agents in laboratory settings. The manual was first published in 1983 and is now in its fourth edition, 2020 (World Health Organization, 2020).

The purpose of the Laboratory Biosafety Manual is to provide laboratories worldwide with expert guidance on assessing biological risks, implementing appropriate biosafety and biosecurity measures, and developing effective biosafety management programs. It aims to help laboratories apply biosafety controls that are practical, sustainable, and tailored to their specific risks, resources, and needs.

Some key milestones in the development of the Laboratory Biosafety Manual include:

- 1st edition (1983): Introduced basic biosafety concepts and encouraged countries to develop national biosafety codes.
- 2nd edition (1993): Incorporated more technical guidance and a microbiological risk assessment approach.
- 3rd edition (2004): Focused on a laboratory biorisk management approach using risk assessments.
- 4th edition (2020): Emphasizes biosafety culture, risk-based approaches, and feasibility in resource-limited settings.

The Laboratory Biosafety Manual has had a significant impact on biosafety and biosecurity practices globally. The manual has helped shape national regulations, institutional biosafety programs, training curricula, facility design, and operational practices for handling biological agents worldwide.

By outlining international biosafety best practices, the manual has helped laboratories improve their capabilities to handle infectious agents and toxins safely and securely. This strengthens biosecurity, enhances research on dangerous pathogens needed to protect public health, and reduces risks of accidental releases.

As the Laboratory Biosafety Manual is periodically updated, it allows national and institutional regulations and practices to evolve based on new biosafety knowledge, technologies, and approaches. The manual continues to provide guidance to help countries and laboratories control biological risks appropriately and improve preparedness against infectious disease outbreaks and bioterrorism threats.

#### 2.3.3.2 CDC Biosafety in Microbiological and Biomedical Laboratories (BMBL)

The CDC's Biosafety in Microbiological and Biomedical Laboratories (BMBL) has become one of the world's most influential guidance documents on biosafety practices. First published in 1984, the BMBL outlines a comprehensive code of practice for laboratories to follow in order to reduce risks from infectious agents and prevent laboratory-associated infections (Centers for Disease Control and Prevention & National Institutes of Health, 2022).

The origins of the BMBL can be traced back to the 1970s, when reports began to surface about scientists globally becoming infected while working with hazardous microorganisms. For example, one seminal study published in 1978 identified over 4000 cases of laboratory-acquired infections between 1930 and 1978, resulting in 168 deaths. It became clear that international guidelines were urgently needed to promote safety in laboratories and protect staff worldwide.

In response, the CDC, NIH, and United States Department of Agriculture (USDA) released early biosafety guidelines in the 1970s that introduced concepts such as risk-based containment levels. Building on these initial US documents, the CDC partnered with international scientists, laboratorians, physicians, and safety experts to create the first comprehensive BMBL code of practice in 1984.

The innovativeness of the BMBL was its inclusion of detailed 'agent summary statements' for specific high-risk pathogens like Ebola virus and *Mycobacterium tuberculosis*. These summaries outlined the risks, recommended precautions, and appropriate containment levels for working with these agents.

Since 1984, the BMBL has been updated every few years to address emerging global issues and incorporate new knowledge and experiences. For example, the latest 6th edition published in 2020 features updated guidance on biosecurity, sustainability, inactivation verification, and clinical laboratories.

The BMBL outlines four ascending biosafety levels (BSL-1 to BSL-4) that describe the containment measures necessary based on a systematic assessment of risks like transmissibility, outbreak potential, and disease severity. The BMBL also emphasizes the critical importance of risk assessments by laboratories to determine the appropriate biosafety controls.

By providing detailed, practical guidance on risk management, the BMBL has helped improve laboratory safety across the world. Countries have adapted and translated the BMBL for their own contexts. Adherence to the recommended practices in the BMBL has reduced laboratory-associated infections globally and accelerated the adoption of engineering controls, proper equipment, and a culture of rigorous biosafety. The BMBL continues to serve as the authoritative international reference that laboratories worldwide consult to ensure the safe handling of infectious materials and protection of their staff and the public.

### 2.3.4 International Standards

International standards represent another category of guidance developed through multi-stakeholder processes to align approaches on biosafety and biosecurity. Standards provide technical specifications, guidelines, or characteristics aimed at ensuring materials, products, processes, or practices meet defined objectives.

In the biosafety and biosecurity sphere, international standards set expectations and evaluate conformity on issues like pathogen risk assessments, laboratory biorisk management, biological containment, decontamination, and transport of infectious substances. Two major standards are ISO 35001 on biorisk management systems and the Canadian Biosafety Standards for facilities handling human and terrestrial animal pathogens.

Developed by accredited standards organizations using global expert input, these standards provide consistent benchmarks and technical guidance to translate complex principles into normalized practices. This facilitates safer research, technology development, and global trade related to high-consequence pathogens. By aligning stakeholders on evidence-based best practices, these standards aim to reduce biorisks and improve preparedness through voluntary adoption.

As scientific capabilities rapidly advance, international standards must continue evolving to address emerging risks and security vulnerabilities requiring collective action. ISO 35001 and the Canadian Biosafety Standards are essential references as we strive to strengthen biosafety and biosecurity worldwide.

#### 2.3.4.1 ISO 35001 Biorisk Management System

In a world of emerging infectious diseases and biosecurity threats, organizations that work with biological agents have a critical responsibility to manage the risks their work may pose. To address this need, the International Organization for Standardization (ISO) implemented the ISO 35001 Biorisk Management System standard in 2019. This milestone represented the culmination of years of effort to develop an international consensus standard for biorisk management.

The origins of ISO 35001 date back to 2011, when the European Committee for Standardization (CEN) published the CEN Workshop Agreement (CWA) 15793—the first standard providing guidance on laboratory biorisk management (CEN Workshop Agreement 15793, 2011). This pioneering document laid the foundation for further standards development. A revised CEN standard was published in 2015, alongside ISO technical requirements for medical laboratory biorisk management. By 2017, ISO established a working group of experts from across the globe to develop ISO 35001.

ISO 35001 provides a comprehensive framework for managing the risks associated with biological materials that can impact human, animal, plant, and environmental health (International Organization for Standardization, 2019). The standard outlines requirements and guidance for policies, objectives, risk assessments, controls, resources, competence, communication, and operational planning to mitigate biorisks. It addresses both laboratory biosafety and biosecurity concerns.

Since its publication, ISO 35001 has been welcomed by the biosafety community as an important step forward for biorisk management practices worldwide. The standard helps organizations align their biosafety and biosecurity activities within a single biorisk system. It brings international consistency while allowing customizable application based on an organization's specific risks. Adoption of ISO 35001 helps organizations demonstrate diligence in protecting workers, the public, and the environment from biological risks.

#### 2.3.4.2 Canadian Biosafety Standards

The landscape of biosafety and biosecurity in Canada was transformed with the release of the Canadian Biosafety Standard (CBS) in 2013. This pivotal national standard ushered in a new era of evidence- and risk-based requirements for facilities handling potentially dangerous human and animal pathogens (Public Health Agency of Canada, 2022). The CBS has become one of the most widely adopted and influential biosafety standards globally. Its graduated, tailored approach focusing on performance over prescription has been replicated across standards worldwide.

This groundbreaking standard consolidated and updated three existing Canadian biosafety standards and guidelines. It aimed to provide harmonized guidance on the design, construction and operation of facilities handling risk group 2–4 human and terrestrial animal pathogens and toxins. By specifying minimum requirements for physical containment, operational practices, and performance testing, the goal was to prevent infections, intoxications and accidental releases. In doing so, the CBS helps safeguard public and animal health, protect facility personnel, and secure sensitive biological assets.

Since its inception, the CBS has continued to evolve, moving toward even greater focus on risk- and performance-based requirements. The 2020 edition reduced prescriptiveness further by encouraging facilities to align requirements with their activities using local risk assessments. By promoting a culture of biosafety and biosecurity, and integrating safety into all aspects of working with dangerous pathogens, the CBS has been transformative for biorisk management in Canada and worldwide. Its impact will continue to be felt as facilities apply these foundational standards to safely advance research, counter emerging infectious diseases, and develop medical countermeasures.

## 2.4 NATIONAL LAWS AND REGULATIONS IN THE PHILIPPINES

While international accords and guidance provide crucial foundations, effective biosafety and biosecurity governance requires detailed implementation at the national level. It is the responsibility of individual countries to transpose global norms and recommendations into binding laws, regulations, and policies tailored to their specific contexts.

As a state party to major international agreements, the Philippines has enacted various legislative and administrative instruments to fulfill its commitments on biosafety, biosecurity, and biorisk management. For example, as part of its obligations under the WHO International Health Regulations (IHR), the Philippines developed the National Action Plan for Health Security (NAPHS) to improve surveillance and response capacity for public health emergencies. The UN Security Council 1540 matrices have also driven local efforts to secure sensitive biological agents and technology that could be misused. Additionally, the Philippines has established governance frameworks like the National Biosafety Framework to meet its commitments under the Cartagena Protocol on Biosafety and its Implementation Plan under the Convention on Biological Diversity.

Nationally, the Philippines has established relevant regulations and policies through executive orders, administrative guidelines, and standards. These include Executive Order No. 430 enacted in 1990 to regulate modern biotechnology and provide a biosafety policy framework. It was further strengthened through EO 514 in 2006, which designated national agencies to oversee biosafety implementation. The Philippine Biosafety Guidelines outline requirements for establishing institutional biosafety committees in facilities handling hazardous biological agents. The Department of Health also released the Laboratory Biosafety and Biosecurity Standards Manual in 2021, incorporating biorisk management principles into clinical and research laboratories nationwide. Some institutions have additionally taken voluntary initiatives such as forming internal oversight committees to strengthen accountability beyond basic compliance.

This section examines key Philippine laws, regulations, and policy frameworks relevant to oversight of high-consequence pathogens and dual-use research. These national measures translate general principles into enforceable controls over activities with significant biological risks.

### 2.4.1 UN 1540 Matrices

The 1540 Matrices track the Philippines' implementation of biosafety and biosecurity measures called for under UN Security Council Resolution 1540 (United Nations Security Council, 2005). The Philippines has enacted national laws prohibiting the manufacture, acquisition, possession, development, transport, transfer, or use of nuclear, chemical, and biological weapons. National legislation which regulates for the biological weapons includes:

- Republic Act 6969 of 1990, Nuclear Wastes and Hazardous Substances
- Republic Act No. 9372 of 2007, Human Security Act
- Republic Act No. 10121 of 2010, Philippine Disaster Risk Reduction and Management Act
- Republic Act No. 10168 of 2012, the Terrorism Financing Prevention and Suppression Act of 2012
- Executive Order No. 39 of 2011 by the President of the Philippines, Section 1 (Designating the Anti-Terrorism Council as the Philippines National Authority on the Chemical Weapons Convention (CWC) and other Disarmament Issues): Philippine National Authority of CWC (PNA-CWC)
- Implementing International Ship and Port Facility Security (ISPS) Code

The regulatory controls account for and secure production, use, and storage of related materials. The Philippines applies border controls, export and import licensing, and transit shipment monitoring to prevent proliferation of WMD dual-use items. National legislation which regulates for biological weapons includes:

- Republic Act 10863 of 2015, Customs Modernization and Tariff Act, Sections 200-202 (powers and functions of the Commissioner and Bureau of Customs) Bureau of Customs/Philippine National Police Control Regulations
- Republic Act No. 10697 of 2015, Act Preventing the Proliferation of WMD by Managing the Trade in Strategic Goods, the Provision of Related Services, and for Other Purposes (Strategic Trade Management Act)

Relevant to biosafety, the matrices show the Philippines has regulations on handling of toxic substances and hazardous wastes. The Department of Environment and Natural Resources (DENR) plays a critical role in overseeing biosafety by regulating the disposal of potentially infectious agents or toxins and other biohazard wastes. For biosecurity, the Philippines has criminalized WMD financing and strengthened policing of illicit trafficking. The matrices demonstrate the Philippines' national systems to control biological agents and toxins while allowing their safe and secure use for legitimate purposes. By consolidating this information, the 1540 Matrices help assess strengths and gaps in the Philippines' implementation of biosafety and biosecurity frameworks.

### 2.4.2 Executive Order 430 National Committee on Biosafety of the Philippines

In 1990, President Corazon Aquino signed Executive Order No. 430 to address concerns over potential risks of biotechnology. It established the National Committee on Biosafety of the Philippines (NCBP) (Aquino, 1990).

The NCBP was tasked with evaluating and recommending biosafety policies on genetic engineering experiments, introducing new species, and releasing genetically engineered organisms. Its goal was to minimize hazards to health, agriculture, environment, and society.

The NCBP is attached to the Department of Science and Technology (DOST). Members include the Chairman, who is the DOST Undersecretary for R&D, four scientist members specializing in biological, environmental, physical, and social sciences, two community members, and representatives from the Department of Agriculture, Department of Environment and Natural Resources, and Department of Health.

The NCBP is responsible for identifying and evaluating genetic engineering hazards, formulating and reviewing national biosafety policies, developing risk assessment protocols, holding public deliberations on proposed policies, and assisting in amending pertinent laws and regulations.

The order directed DOST to provide budget and staff support to the NCBP (Aquino, 1990). Establishing the NCBP enabled oversight to ensure safe development of biotechnology in the Philippines.

### 2.4.3 Philippine Biosafety Guidelines

Under the oversight of the NCBP, the first Philippine Biosafety Guidelines (PBG) were published in 1991 (National Committee on Biosafety of the Philippines, 1991). They were developed by a joint committee of experts from the University of the Philippines Los Baños, International Rice Research Institute, and the Department of Agriculture. The guidelines drew from existing standards in the United States, Australia, and Japan.

The NCBP, tasked with formulating policies and overseeing compliance under EO 430, reviewed and adopted the PBG as the national framework for biosafety (National Committee on Biosafety of the Philippines, 1991). The PBG provides the specific safety protocols and procedures to implement the policies.

The PBG covers laboratory research, field testing and commercial uses of genetically modified organisms, exotic species, and microbial pathogens. They aim to minimize potential risks to human, animal and plant life, and the environment from modern biotechnology applications.

The guidelines establish biosafety levels similar to international frameworks, with requirements for microbial containment, facility safeguards, waste treatment, training, emergency response, etc. Four escalating levels of physical containment are defined, from Biosafety Level 1 for low-risk work to the maximum Biosafety Level 4 for dangerous pathogens like Ebola.

In addition to establishing the overall biosafety framework policies under EO 430, the PBG provides the specific safety protocols and procedures for institutions (National Committee on Biosafety of the Philippines, 1991). Under this system, any institution engaging in genetic engineering work must also set up an Institutional Biosafety Committee, or IBC. Regulated activities require oversight and approvals from the NCBP and Institutional Biosafety Committees (IBCs). IBCs are established in institutions conducting genetic engineering to review proposals and ensure adherence to PBG protocols.

The PBG served as the key biosafety guidelines until the National Biosafety Framework was established under EO 514 in 2006 to expand the governance policies.

### 2.4.4 Executive Order 514 Establishing National Biosafety Framework

In 2006, President Gloria Macapagal-Arroyo signed Executive Order 514 to expand and enhance the national biosafety policies and decision-making framework. This established the National Biosafety Framework (NBF) for the Philippines to keep pace with advances in biotechnology and implement international obligations (Arroyo, 2006).

The NBF appears aligned with the subsequent Implementation Plan for the Cartagena Protocol on Biosafety under the Convention on Biological Diversity. The Implementation Plan, adopted in 2022, aims to guide countries in achieving the Protocol's objectives related to ensuring the safe transfer, handling, and use of living modified organisms (Conference of the Parties to the Convention on Biological Diversity serving as the Meeting of the Parties to the Cartagena Protocol on Biosafety, 2022). It emphasizes having functional national biosafety frameworks, designating competent authorities, incorporating science-based risk assessment and socioeconomic considerations in decision-making, capacity building, public participation, information exchange through biosafety clearing houses, and cooperation.

Similarly, EO 514 establishes the mandate and structure of the NCBP to coordinate and harmonize biosafety policy and standards. It makes risk assessment mandatory and allows factoring socioeconomic, ethical, and cultural aspects into decisions. EO 514 also requires public access to information and participation in policy and risk assessment processes (Arroyo, 2006). In summary, EO 514 puts in place a national biosafety framework for the Philippines that appears consistent with the country's commitments under the Cartagena Protocol and supportive of the Protocol's Implementation Plan.

### 2.4.5 DOH Laboratory Biosafety and Biosecurity Standards Manual

The Philippine biosafety framework established by EO 430, built upon by the PBG, and expanded through EO 514 demonstrates the evolution of national policies to ensure the development, and safe handling and use of genetically modified organisms. However, this framework focused primarily on containment and transport of GMOs, not on implementing a comprehensive biorisk management system for laboratories.

In 2011, the CEN Workshop Agreement (CWA) 15793 outlined a standardized approach to laboratory biorisk management. Building on this, the development of ISO 35001 on Biorisk Management Systems provided further impetus for the Department of Health to create a locally relevant guidance document aligned with international standards.

Thus, in 2017 the DOH published its first Manual of Standards on Laboratory Biosafety and Biosecurity (Health Facility Development Bureau, Department of Health, 2017). This manual prescribed minimum requirements on laboratory design, equipment, waste handling, personal protective equipment, and other topics. It took a conventional approach of basing requirements on biosafety levels and risk groups.

The recently released fourth edition of the WHO Laboratory Biosafety Manual represents a paradigm shift to a flexible, risk-based approach over rigid biosafety levels. Requirements are tailored to the specific risks presented by the agents, procedures, and laboratory environment based on risk assessments. There is also an emphasis on good microbiological practices and procedures as a foundation for safety.

Recognizing the need to adopt this paradigm, the DOH Office for Health Laboratories published a revised Manual of Standards on Laboratory Biosafety and Biosecurity in 2023 (Office for Health Laboratories, Department of Health, 2023). It emphasizes localized risk assessments and sustainable, relevant control measures rather than a rigid risk group/biosafety level structure. Implementation of this revised policy will significantly advance Philippine laboratories' capabilities to manage biological risks and respond to public health emergencies.

The revised manual represents an important step in aligning Philippine policy with the WHO International Health Regulations (IHR). Under the IHR, countries commit to developing core health security capacities. To help countries implement this, the WHO developed the National Action Plan for Health Security (NAPHS) tool which lays out priority actions for building IHR core capacities nationally. The Philippines has adopted its own NAPHS using this WHO tool. One NAPHS priority for the Philippines is developing sustainable biosafety and biosecurity systems across facilities handling biological agents. Another is expanding training to equip laboratory and health facility staff with skills proportionate to their risk levels and responsibilities. Aligning the Philippine laboratory biosafety manual with WHO guidelines helps address these WHO NAPHS priorities for strengthening IHR core capacities in the country.

The 2023 laboratory biosafety and biosecurity manual complies with the WHO NAPHS priorities in several key ways. First, it provides comprehensive guidance adopting the latest WHO recommendations on risk-based biorisk management. This strengthens the national biosecurity system as called for in the NAPHS. Second, by emphasizing localized risk assessments, the manual promotes a tailored approach scalable to diverse facilities and risk levels. This facilitates the expansion of risk-based training opportunities, another NAPHS priority. Additionally, as a common guidance document, the manual helps implement consistent biosafety and biosecurity best practices across all the country's laboratories. Finally, printing and distributing the manual to government and private laboratories ensures these standards reach all facilities handling hazardous biological agents, further aligning with the NAPHS goal of comprehensive national biosecurity. Through these provisions, the revised manual represents an impactful step toward addressing the WHO NAPHS priorities for the Philippines.

#### 2.4.5.1 DOH Administrative Order 2021-0037 on Clinical Laboratory Standards

The Department of Health recently instituted Administrative Order No. 2021-0037 to strengthen biosafety and biosecurity standards for licensed clinical laboratories in the Philippines (Department of Health, 2021). This policy aims to update regulations to better protect healthcare workers and the public.

Specifically, the Order mandates that licensed laboratories comply with biosafety and biosecurity requirements from the 2017 and 2023 DOH manuals. This includes establishing a comprehensive Biorisk Management Program overseen by qualified personnel.

Importantly, laboratory heads must designate a trained Biosafety Officer or Biorisk Officer to lead implementation. These officers should complete specialized training from accredited institutions such as the Research Institute for Tropical Medicine or the National Training Center for Biosafety and Biosecurity.

Having properly qualified personnel lead biorisk management efforts is essential for translating policies into effective safety practices. The Biosafety/Biorisk Officers will spearhead critical functions like conducting risk assessments, developing biosafety and biosecurity plans/protocols, advising on appropriate containment measures, monitoring safe practices and waste management, and fostering a culture of continuous safety improvement.

By requiring laboratory heads to assign trained officers to oversee compliance activities, Administrative Order 2021-0037 promotes rigorous adherence to biosafety and biosecurity standards. This regulatory policy has strengthened protections for healthcare workers, patients, and the public across Philippine clinical laboratories.

## 2.5 INSTITUTIONAL POLICIES AND COMMITTEES

Effective biosafety and biosecurity requires clear policies, procedures, and organizational structures at the institutional level to ensure compliance with international and local laws and regulations. Academic and research institutions that work with biological agents must establish comprehensive biorisk management programs to oversee all activities involving potentially hazardous biological material in accordance with legal requirements and standards. This involves developing institutional biosafety policies and guidelines aligned with national biosafety frameworks, forming safety and ethics review committees, designating responsible officers, and implementing training programs. With appropriate policies and committees in place as mandated by regulations, institutions can promote a culture of biosafety, provide oversight on projects and experiments, ensure transparency and accountability, and mitigate biorisks across the organization. This section examines the key institutional policies and committees needed to enable robust biorisk governance within an academic or research setting that complies with biosafety laws and regulations. Establishing these foundational governance elements is imperative for any institution conducting research and teaching activities involving biological agents in order to safeguard staff, students, the local community, and the environment in accordance with legal obligations.

### 2.5.1 Institutional Biosafety Committees

The Philippine Biosafety Guidelines and the National Biosafety Framework for the Philippines established under Executive Order No. 430 in 1990 and Executive Order 514 in 2006 required government agencies and private institutions undertaking genetic engineering work to establish their own Institutional Biosafety Committees (IBCs). The National Committee on Biosafety of the Philippines (NCBP), which serves as the national regulatory body, registers and maintains a list of active IBCs on their website.

Under this framework, companies and institutions engaging in activities involving genetically modified organisms (GMOs) must constitute an IBC prior to commencing such activities. The IBC is tasked with conducting risk assessments and developing risk management strategies for the applicant's proposed GMO activities. They are responsible for ensuring human health and environmental safety are protected.

The IBC shall have at least five members, three designated as scientist-members and two as community representatives. The scientist-members must have sufficient expertise to properly evaluate and monitor the proposed work with GMOs. The community representatives cannot be affiliated with the applicant and should represent the interests of affected communities. One community representative shall be a local government official. The other shall be selected from residents who are members of civil society organizations. For multi-location field trials, community representatives are designated per site.

The IBC is required to conduct an initial risk assessment using the best available science and identify potential hazards and their mitigation. They must review staff qualifications and ensure competence, good practices, and supervision. The committee must inform and consult the community about planned field testing. IBCs are responsible for submitting project documents for regulatory approval, ensuring compliance with regulations, verifying necessary permits are obtained, developing contingency plans, continuously monitoring activities, conducting facility inspections, limiting access to authorized personnel only, reporting any accidents or illnesses, notifying regulators of any breaches, maintaining records, and submitting activity reports and annual reports.

We highlight that the University of the Philippines Manila (UPM) established its own Institutional Biosafety and Biosecurity Committee (IBBC) as a voluntary initiative to oversee both biosafety and biosecurity in research, teaching, and clinical laboratories. The IBBC was created through a voluntary effort by UPM administration and functions independently from the national regulatory body (University of the Philippines Manila Institutional Biosafety and Biosecurity Committee, 2017).

The IBBC provides UPM-specific policies, standards and procedures for working with potentially hazardous biological materials. It covers risk assessment, classification of laboratories and research projects, commissioning of facilities, waste management, transport of infectious substances, emergency response, and training.

A key unique aspect is the inclusion of biosecurity along with biosafety. The IBBC pays particular attention to dual-use research of concern and potential pandemic pathogens. It is tasked with ensuring proper oversight, control, and accountability for biological materials to prevent their loss, theft, misuse, or intentional unauthorized release. Researchers have a responsibility to protect valuable biological materials and document their storage, use, transfer, or destruction.

For dual-use research of concern, the IBBC guidelines provide oversight for studies involving dangerous pathogens or toxins that could be misapplied to threaten public health or national security. Risk mitigation strategies are implemented.

For potential pandemic pathogens, the IBBC oversees research with new or emerging infectious agents that may cause a future epidemic. Special containment, biosafety, and biosecurity measures are applied when working with these high-risk materials.

The IBBC issued a manual that serves as the main reference for end-users in UPM laboratories regarding requirements, practices, and responsibilities for biosafety and biosecurity, including dual-use and pandemic pathogen research oversight (University of the Philippines Manila Institutional Biosafety and Biosecurity Committee [UPM IBBC], 2017). It demonstrates UPM's commitment to promote safety, security, and responsible conduct of research involving biological agents and toxins.

### 2.5.2 Biorisk Management Committee Example

The Institute of Human Genetics (IHG) at the University of the Philippines Manila provides an excellent example of a healthcare institution proactively developing its biosafety, biosecurity, and biorisk management capacities.

In 2015, the IHG took the initiative to set up the first Institute-based Biorisk Management Committee. The committee was composed of representatives from different IHG laboratories and offices who underwent comprehensive training on biorisk management principles and practices (Institute of Human Genetics, 2015).

The committee developed a tailored biorisk management manual for the Institute aligned with international standards like CWA 15793. This manual serves as the basis for all IHG policies and procedures related to biosafety, biosecurity, and biorisk management.

In the same year, the IHG conducted a 13-day biorisk management training for personnel, funded by the US Biosecurity Engagement Program through Civilian Research and Development Foundation (CRDF) Global. Graduates of the Advanced Biorisk Officers Training facilitated the sessions to build local training capacity.

The Institute also held a series of biosafety and biosecurity awareness seminars in 2015 for all staff. This introduced employees to biorisk concepts and helped them recognize related hazards in their workplace.

The IHG exemplifies a voluntary, proactive approach to implementing biorisk management. Developing in-house expertise through the Biorisk Committee and trained Biorisk Officers was a key strategy. This enhances the Institute's ability to identify, assess, and mitigate biological risks—protecting staff, patients, and the community.

Such voluntary initiatives to build biosafety and biosecurity capabilities demonstrate an organization's commitment to health security. Under the DOH's clinical laboratory licensing framework, healthcare institutions have an opportunity to integrate and align voluntary efforts with mandatory requirements. This creates a robust national system for biorisk management.

## 2.6 CERTIFIED BIORISK OFFICER (CBO)

For a biorisk management program to be successfully implemented within an institution, having competent biosafety or biorisk officers is essential. These qualified professionals lead and oversee the rollout of safety practices, protocols, training, and initiatives to mitigate biological risks. As such, developing specialized training programs to build a strong talent pool of certified biorisk officers has become a priority in the Philippines. The National Training Center for Biosafety and Biosecurity (NTCBB) serves as the pioneering institute to fulfill this important capacity-building need nationwide.

The National Training Center for Biosafety and Biosecurity (NTCBB) serves as the groundbreaking institute for biorisk management education and training in the Philippines. When it was established under the University of the Philippines Manila's National Institutes of Health in 2018, the NTCBB was formally mandated by the UP Board of Regents to serve as the country's primary hub for biosafety and biosecurity expertise (National Training Center for Biosafety and Biosecurity, 2018).

The NTCBB's vision is to be the leading national source for information, training, and collaboration on biosafety and biosecurity principles and protocols. It is tasked with developing and promoting global best practices in biorisk management across Philippine laboratories and institutions that handle hazardous biological agents.

To fulfill this vision, the NTCBB offers targeted training programs. Its flagship course is the Philippines Advanced Biorisk Officer Training (PhABOT), which has proven pivotal for raising biosafety and biosecurity awareness across diverse scientific institutions nationwide. By certifying Biorisk Officers, PhABOT has invigorated biorisk management implementations in academe, clinical laboratories, agriculture, research centers, and more. The officers trained through PhABOT have spearheaded impactful accomplishments including implementing biorisk systems, delivering additional training, conducting research, and shaping policies.

In essence, PhABOT has created a ripple effect by equipping Biorisk Officers with the knowledge and credentials to drive impactful changes locally and nationwide. The officers carry forth the program's capacity building mandates through their leadership roles, and biosafety/biosecurity enhancement initiatives. They help fulfill the NTCBB's vision of ingraining global best practices in biorisk management across Philippine scientific communities.

When the COVID-19 pandemic emerged, the NTCBB rapidly mobilized to offer the Online Biosafety Training Against COVID-19 (BEAT COVID-19) to equip laboratory staff handling coronavirus specimens. This agile response was crucial for disseminating biosafety fundamentals during a public health crisis. The NTCBB also collaborated with the Department of Health to provide Basic Biosafety Officer Training, which helps clinical laboratory staff meet evolving regulatory requirements such as the AO 2021-007.

By responding to urgent needs while continuing flagship programs, the NTCBB fulfills its mandate of ingraining global best practices to protect laboratories, workers, and the public. As the pioneering institute for biorisk management in the Philippines, the NTCBB will continue serving an integral role in advancing nationwide capacity.

## 2.7 CONCLUDING REMARKS AND FUTURE DIRECTIONS

The Philippines has made significant strides in developing a governance framework for biosafety and biosecurity, guided by binding international accords and voluntary guidance from bodies like the WHO, ISO, and BWC. Domestically, measures like Executive Orders 430 and 514 established oversight foundations, while instruments such as the DOH laboratory standards manual, biosafety guidelines for genetic engineering, and healthcare regulations mandating biorisk programs continue strengthening the scaffolding.

However, work remains to fully erect the structure. Aspects like unified legislation focused exclusively on biosafety and biosecurity, along with enhanced implementation capacities and inter-agency coordination, demand reinforcement. Competing priorities and resource constraints impede complete actualization of sophisticated global frameworks nationally. But the sincere efforts thus far, despite limitations, signal promise.

With biological risks inexorably rising, the need for robust, coherent governance is urgent. The Philippines must summon the resolve to address persistent gaps and inconsistencies through consolidated legislation, capacities, and funding. The country's demonstrated willingness to engage in strengthening legal scaffolds, though imperfect, can serve as a foundation. The essential elements instituted to date must be built upon.

Importantly, the absence of a dedicated biosafety and biosecurity law does not excuse laboratories and other entities from implementing safety and security measures with biological agents. All those working with hazardous biological materials have an ethical and moral obligation to follow biosafety and biosecurity best practices, whether mandated by law or not. Preventing harm to human health and the environment must be the paramount consideration.

Sustained progress relies on investments and political commitment to implement policies fully. But the Philippines' earnest commitment to biosafety and biosecurity thus far, guided by binding accords like the BWC and WHO IHR, illustrates understanding of the grave threats posed by biological agents and toxins. We must collectively avert the devastating consequences of misuse and accidental releases. With cooperation, resources, and vigilance, the country can construct an enduring governance framework that secures society from biorisks, known and unknown. The stakes are too high to settle for half-measures.

However, some gaps and limitations persist that provide opportunities for further strengthening the biosafety governance framework in the Philippines (Destura et al., 2021). While the policy framework has advanced, operationalizing regulations fully and consistently across the country remains challenging. Regulators need expanded resources and training on cutting-edge biotech oversight. Fragmented governance and inadequate coordination among agencies hamper coherent nationwide implementation of biosafety measures. Consolidated legislation could strengthen this. Balancing precautionary regulation while enabling innovation is an ongoing challenge. Guidelines must remain adaptive and science-based. Monitoring industry compliance and laboratory adherence to guidelines across dispersed sites is difficult with limited inspectors, raising biosafety risks. Administrative hurdles in proposal reviews and permit processing should be streamlined so they do not stifle beneficial research and development. More initiatives to increase public awareness and participation in policy processes could help address ethical concerns and build confidence in the biosafety system. Regional collaboration and information exchange would allow the Philippines to leverage experiences and expertise from neighboring countries to overcome common challenges.

Importantly, by adopting the latest international guidance and standards on biorisk management, the Philippines is better prepared to detect and respond to emerging and re-emerging infectious diseases. Implementing robust, flexible and sustainable biosafety systems based on WHO recommendations improves pandemic preparedness and health security.

## 2.8 ACKNOWLEDGMENTS

I gratefully acknowledge the many individuals, institutions,and organizations whose published works have contributed valuable knowledge and insights to this book chapter through their research, analysis, and guidance documents on biosafety, biosecurity, and biorisk management.

I would like to express my sincere gratitude to the following individuals for their important contributions: Broughton, Destura, Marks, Roos, Schnirring, and Tuzmukhamedov. Their work has significantly enriched the knowledge base in this field.

I also extend my appreciation to the following institutions and organizations: the Centers for Disease Control and Prevention, the Conference of the Parties to the Cartagena Protocol on Biosafety, the Department of Health (Philippines), the Health Facility Development Bureau (Philippines), the Institute of Human Genetics (Philippines), the International Organization for Standardization, the National Committee on Biosafety of the Philippines, the National Institutes of Health, the Office for Health Laboratories (Philippines), the Office of the President of the Philippines, the Public Health Agency of Canada, the Secretariat of the Convention on Biological Diversity, the United Nations Security Council, the University of the Philippines Manila Institutional Biosafety and Biosecurity Committee, and the World

Health Organization. These entities have provided invaluable resources,guidance, and support in the field of biosafety, biosecurity, and biorisk management.

I am grateful for the collective efforts of these individuals, institutions, and organizations to elucidate the scientific, technical, legal, and policy aspects of biosafety and biosecurity through research, analysis, and normative guidance. Their work provides an essential foundation for ongoing efforts to strengthen biorisk management globally and in the Philippines.

## REFERENCES

Aquino, C. C. (1990, October 15). Constituting the National Committee on Biosafety of the Philippines (NCBP) and for other purposes (Executive Order No. 430). Manila, Philippines: Office of the President of the Philippines. Retrieved from https://www.officialgazette.gov.ph/1990/10/15/executive-order-no-430-s-1990/

Arroyo, G. M. (2006, March 17). Establishing the national biosafety framework, prescribing guidelines for its implementation, strengthening the National Committee on Biosafety of the Philippines, and for other purposes (Executive Order No. 514). Manila: Office of the President of the Philippines. Retrieved from https://www.officialgazette.gov.ph/2006/03/17/executive-order-no-514-s-2006/

Broughton, E. (2005). The Bhopal Disaster and its aftermath: A review. *Environmental Health*, 4(1), 1–6.

CEN Workshop Agreement 15793:2011. (2011). Laboratory biorisk management (CWA 15793:2011). Retrieved from https://internationalbiosafety.org/wp-content/uploads/2019/08/CWA-15793-English.pdf

Centers for Disease Control and Prevention & National Institutes of Health. (2022). *Biosafety in Microbiological and Biomedical Laboratories* (6th ed.). Retrieved from https://www.cdc.gov/labs/BMBL.html

Conference of the Parties to the Convention on Biological Diversity serving as the Meeting of the Parties to the Cartagena Protocol on Biosafety. (2022). Implementation plan for the Cartagena Protocol on Biosafety (Decision CP-10/3). Secretariat of the Convention on Biological Diversity. Retrieved from https://bch.cbd.int/protocol/post2020/Plan.shtml#:~:text=The%20Implementation%20Plan%20is%20a,objectives%20and%20indicators%20are%20outlined

Department of Health (Philippines). (2021). New rules and regulations governing clinical laboratories (Administrative Order No. 2021-0037). Retrieved from https://hfsrb.doh.gov.ph/clinical-laboratory/

Destura, R. V., Lam, H. Y., Navarro, R. C., Lopez, J. C. F., Sales, R. K. P., Gomez, M. I. F. A., … Ulanday, G. E. (2021). Assessment of the biosafety and biosecurity landscape in the Philippines and the development of the national biorisk management framework. *Applied Biosafety*, 26(4), 232–244.

Health Facility Development Bureau, Department of Health (Philippines). (2017). Manual of Standards on Laboratory Biosafety and Biosecurity.

International Organization for Standardization (ISO). (2019). Biorisk management for laboratories and other related organisations (ISO 35001:2019). Retrieved from https://www.iso.org/standard/71270.html

Institute of Human Genetics. (2015). IHG launches the first Institute-Based Biorisk Management. Retrieved from https://ihg-old.upm.edu.ph/node/138

Marks, J. D. (2011). Forensic aspects of biological toxins. In B. Budowle, S.E. Schutzer, R.G. Breeze, P.S. Keim, & S.A. Morse (Eds.), *Microbial Forensics* (2nd ed., pp. 327–353). Academic Press. https://doi.org/10.1016/B978-0-12-382006-8.00020-7

National Committee on Biosafety of the Philippines. (1991). *Philippine Biosafety Guidelines (PBG)* (1st ed.). Department of Science and Technology. Retrieved from https://dost-bc.dost.gov.ph/download/guidelines/category/13-philippine-biosafety-guidelines-series-of-1991

National Training Center for Biosafety and Biosecurity. (2018). About the NTCBB. University of the Philippines Manila. Retrieved from https://nih.upm.edu.ph/institute/national-training-center-biosafety-and-biosecurity

Office for Health Laboratories (OHL), Department of Health (Philippines). (2023). Laboratory biosafety and biosecurity standards manual. Retrieved from https://sites.google.com/view/doh-hfdb/2023-updates/dc-2023-0454

Public Health Agency of Canada. (2022). *Canadian Biosafety Standard* (3rd ed.). Ottawa, ON: Government of Canada. Retrieved from https://www.canada.ca/en/public-health/services/canadian-biosafety-standards-guidelines/third-edition.html

Roos, R. & Schnirring, L. (2011, September 1). Public health leaders cite lessons of 2001 anthrax attacks. Center for Infectious Disease Research and policy (CIDRAP). Retrieved from https://www.cidrap.umn.edu/anthrax/public-health-leaders-cite-lessons-2001-anthrax-attacks

Secretariat of the Convention on Biological Diversity. (2000). *Cartagena protocol on biosafety to the convention on biological diversity: Text and annexes.* Montreal: Secretariat of the Convention on Biological Diversity.

Secretariat of the Convention on Biological Diversity. (2011). *Convention on biological diversity: Text and annexes.* Secretariat of the Convention on Biological Diversity. https://www.cbd.int/doc/legal/cbd-en.pdf

The Wassenaar Arrangement & The Australia Group. (1996). Retrieved from https://www.hsgac.senate.gov/wp-content/uploads/imo/media/doc/reinsch.pdf

Tuzmukhamedov, B. (2021). *Convention on the prohibition of the development, production and stockpiling of bacteriological (biological) and toxin weapons and on their destruction.* United Nations Audiovisual Library of International Law. Retrieved from https://legal.un.org/avl/pdf/ha/cpdpsbbtwd/cpdpsbbtwd_e.pdf

United Nations Security Council. (2004). Resolution 1540 (2004). Retrieved from https://undocs.org/S/RES/1540(2004)

United Nations Security Council. (2005). 1540 matrices. Retrieved from https://www.un.org/en/sc/1540/national-implementation/1540-matrices.shtml

University of the Philippines Manila Institutional Biosafety and Biosecurity Committee (UPM IBBC) Guidelines on Biosafety and Biosecurity Committee. (2017). *Laboratory on Biosafety and Biosecurity Guidelines* (1st ed.).

World Health Organization. (2005). Enhancement of laboratory biosafety (WHA58.29). Retrieved from https://apps.who.int/gb/ebwha/pdf_files/WHA58-REC1/english/A58_2005_REC1-en.pdf

World Health Organization. (2016). *International Health Regulations (2005)* (3rd ed.). Retrieved from https://www.who.int/publications/i/item/9789241580496

World Health Organization. (2020). *Laboratory Biosafety Manual* (4th ed.). Geneva: World Health Organization. Retrieved from https://www.who.int/publications/i/item/9789240011311

# 3 Principles of Biosafety in Resource Limited Settings

*John Mark Velasco*
Institute of Molecular Biology
National Institutes of Health and Department of Clinical Epidemiology
College of Medicine
University of the Philippines
Manila, Philippines

## 3.1 INTRODUCTION

### 3.1.1 Importance of Biosafety in Low-Middle Income Country Settings (LMICs)

Biosafety is the set of measures and practices that are designed to prevent or minimize the potential risks associated with the handling and use of biological agents, which can include everything from bacteria and viruses, genetically modified organisms, toxins, and hazardous biological waste. In developing countries, where healthcare infrastructure may be limited and access to medical care is often restricted, the risks associated with infectious diseases are even greater. Furthermore, the use of biological agents in research and industry can pose a threat to the environment, wildlife, and agricultural production. The importance of biosafety is further emphasized by the higher risk of accidental release of biological agents in developing countries where resources for biosafety may be limited and regulatory oversight may be weaker. Accidental release can occur during research, manufacturing, and transport of biological agents, and can result in the spread of infectious diseases or the release of genetically modified organisms into the environment.

This chapter provides an overview of the basic principles of biosafety as applied to a Low- and Middle-Income Country (LMIC) setting. It explores the unique challenges faced by LMICs in implementing biosafety practices and offers practical guidance for establishing and maintaining effective biosafety programs. LMICs face a number of public health challenges, including infectious diseases, environmental pollution, and natural disasters. We will provide an overview of the principles of biosafety, including the hierarchy of controls, risk assessment, containment, and measures, which can be used to mitigate these risks with a specific focus on the challenges and opportunities associated with implementing these principles in a developing country context. In LMICs, biosafety is an especially

DOI: 10.1201/9781003426219-3

important concern, as the potential risks associated with biological materials can be more significant due to factors such as limited resources, inadequate infrastructure, and lack of training and education. We will examine the importance of biosafety, the risks associated with biological agents, and the measures that can be taken to minimize those risks. We will also discuss the core principles of biosafety, including risk assessment, containment, and management, as well as the importance of training and education for all those who handle or work with biological materials.

### 3.1.2 Chapter Objectives

By focusing on risk assessment, risk management, laboratory infrastructure, training, and regulatory frameworks, this chapter aims to (1) provide a comprehensive overview of the basic principles of biosafety in the context of LMICs; (2) highlight the unique challenges faced by LMICs in implementing biosafety practices and the importance of addressing these challenges; (3) equip healthcare workers with the knowledge and tools necessary to ensure the safe handling of biological agents and protect public health; (4) explain the fundamental concepts and terminology of biosafety to readers; (5) explore the process of risk assessment and guide readers in identifying and evaluating biological hazards specific to their local context; (6) offer practical guidance on risk management strategies tailored to the limited resources and infrastructure typically found in LMICs; (7) emphasize the significance of biosafety training and capacity building programs; and (8) discuss future directions and recommendations for improving biosafety practices, including integration with public health systems.

### 3.1.3 Basic Concepts and Terminology in Biosafety

*Administrative Controls:* Administrative controls refer to policies, procedures, and protocols implemented to minimize risks in a biosafety setting. They include practices such as standard operating procedures (SOPs), training programs, access control, and waste management procedures.

*Biological restricted area:* Area where access to Biological Select Agent and Toxins (BSAT) is possible. Entry will be subject to special access restrictions. Physical security controls will be used to control access and secure property and materials. Biological Restricted Areas may be of different types depending on the nature and varying degree of access to BSAT, or other relevant matter contained in the area. An organization may extend this definition to include any area where access to Valuable Biological Material (VBM) is possible.

*Biological Select Agents and Toxins (BSAT):* Biological agents and toxins that present a high bioterrorism risk to national security and have the greatest potential for adverse public health impact with mass casualties of humans and/or animals or that pose a severe threat to plant health or to plant products.

*Biological Materials of Concern (BMC):* Any material composed of, containing, or that may contain biological agents and/or their harmful products, such as toxins and allergens. Biological materials may be blood, secretions, or tissues of human or animal origin. Other biological materials include debris or organic material from nature, culture, or preservation media, and/or cell cultures from human, animal, and plants.

*Biosafety:* Biosafety refers to the set of principles, practices, and measures implemented to prevent or minimize risks associated with the handling, storage, transport, and disposal of biological agents. It encompasses strategies to protect individuals, communities, and the environment from the potential harmful effects of biological hazards.

*Biosafety Level 2 (BSL-2):* Laboratory standard practices, safety equipment, and facility specifications are applicable to laboratories in which work is performed using a broad-spectrum of biological agents and toxins that are associated with causing disease in humans of varying severity. With good practices and procedures, these agents and toxins can generally be handled safely on an open bench, provided the potential for producing splashes and aerosols is low. Work done with any human, animal, or plant-derived specimens (e.g., blood, body fluids, tissues, or primary cell lines), where the presence of a biological agent or toxin may be unknown, can often be safely conducted under conditions typically associated with BSL-2.

*Biosafety Level 3 (BSL-3):* Laboratory standard practices, safety equipment, and facility specifications are applicable to laboratories in which work is performed using indigenous or exotic biological agents with a potential for respiratory transmission and those that may cause serious and potentially lethal infection.

*Biosafety Level 4 (BSL-4):* Laboratory standard practices, safety equipment, and facility specifications are applicable primarily for laboratories working with dangerous and exotic biological agents that pose a high individual risk of life-threatening disease that may be transmitted via the aerosol route and for which there is no available vaccine or therapy. Agents with a close or identical antigenic relationship to agents requiring BSL-4 containment must be handled at this level until sufficient data are obtained either to confirm continued work at this level or to re-designate the level.

*Biosafety Regulations:* Biosafety regulations are legal frameworks established by governments to ensure the safe handling of biological agents and the protection of public health and the environment. These regulations define the requirements for facilities, equipment, training, waste management, and compliance monitoring.

*Biosafety Training and Education:* Biosafety training and education programs provide individuals with the knowledge, skills, and awareness necessary to work safely with biological agents. Training covers topics such as risk assessment, use of personal protective equipment (PPE), laboratory practices, emergency response, and waste management.

*Engineering Controls:* Engineering controls are physical measures designed to reduce or eliminate exposure to biological agents or biological hazards. Examples include biological safety cabinets, ventilation systems, containment devices, and barrier systems.

*Hazard*: A danger or source of danger; the potential to cause harm.

*Hierarchy of Controls:* The hierarchy of controls is a framework that guides the implementation of risk management strategies. It includes five levels: elimination, substitution, engineering controls, administrative controls, and personal protective equipment (PPE).

*Laboratory-Acquired Infections (LAIs)*: Laboratory personnel working with biological hazards are at risk of contracting infections due to accidental exposure. For example, these can occur through needle stick injuries, cuts, splashes, inhalation of aerosols, or ingestion. Inadequate personal protective equipment (PPE), poor laboratory practices, or breaches in containment protocols can increase the likelihood of laboratory-acquired infections.

*Laboratory Biosafety Levels (BSLs):* Laboratory biosafety levels categorize laboratories based on the specific risks associated and appropriate to the work being conducted. BSLs range from BSL-1 (low risk) to BSL-4 (highest risk), with each level having specific requirements for facility design, practices, and containment measures.

*Personal Protective Equipment (PPE):* PPE comprises specialized garments, equipment, and devices worn to protect individuals from potential exposure to biological hazards. Examples of PPE include gloves, masks, goggles, face shields, gowns, and respirators.

*Risk Assessment:* Process of evaluating the risk(s) arising from hazards, taking into account the adequacy of any existing controls and deciding whether the risks are acceptable or not. It is the systematic process of identifying and evaluating potential hazards and associated risks posed by biological agents and involves assessing the likelihood of exposure, potential consequences, and the vulnerability of individuals and the environment.

*Risk Communication:* Risk communication is the process of sharing information about the potential risks associated with the use of biotechnology. Effective risk communication involves the sharing of information about the potential risks associated with work involving biological agents in a way that is clear and easy to understand. This information should be communicated to stakeholders, including regulators, scientists, farmers, and the public.

*Risk Management:* Risk management refers to the implementation of strategies and measures to mitigate or control identified risks. This process involves identification of the hazards and risks, evaluating the risks, implementing a risk mitigation plan or establishing protocols to minimize or eliminate potential hazards, as needed and evaluating effectiveness of controls.

*Valuable Biological Materials (VBM):* Biological materials that require (according to their owners, users, custodians, caretakers or regulators) administrative oversight, control, accountability, and specific protective and monitoring measures in laboratories to protect their economic and historical (archival) value, and/or the population from their potential to cause harm. VBM may include pathogens and toxins, as well as non-pathogenic organisms, vaccine strains, foods, genetically modified organisms (GMOs), cell components, genetic elements, and extraterrestrial samples.

*Waste Management:* Waste management in biosafety refers to the proper handling, treatment, and disposal of waste materials generated in biosafety facilities, such as laboratories, research institutions, and healthcare settings, where biological materials are handled. It aims to prevent the release of hazardous biological agents into the environment. Implementation of proper and effective waste management practices is crucial to ensure the safety of personnel, mitigate environmental impact, and prevent the spread of infectious diseases.

## 3.2 RISK ASSESSMENT

### 3.2.1 Overview of Risk Assessment Process

The Assessment, Mitigation, and Performance (AMP) model of biorisk management is a framework widely used in biosafety and biosecurity. This model was first articulated by the World Health Organization, Advanced Trainer Program, developed and executed in 2010. It is a systematic approach to manage risks associated with biological agents, in order to protect and secure human health and the environment. The AMP model provides a structured approach to assess, mitigate, and prepare for potential biorisks. It helps ensure the effective management of biological agents, promotes biosafety and biosecurity, and minimizes the risks associated with the handling, storage, transportation, and disposal of such agents.

Assessment: This phase involves conducting a comprehensive assessment of potential biorisks. It includes identifying and characterizing biological agents, evaluating the associated potential hazards and risks, and understanding the vulnerabilities and exposure pathways. The assessment may involve risk assessments, vulnerability assessments, and threat assessments to determine the likelihood and consequences of potential biological risks.

Mitigation: In this phase, appropriate risk mitigation measures are implemented to reduce or eliminate identified biorisks. This phase focuses on the development and implementation of risk management strategies and controls and includes the implementation of engineering controls, administrative controls, and the use of personal protective equipment (PPE). Mitigation efforts may also involve biocontainment measures and waste management protocols.

Performance: This phase focuses on building capacity, resilience, and preparedness to effectively respond to biological risks or hazards. This includes developing emergency response plans, establishing protocols for incident reporting and investigation, and conducting drills and exercises. This also emphasizes training and education programs for personnel involved in biorisk management, raising awareness among stakeholders, and fostering an institutional culture of safety.

### 3.2.2 Risks Associated with Biological Hazards and Agents

Biological agents pose a number of risks to public health and the environment. These risks can be classified into three categories: infectious diseases, genetically modified organisms, and recombinant DNA. Infectious disease agents can cause a wide range of infectious diseases, from the common cold to more severe diseases like tuberculosis, HIV/AIDS, and Ebola. One should take into careful consideration various properties of infectious agents such as virulence, severity, case fatality rate, transmission routes, infectiveness or transmission potential, infectious dose, and availability of effective treatment or vaccines for a particular pathogen. In developing countries, where healthcare infrastructure and health resources may be limited and access to medical care is often restricted, the risks associated with infectious diseases are even greater. Genetically modified organisms (GMOs) are organisms whose genetic material has been altered using genetic engineering techniques. GMOs are used in a wide range of applications, including agricultural production, pharmaceutical manufacturing, and research. However, the use of GMOs can pose a risk to the environment and biodiversity, as well as to public health. Recombinant DNA technology involves the manipulation of genetic material to create new genetic combinations. This technology is used in a wide range of applications, including pharmaceutical manufacturing, research, and genetic engineering which can in turn be weaponized by modifying the properties of pathogens to make them more virulent or more transmissible (e.g., Dual-Use Research of Concern). The use of recombinant DNA can pose a risk to public health and the environment as the new genetic combinations may have unpredictable effects.

### 3.2.3 Assessing Risks and Vulnerabilities

Risk assessment is the process of identifying, evaluating, and prioritizing potential risks associated with biological agents. This process involves identifying the hazards associated with the agent, assessing the likelihood and consequences of exposure, determining the level of risk, and implementing strategies to minimize and mitigate the potential risks associated with work involving biological agents. In LMICs, risk assessment is particularly important because of the limited resources available for monitoring and surveillance. For instance, the lack of infrastructure and expertise in the field of biotechnology can make it difficult to identify potential risks and develop effective risk management strategies. Risk management strategies can include but are not limited to the development of best practices for work involving biological agents, the use of biotechnology, the establishment of biosafety training programs, and the development of emergency response plans in the event of a biosafety incident.

### 3.2.4 Prioritizing Risks and Allocating Resources

Strategies to prioritize risks and resource allocation for biosafety programs can include the following methods:

a) Conduct a Comprehensive Risk Assessment: Start by conducting a comprehensive risk assessment to identify and evaluate the potential hazards and associated risks within the LMIC's biosafety context. Assess the likelihood of occurrence and potential consequences of different risks to prioritize them effectively.

b) Engage Stakeholders: Involve key stakeholders such as government agencies, regulatory bodies, research institutions, institutional managers/ supervisors, leadership and relevant experts in the prioritization process. Seek their input and insights on risk perception, local priorities, and resource availability.
c) Use Risk Scoring and Ranking Systems: Develop a risk scoring and ranking system to objectively assess and prioritize identified risks. Assign scores based on the likelihood and severity of risks, considering factors such as potential impact on individual and public health, the environment, and socioeconomic factors. There are many biorisk assessment methods including manual as well as free software, which can assist in creating the risk scoring and ranking systems. Some examples of risk scoring and ranking systems include the Likelihood and Consequence of Risk Model, Biosafety and Biosecurity Risk Assessment Model (BioRAM) developed by Sandia National Laboratories, the 5x5 Risk Matrix, and the Deliberate Risk Assessment Matrix being used by the US military (Murnyak et al., 2003).
d) Consider Local Epidemiology and Public Health: Take into account the local epidemiological data, the diseases that are endemic or prevalent, and the public health priorities to guide risk prioritization. Focus on risks that have a significant impact on the local population and align this with national health strategies. One can use as a basis for prioritization, the top ten infectious diseases causing significant morbidity and/or mortality in the country. High-consequence diseases or pathogens, which have the potential to cause local outbreaks or pandemics and especially those associated with a high case fatality rate or those, which are very contagious, should be prioritized.
e) Evaluate Potential Consequences and Impacts: Assess the potential consequences and impacts of each risk, including the potential for morbidity, mortality, economic and health burden, and ecological damage. Prioritize risks that have higher potential consequences to human and animal health and the environment.
f) Evaluate Risk Management Feasibility: Consider the feasibility and effectiveness of risk management measures for each identified risk. Assess the available resources, infrastructure, and technical capacity required to address each risk effectively.
g) Conduct Cost-Benefit Analysis: Perform a cost-benefit analysis to evaluate the potential costs and benefits of implementing risk management strategies for each identified risk. Prioritize risks where the benefits of mitigation outweigh the costs, considering long-term impact and sustainability.
h) Review International and Local Guidelines and Best Practices: Refer to international and local guidelines, standards, and best practices in biosafety to inform risk prioritization and resource allocation decisions. Adapt and contextualize these guidelines to the specific needs and resources of the LMIC.
i) Seek External Support and Funding: Explore opportunities for external support and funding through partnerships, grants, and donor organizations

to augment available resources for high-priority risks. Collaborate with international organizations and research institutions to leverage expertise and resources.

j) Monitor and Reevaluate: The risk assessment and mitigation process is continuous and one should regularly monitor and reevaluate the risk prioritization and resource allocation decisions. Periodically review the effectiveness of implemented risk management strategies and reallocate resources based on emerging risks and changing priorities.

By employing a combination of these strategies, biosafety programs in LMICs can prioritize risks effectively and ensure optimal allocation and usage of limited resources in a manner that maximizes the protection of human health, the environment, and public safety.

## 3.3 RISK MANAGEMENT STRATEGIES: HIERARCHY OF CONTROLS

The hierarchy of controls is a widely accepted framework used in biosafety to prioritize and implement control measures to minimize risks associated with biological agents. It provides a systematic approach to risk management by identifying and implementing control measures in a specific order of preference, starting with the most effective and protective controls. The hierarchy consists of five levels, each building upon the previous one, namely: (1) Elimination, (2) Substitution, (3) Engineering Controls, (4) Administrative Controls, and lastly, (5) Personal Protective Equipment.

Elimination: The highest level of control is elimination, which involves completely removing the hazard or the need for exposure to it. In biosafety, elimination can be achieved by substituting hazardous biological agents or processes with safer alternatives, such as using non-pathogenic strains or non-biological methods. Examples include the use of engineered non-replicative and non-infective pseudoviruses in neutralization experiments in lieu of live and highly pathogenic viruses (e.g., SARS-CoV-2, Ebola) (Cao et al., 2021).

Substitution: If elimination is not feasible, the next level is substitution, where the hazardous biological agent or process is replaced with a less hazardous one. Substitution can involve using less virulent strains, attenuated variants, or alternative materials or procedures that pose a lower risk. An example is the use of *Bacillus anthracis* surrogates, which can be phylogenetic relatives of *B. anthracis* (e.g., *B. cereus*, *B. thuringiensis*) in experiments, which require lower biosafety levels to work with instead of using the select agent *B. anthracis* (Greenberg et al., 2010), which requires biosafety level 3.

Engineering Controls: Engineering controls are physical or mechanical measures designed to isolate or minimize exposure to biological agents. These controls include the design, installation, and maintenance of biosafety facilities, ventilation systems, biological safety cabinets, and containment equipment. Engineering controls aim to prevent the release or spread of biological agents and provide a barrier between personnel and the hazards. During the initial phase of the SARS-CoV-2 pandemic, when the pathogen was still relatively unknown, it was generally recommended to conduct work with the virus in biosafety level 3 laboratories with negative pressure rooms. In resource-limited settings, this would have required significant creation or modification of infrastructure and high capital outlay to construct negative pressure laboratories at short notice. Instead of investing in the installation of costly centralized negative pressure ventilation and air conditioning systems with HEPA filtration, some laboratories used Class II type B2 total exhaust biosafety cabinets, which protected the laboratory worker and at the same time provided the required negative pressure during SARS-CoV-2 specimen processing.

Administrative Controls: Administrative controls are organizational and administrative measures that establish policies, procedures, and safe work practices to reduce exposure to biological agents. They include the development and implementation of standard operating procedures (SOPs), training programs, access control, work authorization procedures, and incident reporting systems. Administrative controls help to ensure that safe practices are followed, personnel are properly trained, and biosafety protocols are consistently enforced.

Personal Protective Equipment (PPE): PPE is the last line of defense in the hierarchy of controls and provides protection for individuals against exposure to biological hazards. It includes items such as gloves, masks, gowns, respirators, and eye protection. PPE should be properly selected, used, maintained, and disposed of, and personnel should receive appropriate training on its correct usage such as donning and doffing. It is important to note that the hierarchy of controls is not a strict linear progression. In practice, a combination of controls from different levels may be necessary to effectively manage risks. The hierarchy serves as a guideline to prioritize control measures, with the goal of reducing exposure to biological hazards and protecting the health and safety of individuals working with or in proximity to biological agents. An example of this is the practice of layering and combining different controls to increase the level of protection afforded to laboratory workers. The most common PPEs used during the start of the COVID-19 pandemic were gloves, disposable full coveralls with hood, safety goggles, face shields, N95 masks or powered air purifying respirators (PAPR), and boots and booties. Many PPEs were in short supply due to the global demand and many laboratories had to rely on disposable gowns, ordinary surgical masks, makeshift face shields, etc., but these were combined with other controls such as the use of engineering controls via biosafety cabinets, administrative controls/SOPs which still were able to prevent an increase of laboratory-acquired infections from SARS-CoV-2 despite its high infectivity.

## 3.4 LABORATORY INFRASTRUCTURE LMICS

### 3.4.1 Biosafety Levels, Facility Design, and Containment

Biosafety levels and facility design are crucial aspects of laboratory safety in biological research. They are designed to protect laboratory personnel, the surrounding environment, and public health from potential risks associated with working with biological agents. This section provides an overview of biosafety levels and their corresponding facility design considerations, highlighting their importance in ensuring safe and responsible research and laboratory practices. This section will also discuss the principle of containment which is a fundamental aspect of biosafety and which focuses on preventing the unintentional release of biological agents or hazardous materials from laboratory or research settings. This section will further elucidate on implementation of measures to control and minimize the risk of exposure to biological agents, protecting laboratory personnel, the environment, and public health.

Biosafety Levels (BSLs): Biosafety levels, commonly referred to as BSLs, are a set of containment principles and practices that define the specific safety requirements for laboratories working with different types of biological agents. The BSL system classifies laboratories into four levels, ranging from BSL-1 to BSL-4, with each level associated with increasingly stringent safety measures.

BSL-1 laboratories handle low-risk biological agents that pose minimal threats to human health and the environment. They require basic standard laboratory practices, such as hand hygiene, personal protective equipment (PPE), and proper waste management. BSL-2 laboratories work with moderate-risk agents that can cause human disease, but have effective treatments or preventive measures. They have additional safety features, including controlled access, dedicated laboratory equipment, and enhanced personnel training. BSL-3 laboratories deal with indigenous or exotic agents that may cause serious or potentially lethal diseases through inhalation. They incorporate stringent engineering controls, such as specialized ventilation systems, to prevent the release of these agents into the environment. Personnel must adhere to strict containment practices and wear appropriate respiratory protection. BSL-4 laboratories handle high-risk and exotic agents that pose severe threats to human health and lack effective treatments or preventive measures. These laboratories employ maximum containment measures, including complete personal protective suits, sophisticated heating, ventilation, air conditioning (HVAC) handling systems, and multiple physical barriers, to ensure the highest level of protection.

Facility Design Considerations: The design and layout of a laboratory facility play a critical role in implementing and maintaining biosafety measures. Key considerations include: (1) risk assessment and facility classification; (2) physical structure and barrier systems; (3) ventilation and airflow control; (4) waste management and decontamination; (5) safety and emergency response planning; and (6) containment strategies. Renovation of existing structures or facilities is generally more difficult and more expensive to perform versus creation of a new facility from the ground up. In some cases especially if time is of the essence and resources are limited, utilizing and retrofitting existing facility structures and making minor modifications is more appropriate. If facility design and funding are critical limitations, these deficiencies and the associated risks can be mitigated by instituting and strengthening other control measures so that the overall biosafety residual risk remains low. An example is if the airflow of an existing air conditioner directly blows on the opening of the biosafety cabinet and affects the laminar airflow of the biosafety cabinet, if you cannot move the position of the biosafety cabinet due to the limited space, instead of destroying a wall to enlarge the laboratory room space or moving the position of the air conditioner, one can just redirect the airflow of the air conditioner by installing an air vent deflector, which is inexpensive and very easy to install.

Risk Assessment and Facility Classification: A thorough risk assessment is essential to determine the appropriate BSL classification and associated facility requirements. This assessment considers the properties of the biological agents, their routes of transmission, potential consequences of exposure, and the level of expertise of laboratory personnel. The level of risk aversion or risk tolerance of a specific institution should be taken into consideration but without compromising minimum biosafety requirements. Risk assessments, as typically performed within institutions, inherently give subjective and arbitrary results and largely depend on the experience and expertise of the one conducting the risk assessment. The important thing is to focus on mitigating the risks identified. A rigid and inflexible approach to risk assessment will result in overkill and overdesign which results in sub-optimal utilization of limited resources. Input, consultation and dialogue with scientists, laboratory subject matter experts and laboratory end-users are critical in conducting risk assessments and reaching a recommendation, which will allow the effective and optimal use of limited resources.

Physical Structure and Barrier Systems: The laboratory facility should be designed with physical barriers, such as solid walls, airtight doors, and windows, to prevent unauthorized access and limit the spread of infectious agents. The structural integrity of the facility is crucial in maintaining containment.

Ventilation and Airflow Control: Proper airflow control and ventilation systems are crucial in minimizing the risk of airborne transmission. Negative pressure systems are commonly used in BSL-3 and BSL-4 laboratories to ensure inward airflow, preventing the unintentional release of hazardous agents into surrounding areas and the environment.

Waste Management and Decontamination: Adequate waste management procedures, including segregation, disinfection, and proper disposal, are essential to prevent the spread of biological agents. Decontamination protocols for equipment, surfaces, and waste materials should be implemented to ensure a safe working environment.

Safety and Emergency Response Planning: The facility should have comprehensive safety plans in place, including emergency response protocols, evacuation procedures, and personnel training on responding to accidents or exposures. Regular drills and exercises help reinforce preparedness and ensure effective response capabilities.

Containment Strategies: The principle of containment is essential in minimizing the risk of accidental exposure to biological agents, preventing their release into the environment, and safeguarding the health and well-being of laboratory personnel, the public, and ecosystems. Containment strategies help prevent accidental releases and protect personnel from exposure. The principle of containment is guided by the following key principles, namely: physical barriers, engineering controls, facility design, personal protective equipment, standard operating procedures, and training and education. The use of physical barriers, such as biosafety cabinets, sealed containers, and specialized laboratory equipment, is crucial in preventing the escape of biological agents. These barriers create a controlled environment that isolates the agents and prevents their release into the surrounding areas. Engineering controls include ventilation systems, air pressure differentials, and filtration systems, which ensure proper airflow and containment within the laboratory. These controls help to limit the spread of airborne contaminants and maintain a safe working environment. By implementing effective containment measures and unidirectional workflows, laboratories can maintain a controlled environment that allows for safe and responsible research with biological agents.

Biosafety considerations should be incorporated into the design and layout of laboratory facilities. The facility should have designated containment areas with controlled access, certified biosafety cabinets, proper storage facilities for hazardous materials, separate zones for clean and contaminated activities, and specialized workstations or dedicated work areas where activities involving biological agents are conducted. The facility design should support efficient and, if applicable, unidirectional workflow which can minimize the risk of cross-contamination. The use of appropriate PPE, such as gloves, lab coats, goggles, and respiratory protection, is essential in reducing the risk of exposure to biological agents since PPE acts as a physical barrier between laboratory personnel and hazardous materials, preventing direct contact or inhalation. The development and implementation of SOPs can also provide detailed guidelines for laboratory practices and procedures. SOPs outline specific containment measures, handling protocols, decontamination procedures, waste management practices, and emergency response plans. They ensure consistent and safe practices within the laboratory. Lastly, proper training and education of laboratory personnel are critical in ensuring adherence to containment principles. Training should cover aspects such as biosafety practices, proper use of equipment, handling of hazardous materials, waste management, and emergency procedures. Ongoing training and refresher courses help reinforce biosafety practices and promote a culture of safety.

### 3.4.2 Adapting Laboratory Infrastructure to LMIC Settings

Laboratory infrastructure plays a vital role in ensuring biosafety in LMICs. However, limited resources, infrastructure challenges, and specific contextual factors may pose unique barriers to achieving optimal biosafety standards. Here we present strategies for adapting laboratory infrastructure in LMICs to enhance biosafety practices and mitigate risks associated with working with biological agents.

Risk Assessment and Prioritization: Conduct a comprehensive risk assessment to identify the most significant hazards and prioritize the allocation of resources based on the level of risk. Consider the specific challenges and prevalent biological agents in the LMIC setting to guide the design and adaptation of laboratory infrastructure. We recommended that countries have their own specific prioritization list of biological materials of concern, which takes into consideration the endemicity of certain pathogens and the immunity of the population. Some countries require manipulation of some pathogens (e.g., chikungunya, Japanese encephalitis virus) to be done under biosafety level 3, but the biosafety level requirement may be lower in countries where these pathogens are highly endemic.

Facility Design and Engineering Controls: Optimize the existing laboratory infrastructure by implementing basic engineering controls such as proper ventilation, adequate lighting, and separate work areas to minimize cross-contamination. Laboratory layouts can be modified to ensure efficient workflow and minimize unnecessary movement of personnel, samples, and equipment. One should also utilize cost-effective strategies such as physical barriers, designated storage areas, and separate zones for clean and contaminated activities. Instead of creating new facilities or extensive modification of existing infrastructure, which are usually very expensive and time-consuming, modifying and adapting existing facilities can instead be done. Sometimes regulatory requirements may be detrimental, may not be appropriate or even exceed the biosafety requirements. An example is the use of the Cepheid GeneXpert equipment for SARS-CoV-2 testing during the early phase of the COVID-19 pandemic. Some countries still require that the GeneXpert equipment be located and used inside laboratory rooms with negative pressure even though the specimens being loaded into the equipment have already been inactivated and pose minimal risk of infection to the laboratory workers. Placing the equipment in a well-ventilated room with adequate air exchange and implementing a unidirectional workflow would have theoretically sufficed instead of requiring construction of negative pressure systems, which entailed a significant investment of time and money. This resulted in delays in the rollout and utilization of numerous existing GeneXpert machines used by the country tuberculosis program for SARS-CoV-2 testing.

Biosafety Cabinets and Equipment: Assess the need for biosafety cabinets and other specialized equipment based on the specific activities and risks involved in the laboratory. Consider alternative options such as local fabrication or refurbishment of equipment to reduce costs and adapt to local conditions. Regular maintenance and calibration programs should be implemented for equipment to ensure their continued effectiveness. Non-hazardous work, low-risk procedures that do not generate aerosols, splashes, or handling of non-infectious agents or non-biological materials, can usually be performed on laboratory countertops. Laminar flow hoods can be used to provide a sterile area to protect the work product or chemical fume hoods may be used if the work involves handling chemicals emitting fumes or vapors but cannot be used in place of biosafety cabinets. The BSC class and type used should be appropriate to the need and level of protection required for personnel, environment, and the product/agent to conduct laboratory activities safely. BSCs are deemed absolutely necessary if one is working with or handling infectious agents, conducting aerosol or splash generating procedures such as during centrifugation or performing cell cultures, high risk experiments involving recombination or genetic modification experiments, gain-of-function experiments, or animal model studies involving agents infectious to humans to name a few.

Personal Protective Equipment (PPE): Develop PPE protocols that are practical and suitable for the local context, taking into account availability, affordability, and cultural acceptability. One should be careful not to forget the proper donning and doffing procedures for PPE usage since even if you have access to high-quality PPEs, self-contamination can still occur and render the PPE ineffective if personnel are not properly trained on how to put on and remove PPE. In cases where PPEs are unavailable or unaffordable, one can combine and substitute these with less expensive PPEs and still afford similar levels of protection for the laboratory personnel. A careful risk assessment of the potential hazards associated with a specific experiment is needed in order to select the appropriate PPE to use as well as compatibility of the PPEs to the chemicals or materials, which will be used in the conduct of the experiment. Always refer to safety data sheets, understand the procedures of the experiment, and adhere to local and/or international regulatory standards and guidelines. Ensure that laboratory personnel are properly trained on the correct usage of the PPE and specific PPEs should fit well specific laboratory personnel (i.e., fit testing for respirators) and that PPEs are regularly inspected to ensure it retains its effectiveness. Lastly, emergency procedures should be in place including decontamination and biological spill procedures in case of a contamination or spillage. Keep in mind that PPEs are considered as the last line of defense and one should prioritize control measures, which are placed higher in the hierarchy of controls (e.g., engineering and administrative controls). Provide training on proper usage, maintenance, and disposal of PPE to laboratory personnel.

Training and Capacity Building: Develop tailored training programs on biosafety practices and procedures, emphasizing the importance of risk awareness and mitigation in resource-limited settings. Collaborate with local and international partners to provide technical support, mentorship, and knowledge transfer to strengthen local capacity and expertise. Foster a culture of safety through regular communication, refresher training sessions, and regular monitoring and evaluation of biosafety practices.

Waste Management: Establish appropriate waste management protocols to handle and dispose of hazardous materials, including sharps, biological waste, and chemical reagents. Explore local solutions such as autoclaving, chemical disinfection, or on-site treatment to minimize reliance on costly external waste management services. Implement waste segregation practices and provide training on safe handling and disposal to minimize environmental contamination and public health risks.

Regulatory Compliance and Standards: Align laboratory practices with national and international biosafety guidelines and regulations. Advocate for the development and enforcement of biosafety standards specific to the LMIC context, considering the unique challenges and resources available. Engage with regulatory authorities to seek their support in providing technical assistance, guidance, and oversight.

Collaboration and Resource Sharing: Foster collaboration and knowledge-sharing networks among laboratories, both within the LMIC and globally, and to exchange best practices, experiences, and resources. Leverage partnerships with academic institutions, research organizations, and international agencies to access funding opportunities, technical expertise, and shared resources.

It is essential to strike a balance between overdesigning and cost-effectiveness, as excessive design measures may lead to unnecessary financial burdens and resource utilization. Laboratory design and safety considerations should be based on a comprehensive and accurate risk assessment taking into account the specific hazards, operational requirements, and regulatory guidelines applicable to the laboratory setting. Adapting laboratory infrastructure in LMICs for biosafety requires a pragmatic and context-specific approach. By implementing these strategies, LMICs can enhance biosafety practices, mitigate risks, and protect laboratory personnel, the environment, and public health. Although challenges exist, creative and adaptive solutions can be implemented to promote safe and responsible laboratory practices within the resource constraints of LMICs.

## 3.5 GOOD LABORATORY PRACTICES AND TECHNIQUES

Good laboratory practices and techniques are essential for maintaining safety, accuracy, and reliability in laboratory settings. These practices include a range of guidelines and procedures that promote consistent and standardized approaches to laboratory work. By adhering to these practices, laboratories can minimize the risk of contamination, ensure the integrity of results, protect the health and safety of personnel, and maintain compliance with regulatory requirements. One fundamental aspect of good laboratory practices is proper documentation. Accurate and detailed record keeping is crucial for maintaining a traceable and transparent record of laboratory activities. This includes documenting experimental procedures, observations, data, and results. Proper documentation facilitates reproducibility, allows for the identification and resolution of any discrepancies, and provides a historical record of laboratory work. Maintaining a clean and organized laboratory environment is another critical practice. Regular cleaning and disinfection of surfaces, equipment, and work areas help prevent cross-contamination and the build-up of hazardous substances. Proper waste management, including the segregation, labeling, and disposal of different types of waste, is also essential to minimize risks and ensure compliance with environmental regulations. Adherence to proper sample handling protocols is vital for maintaining the integrity of samples and avoiding contamination. This includes correct labeling, storage, and transportation of samples, as well as appropriate use of personal protective equipment (PPE) during sample handling. Following established protocols and guidelines for sample preparation and analysis, helps ensure accurate and reliable results. Regular calibration and maintenance of laboratory equipment are critical for obtaining accurate measurements and reliable data. This includes routine inspection, calibration, and verification of instruments to ensure they are functioning correctly. Proper training of laboratory personnel on equipment operation and maintenance is necessary to prevent errors and optimize instrument performance. Effective communication and teamwork are essential aspects of good laboratory practices. Clear communication among laboratory personnel promotes efficient coordination of tasks, reduces the risk of errors, and enhances safety. Collaboration and teamwork encourage sharing of knowledge and expertise, facilitating problem-solving and the exchange of best practices. Proper training and education of laboratory personnel are crucial for the implementation of good laboratory practices. Training programs should cover topics such as biosafety, chemical safety, equipment operation, sample handling, and waste management. Ongoing training ensures that laboratory personnel are updated on new techniques, regulations, and safety protocols. Risk assessment and management are integral components of good laboratory practices. Identifying potential hazards, assessing associated risks, and implementing appropriate control measures are essential for maintaining a safe working environment. This includes conducting regular risk assessments, implementing risk mitigation strategies, and establishing emergency response plans. Quality control measures are crucial to ensure accuracy and reliability in laboratory results. This involves implementing quality assurance programs, participating in proficiency testing, and conducting internal audits to monitor and improve laboratory performance. Regular review of procedures, protocols, and data helps identify areas for improvement and ensures compliance with quality standards. Continual improvement is a key principle of good laboratory practices. Regular evaluation and feedback mechanisms help identify areas for improvement, address shortcomings, and promote a culture of continuous learning and development. By embracing a commitment to ongoing improvement, laboratories can enhance their capabilities, maintain high-quality standards, and contribute to scientific advancements.

## REFERENCES

Cao, Zengguo, Hongli Jin, Gary Wong, Ying Zhang, Cuicui Jiao, Na Feng, Fangfang Wu, Shengnan Xu, Hang Chi, Yongkun Zhao, Tiecheng Wang, Weiyang Sun, Yuwei Gao, Songtao Yang, Xianzhu Xia and Hualei Wang. (2021). "The Application of a Safe Neutralization Assay for Ebola Virus Using Lentivirus-Based Pseudotyped Virus." *Virologica Sinica* 36(6): 1648–1651.

Centers for Disease Control and Prevention. (2020). *Biosafety in Microbiological and Biomedical Laboratories (BMBL)* (6th ed.). U.S. Department of Health and Human Services. Atlanta, GA.

Greenberg, D. L., J. D. Busch, P. Keim and D. M. Wagner (2015). "Identifying Experimental Surrogates for Bacillus anthracis Spores: A Review." *Investigative Genetics* 1(1) (2010): 4.

*Laboratory Biorisk Management: Biosafety and Biosecurity.* Edited By Reynolds, M., Salerno Jennifer, Gaudioso. Boca Raton: CRC Press.

Murnyak et al. (2003). "The Risk Assessment Process Used in the Army's Health Hazard Assessment Program. Acquisitions Review Quarterly." https://apps.dtic.mil/sti/tr/pdf/ADA423528.pdf.

World Health Organization. (2020). *Laboratory Biosafety Manual* (4th ed.). World Health Organization. Geneva, Switzerland.

# 4 Principles and Concepts of Biosecurity

*Rodel Jonathan S. Vitor II*
Department of Biology, College of Science
De La Salle University
Manila, Philippines
National Training Center for Biosafety and Biosecurity
National Institutes of Health
University of the Philippines
Manila, Philippines

*Fedelyn P. Estrella*
Lyceum of the Philippines University Cavite and
National Training Center for Biosafety and Biosecurity
National Institutes of Health
University of the Philippines
Manila, Philippines

## 4.1 INTRODUCTION TO BIOSECURITY AND RELATED CONCEPTS

Laboratory biosecurity refers to institutional and personnel security measures designed to prevent the loss, theft, misuse, diversion, or intentional release of biological agents being handled in the laboratory (WHO, 2006). Addressing laboratory biosecurity risks in many ways parallels and complements that of biosafety risk management. Effective biosafety practices are the foundation of laboratory biosecurity, and biosecurity risk control measures must be performed as an integral part of an institution's biosafety program management (WHO, 2020b). Risk assessment and management techniques, personnel expertise and accountability, control and accountability for research materials such as microorganisms and culture stocks, access control components, material transfer documentation, training, emergency planning, and program management are just a few areas where biosafety and biosecurity have similar viewpoints (WHO, 2020a).

To regulate the movement of a specific list of biological agents from one institution to another, the US government passed the Select Agent Regulations in 1996 (Ryan, 2016). The US government's viewpoint evolved a little following the 9/11 attacks and the 2001 anthrax attacks, commonly known as Amerithrax. Any facility in the United States that handled or housed one or more agents from the new, longer list of agents was then required to take additional security precautions under the amended Select Agent Regulations (Burnette, 2013). The creation of two levels of select agents was addressed by the 2012 modification of the Select Agent Regulations (CDC, 2013).

DOI: 10.1201/9781003426219-4

Materials with the highest potential for intentional misuse are classified as Tier 1 agents, along with the remaining select agents. By requiring more stringent security measures for Tier 1 agents, this revision aimed to make the rules more risk-based.

For bioscience institutions, several nations have likewise gradually adopted and prescribed biosecurity norms. The Biological Agents and Toxins Act in Singapore has a comparable scope to the laws in the US, but it has harsher penalties for breaking the rules (The Republic of Singapore, 2015). The Act on Prevention of Infectious Diseases in South Korea was revised in 2005 to mandate that organizations that deal with the 'highly dangerous pathogens' mentioned in the act implement laboratory biosafety and biosecurity standards to prevent these agents from being lost, stolen, diverted, released, or used improperly (Republic of Korea, 2015). The Ministry of Health, Labor, and Welfare of Japan has likewise modified the Infectious Disease Control Law. Additionally, four schedules of chosen agents were formed, each of which had its own reporting and handling guidelines for possession, transportation, and other operations (Japan, 1998). While other countries have established their own list of valuable biological agents and toxins, the Philippines has not yet finalized the list, and it is hoped that in the near future, the list will be available for researchers.

This chapter will not focus on agricultural biosecurity, which is critical in safeguarding crops, livestock, and agricultural ecosystems from the introduction and spread of pests, diseases, and invasive species (Dewulf & Immerseel, 2019). In brief, it involves implementing strict control measures, including quarantine procedures, surveillance, and risk assessments, to prevent the entry of harmful agents that could devastate agricultural production. Proactive agricultural biosecurity is essential to maintain food security, preserve biodiversity, and ensure the resilience of farming systems in the face of evolving threats.

## 4.2 COMPONENTS OR PILLARS OF LABORATORY BIOSECURITY

The World Health Organization has published the Laboratory Biosecurity Guidance to guide laboratorians with regards to biosecurity (WHO, 2006), however, changes in the landscape of laboratory biosecurity especially the concept of the pillars of biosecurity have soon arise. Before the publication of the fourth edition of the Laboratory Biosafety Manual (WHO, 2020b), there were only five pillars: (1) inventory control (materials control and accountability), (2) information control, (3) personnel control, (4) physical security control, and (5) transport control. The fourth edition added three more, which are the following: (1) emergency and incident response, (2) emerging biological risks, and (3) dual use research of concern. The three added pillars highlight the need for better oversight, especially in the evolving field of biosecurity amidst the increasing emerging and re-emerging diseases, as well as the recent COVID-19 pandemic.

Low- to middle-income class countries are usually more focused on biosafety principles as they focus more on protecting the users from the biological materials that they are working with. However, in the ever-changing landscape of laboratory biosecurity, the implementation of the pillars of biosecurity to ensure that valuable biological materials as well as data and information produced within the laboratory are protected. This will also prevent the possible misuse of biology and biotechnology as a tool of destruction to create havoc and distress among the general population.

### 4.2.1 Inventory Control

To establish effective control of biological agents that are at risk and to deter theft and/or misuse, a thorough program of accountability is required. A thorough inventory that includes information on the biological agent(s), its quantities, storage location and use, the person responsible, documentation of internal and external transfers, and an inactivation and/or disposal of the materials can be put together as a method to accomplish this. It is advised to conduct periodic reviews, and any discrepancies should be looked into and fixed.

### 4.2.2 Information Control

The confidentiality and integrity of sensitive information stored in the laboratory that might be exploited maliciously must also be protected through processes and procedures. It is crucial to recognize, mark, and safeguard sensitive information within the framework of the biosecurity program against illegal access. Research data, test findings, information on animal trials, important personnel lists (such as IT and biosafety contacts), security plans, access codes, passwords, storage sites, and inventories of biological agents are all examples of sensitive information. Sensitive information must not be disclosed to unauthorized parties under any circumstances.

### 4.2.3 Personnel Control

The personnel's training, capability, dependability, and integrity ultimately determine the efficacy of any procedural controls for biosecurity. An effective administration of employees is necessary for a laboratory to operate. It guarantees that regular work practices and procedures are carried out by qualified staff who act in a dependable and trustworthy manner. To make sure that there is a valid need for access and that the proper vetting and escorting procedures are followed, it is necessary to develop laboratory access request and approval processes for visitors and other outside individuals in addition to laboratory staff.

Depending on the results of the risk assessment, all staff members should get laboratory biosecurity training in addition to biosafety training. The staff should benefit from this training by having a better understanding of the necessity to safeguard biological agents as well as the justification for the particular biosecurity measures that have been put in place. Reviewing pertinent national standards and institution-specific protocols should also be part of it. It is also important to clarify the duties and responsibilities of personnel in terms of security in both normal and emergency situations. Training and standards should be proportionate to the hazards involved because not all positions pose the same amount of biosecurity risk.

### 4.2.4 Physical Security Control

Physical security countermeasures are used to both reduce the threat from insiders (those who have a legitimate presence in the facility such as employees and approved visitors) who do not need access to a particular asset and to prevent unauthorized access of outside adversaries (i.e., those who do not have a legitimate presence in the facility and have malicious intent such as criminals, terrorists, and extremists/activists). By restricting access to the laboratory and other potentially dangerous places, physical security systems not only serve biosafety objectives but also advance their biosecurity goals. In order to increase a facility's ability to prevent, detect, assess, postpone, respond to, and recover from a security incident, an effective physical security system integrates a number of different components. These components, which are often graded, comprise borders, access controls, intrusion detection, alert evaluation, and reaction. A system of graded protection builds up layers of risk-based security around the facility's assets while gradually increasing security. The assets that will have the greatest impact on national and possibly worldwide security, as well as the health and safety of workers, the general public, and the environment, should be given the highest level of protection. These components should also be chosen and put into place following a site-specific biosecurity risk assessment to make sure they are all workable, long-lasting, and appropriate for the dangers that have been identified. By performing a robust risk assessment, institutions will be able to prioritize resources in low- to middle-income class countries, thereby ensuring that there is a balance between the security controls chosen versus the perceived or actual risks evaluated.

### 4.2.5 Transport Control

The transfer of biological agents must adhere to national and international regulations for packing, marking, labeling, and documentation. To guarantee adequate monitoring within the biosecurity program, this process should be controlled to a degree proportionate to the biological agent's identified biosecurity threats. Biological agents may be shipped using authorized couriers and ordered from reliable suppliers as part of procedures to ensure their safety. It is important to write down and execute any necessary procedures for the shipper, carrier, and receiver's obligations to control biosecurity risks. As soon as goods are taken out of safe places, there are vulnerabilities because more people may now have access to them. In accordance with the findings of a biosecurity risk assessment, transfers should be prepared, pre-approved by responsible parties, and use chain of custody paperwork (or an equivalent) for adequate record-keeping. Inventory records must be updated to reflect all internal and external transfers, as well as arriving and exiting specimens.

### 4.2.6 Emergency and Incident Response

To practice the responses to simulated incidents or emergencies, drills and exercises can also be employed throughout the planning and preparation phases. They can aid in finding areas for improvement and other weaknesses. At least once a year, plans should be reviewed and updated, and any necessary alterations and improvements should be made using the data gathered from drills, incident reports, and investigations. Despite existing prevention or risk control procedures, even the most well-prepared laboratory may have unintended or purposeful mishaps or emergencies. Through planning and preparing for potential incidents (like discrepancies found in inventories, missing biological agents, or unauthorized individuals in the laboratory), effective incident response is a risk control strategy that can lessen the effects of these unknown events. It may also help detect, communicate, assess, respond to, and recover from actual events. To ensure accurate reporting, as well as to allow investigation, root-cause analysis, corrective action, and process improvement, an incident response protocol should be created and implemented.

### 4.2.7 Emerging Biological Risks

Genetically modified microbes, synthetic biology, gain-of-function research, stem cell research, gene editing, and gene drives are examples of emerging biological dangers. Improvements in the health of people, plants, and animals are intrinsically related to advances in the life sciences. Promoting high-quality life sciences research that is carried out ethically, securely, and responsibly can increase the security of global health as well as aid in economic growth, evidence-based policymaking, and public faith in science. However, in order to perform required and advantageous life sciences research, nations, laboratories, and scientists must also take into account the dangers posed by events and/or the potential for deliberate misuse of the research. They must then choose the proper risk control procedures to reduce those risks. The scientific community must follow these principles while working with novel technology about which there is currently little knowledge (WHO, 2020):

1. Promote a culture of integrity and excellence, distinguished by openness, honesty, accountability, and responsibility; such a culture is the best protection against the possibility of accidents and deliberate misuse, and the best guarantee of scientific progress and development;
2. Provide direction for biosafety/biosecurity oversight and the risk assessment process for emerging technologies in the life sciences, and as additional information is obtained over time, contribute to better understanding of their risks and biosafety/ biosecurity needs; and
3. Monitor and assess the scientific, ethical, and social implications of certain biotechnologies and, as warranted, monitor the development of those technologies and their integration into scientific and clinical practice.

### 4.2.8 Dual Use Research of Concern

Life sciences research that has the potential to provide knowledge, information, products, or technologies that could be directly misapplied to create a serious threat with potential repercussions to the public's health and safety, agricultural species, other plants, animals, and the environment is considered dual use research of concern. Where appropriate, awareness of the dual use of agents, tools, and technologies should be considered while developing laboratory biosecurity programs. In order to choose the best biosecurity measures to guard against unauthorized access, loss, theft, misuse, diversion, and intentional release, laboratories should accept responsibility for the dual-use nature of such agents and experiments, such as genetic modification, and follow national guidelines. A balanced approach to laboratory biosecurity is necessary in light of the possible misuse of biosciences as a worldwide hazard in order to protect legitimate access to crucial research and clinical materials.

## 4.3 FIELD BIOSECURITY

Biosecurity not only begins and ends at the laboratory level, but that it has direct implications into other facets of concern, magnitude, and consequence. These include global health, defense, and developing technologies (Burnette, 2021). Biosecurity is fundamentally about dealing with risk; that is, dealing with problems before they are actualized. When dealing with threats to human, animal, and plant health the motif of 'prevention is better than cure' dominates. This involves the identification of pathways and vectors by which biosecurity threats might arrive in a country, the likelihood of such events occurring, and their potential impacts (Dobson et al., 2013). The global expansion of biosecurity initiatives is fueled by one factor in particular and has to do with rising biological danger. Growing consequences of climate change on the ranges of species have resulted in an increase in invasive species incursions and the increased distribution of disease-carrying vectors, such as the midge that spreads bluetongue. Northern Europe has already experienced the spread of a virus affecting ruminants (Mayo et al., 2020). Governments in the UK have voiced concern about climate change because of the size of recent outbreaks; there is growing worry about the cost of biosecurity issues with rare diseases and growing endemic illness challenges. On the other hand, the Cartagena Protocol on Biosafety to the Convention on Biological Diversity sets out further measures relating to regulating living-modified organisms (LMOs). The protocol requires 'advance informed agreement' between exporting and importing states for the movement of LMOs intended for release into the environment and makes it clear that the precautionary approach should be adopted (Dobson et al., 2013).

Foreign agents used in bioterrorism are by definition invasive alien species (IAS). While the IAS released with intent to cause harm have so far been limited to microbes or any organism that causes direct or indirect harm to humans could become a terrorist tool. Vectors of human disease, from mosquitoes to migratory birds, may be of particular concern. However, harmful impacts (more subtle but cumulative) could also occur if terrorists choose to introduce organisms that damage infrastructure or compete with or prey upon native plants, wildlife, or domestic stock. Serious consideration of biosecurity in this broader perspective will highlight many challenges, not least of which is that the overwhelming majority of attention and resources are devoted to military threats, while non-military threats are often ignored (Meyerson & Reaser, 2003).

## 4.4 BIOLOGICAL WEAPONS AND BIOTERRORISM

The origins of biological weapons can be traced back centuries, where mankind's earliest confrontations with infectious diseases inadvertently laid the foundation for this dark facet of warfare. Ancient armies would often resort to tactics like catapulting infected corpses over city walls to spread disease within enemy ranks. The deliberate use of smallpox-contaminated blankets as a biological weapon during the French and Indian War in the 18th century is a grim testament to the historical use of biological agents in warfare (Ryan, 2016). However, it was during the 20th century that biological weapons assumed a more sinister and systematic form. The horrors of World War I witnessed the deployment of chemical agents, but biological agents were largely unused due to the unpredictability of their effects and the difficulty in delivery. Nevertheless, the lessons learned during this period laid the groundwork for more sophisticated biological weapons programs in subsequent decades.

Biological weapons are distinct from conventional and chemical weapons in that they utilize living organisms, such as bacteria, viruses, or toxins, to inflict harm. The science behind these weapons is rooted in microbiology, genetics, and the understanding of pathogenic microorganisms (Burnette, 2021).

1. Pathogens: Pathogenic microorganisms, such as anthrax bacteria or smallpox virus, are the primary agents used in biological weapons. These organisms can cause disease in humans, animals, or plants.
2. Delivery Mechanisms: The delivery of biological agents can take various forms, including aerosols, contaminated food or water, and vectors like insects. Aerosols, which allow for the inhalation of pathogens, are particularly effective in spreading disease.
3. Genetic Engineering: Advances in genetic engineering have enabled the manipulation of pathogens to enhance their virulence or make them resistant to treatment. This represents a concerning development in the evolution of biological weapons.

The proliferation of biological weapons and the emergence of bioterrorism as a viable threat have grave implications for global security and public health (Macintyre, 2022). Several key factors contribute to the threats posed by these sinister tools:

1. Mass Casualties: Biological weapons have the potential to cause mass casualties, as infectious diseases can rapidly spread through densely populated areas. A single attack can result in thousands of fatalities and incapacitated individuals.
2. Stealth and Subterfuge: Biological weapons are notoriously difficult to detect and attribute, making them attractive to both state and non-state actors. The anonymity and ambiguity surrounding bioterrorism incidents can complicate response efforts.
3. Globalization: In an era of rapid globalization, the interconnectedness of nations and the ease of travel can facilitate the rapid spread of infectious diseases, amplifying their impact.
4. Dual-Use Nature: Many biological agents used in weapons have legitimate research applications, making it challenging to monitor and control access to these materials.

5. Economic Disruption: Bioterrorism can disrupt economies by affecting critical infrastructure, agriculture, and public services. The economic consequences can be severe and long-lasting.
6. Psychological Impact: Beyond the immediate physical harm, the fear and panic generated by bioterrorism can have enduring psychological effects on affected populations.

While the use of biological weapons in armed conflicts has been relatively limited in recent history, instances of bioterrorism and clandestine programs demonstrate the persistent threat (Wheelis et al., 2006). Notable historical incidents include:

1. Aum Shinrikyo: The Aum Shinrikyo cult in Japan attempted several bioterrorism attacks in the 1990s, including releasing anthrax in Tokyo. Their attacks were limited in scope but highlighted the potential for non-state actors to engage in bioterrorism.
2. Sverdlovsk Anthrax Outbreak: In 1979, an accidental release of anthrax spores from a Soviet military facility in Sverdlovsk (now Yekaterinburg) resulted in a significant outbreak, exposing the world to the dangers of accidental releases and covert biological weapons programs.
3. Iraq's Biological Weapons Program: Saddam Hussein's regime in Iraq pursued an active biological weapons program in the 1980s and 1990s, which included the development of anthrax and botulinum toxin as potential agents.
4. The Amerithrax Attacks: In 2001, anthrax-laced letters were mailed to several US government officials and media outlets, resulting in five deaths and multiple injuries. The perpetrator was eventually identified as a US government scientist, highlighting the potential for insider threats.

The realization of the devastating consequences of biological weapons has led to a series of international efforts aimed at preventing their use and proliferation (Enemark, 2017).

1. Biological Weapons Convention (BWC): The BWC, which came into force in 1975, is an international treaty aimed at prohibiting the development, production, and acquisition of biological weapons. It also promotes the peaceful use of biotechnology.
2. United Nations Security Council Resolutions: The United Nations Security Council has passed resolutions emphasizing the need to prevent the proliferation of biological weapons and to respond to bioterrorism threats.
3. Global Health Security Agenda (GHSA): The GHSA is an international partnership that seeks to enhance global health security by preventing, detecting, and responding to infectious disease threats, including those posed by bioterrorism.
4. International Health Regulations (IHR): The IHR, overseen by the World Health Organization (WHO), provides a framework for the international response to public health emergencies, including outbreaks of infectious diseases.

Biological weapons and bioterrorism represent a sobering and complex challenge to global security and public health. Their historical roots, rooted in the malevolent use of pathogens, have evolved alongside advances in microbiology and genetics. The threats posed by biological weapons are characterized by their potential for mass casualties, stealth, and economic disruption. The international community has responded with a range of efforts aimed at preventing the proliferation and use of biological weapons, including the Biological Weapons Convention, United Nations Security Council resolutions, and partnerships like the Global Health Security Agenda. However, the ongoing threat of bioterrorism and the potential for covert biological weapons programs necessitate ongoing vigilance and preparedness. In the following sections of this essay series, we will delve deeper into the science of biological weapons, the challenges of detection and attribution, the implications for public health, and strategies for countering these threats in an ever-evolving world.

Even though the misuse of biological materials for bioweapons and bioterrorism is perceived to be lower in low- to middle-income countries, it cannot be discounted that these countries should understand that having security measures in place will thwart the possibility of the materials being used to deliberately cause havoc. This will ensure that countries will put measures vis-à-vis biosafety to ensure a safe and secure environment for laboratorians and scientists.

## 4.5 CYBERBIOSECURITY

The convergence and confluence of biology with the digital world have given rise to the new field of cyberbiosecurity. Cyberbiosecurity sits at the crossroads of cybersecurity, cyber-physical security, and biological security (Murch et al., 2018). The increase in interest in cyberbiosecurity lies in its role as a safeguard against potential catastrophic events that may result from cyberattacks on biological systems. Most institutions have become increasingly reliant on digitalization and the internet, resulting in the majority of laboratory processes being exploited. Vulnerabilities within the laboratory ecosystem are vast, which may include manipulation of genomic and engineering data, theft of sensitive and personalized data, compromising digital assistive technologies, and even accessing sensitive government records which may be of public health, wildlife, and agricultural significance. These breaches have far-reaching consequences, threatening public health security and leading to economic losses and compromises on national security (Richardson et al., 2019). For example, given that personalized medicine is now increasing in popularity, manipulation of patient genetic data may lead to misdiagnosis or inappropriate treatment. Thus, without cyberbiosecurity, lives could be endangered.

Understanding cyber biosecurity vulnerabilities is crucial for effective risk mitigation. These vulnerabilities exist in various aspects of the biotech ecosystem, including laboratory automation, data storage, communication systems, and cyber-physical systems like biomanufacturing facilities (Schabacker et al., 2019). For instance, the reliance on Internet of Things (IoT) devices in laboratories presents opportunities for cyberattacks to infiltrate and manipulate experiments. Similarly, the vast amount of sensitive genetic data stored in digital databases makes them attractive targets for data breaches. To assess these vulnerabilities, organizations must conduct comprehensive risk assessments. This entails identifying potential threats, evaluating their likelihood and consequence, and prioritizing mitigation strategies. The NIST Cybersecurity Framework provides a useful model for organizations to assess and manage their cyberbiosecurity risks. By categorizing threats, protecting against them, detecting any breaches, responding effectively, and recovering from incidents, organizations can develop a resilient cyberbiosecurity posture (National Institute of Standards and Technology, 2018).

Mitigating cyberbiosecurity risks requires the implementation of best practices across the different interfaces (Mueller, 2021). Access control, for instance, plays a pivotal role in preventing unauthorized access to biological data and systems. Implementing multi-factor authentication, least privilege access, and robust password policies can bolster defenses. Encryption techniques ensure the confidentiality of data both in transit and at rest, while intrusion detection systems continuously monitor for malicious activities. Personnel training and creating a security-conscious organizational culture are equally essential (Reed & Dunaway, 2019). Human errors are still one of the main problems regarding cyber incidents. Organizations can significantly decrease the risk resulting from insider threats by educating staff on the importance of cyberbiosecurity, conducting regular training, and encouraging a culture of reporting suspicious activities.

In summary, cyberbiosecurity is an increasingly critical field that safeguards the convergence of biology and technology. Its significance lies in protecting against potentially catastrophic cyberattacks that could compromise public health, economic stability, and national security. With everything being connected through the internet, low- to middle-income countries will benefit from understanding the principles and instigating cybersecurity measures that will not only protect the valuable information being stored in their institutions but rather also protect the institution's integrity. As the threat landscape evolves, understanding vulnerabilities, implementing best practices, and embracing emerging technologies will be essential in building a resilient cyberbiosecurity framework for the future. International collaboration and policy development will play a pivotal role in addressing the global challenges posed by cyberbiosecurity threats.

## REFERENCES

Burnette, R. (Ed.). (2013). *Biosecurity: Understanding, assessing, and preventing the threat.* John Wiley & Sons, Inc.

Burnette, R. N. (Ed.). (2021). *Applied biosecurity: Global health, biodefense, and developing technologies.* Springer Nature Switzerland.

CDC. (2013). *Security guidance for select agent or toxin facilities.* CDC.

Dewulf, J., & Immerseel, F. V. (Eds.). (2019). *Biosecurity in animal production and veterinary medicine: From Principles to practice.* CABI.

Dobson, A., Barker, K., & Taylor, S. L. (Eds.). (2013). *Biosecurity: The socio-politics of invasive species and infectious diseases.* Routledge.

Enemark, C. (2017). *Biosecurity dilemmas: Dreaded diseases, ethical responses, and the health of nations.* Georgetown University Press.

Japan. (1998). *Act on prevention of infectious diseases and medical care for patients with infectious disease* (Act No. 114). Retrieved from https://www.japaneselawtranslation.go.jp/en/laws/view/2830/en

Macintyre, R. (2022). *Dark Winter: An insider's guide to pandemics and biosecurity.* New South Publishing.

Mayo, C., McDermott, E., Kopanke, J., Stenglein, M., Lee, J., Mathiason, C., Carpenter, M., Reed, K., & Perkins, T. A. (2020). Ecological dynamics impacting bluetongue virus transmission in North America [review]. *Frontiers in Veterinary Science, 7.* https://doi.org/10.3389/fvets.2020.00186

Meyerson, L. A., & Reaser, J. K. (2003). Bioinvasions, bioterrorism, and biosecurity. *Frontiers in Ecology and the Environment, 1*(6), 307–314. https://doi.org/10.1890/1540-9295(2003)001[0307:BBAB]2.0.CO;2

Mueller, S. (2021). Facing the 2020 pandemic: What does cyberbiosecurity want us to know to safeguard the future? *Biosafety and Health, 3*(1), 11–21. https://doi.org/10.1016/j.bsheal.2020.09.007

Murch, R. S., So, W. K., Buchholz, W. G., Raman, S., & Peccoud, J. (2018). Cyberbiosecurity: An emerging new discipline to help safeguard the bioeconomy [perspective]. *Frontiers in Bioengineering and Biotechnology, 6.* https://doi.org/10.3389/fbioe.2018.00039

National Institute of Standards and Technology. (2018). *Framework for improving critical infrastructure cybersecurity.* Retrieved June 18 from https://nvlpubs.nist.gov/nistpubs/CSWP/NIST.CSWP.04162018.pdf.

Reed, J. C., & Dunaway, N. (2019). Cyberbiosecurity implications for the laboratory of the future [review]. *Frontiers in Bioengineering and Biotechnology, 7.* https://doi.org/10.3389/fbioe.2019.00182

Republic of Korea. (2015). *Infectious disease control and prevention act.* (Act No.17067). Retrieved from https://www.law.go.kr/LSW/lsInfoP.do?nwJoYnInfo=N&ancYnChk=0&ancNo=13392&chrClsCd=010202&efYd=20160107&lsiSeq=172762&efGubun=Y&ancYd=20150706#0000

Richardson, L. C., Connell, N. D., Lewis, S. M., Pauwels, E., & Murch, R. S. (2019). Cyberbiosecurity: A call for cooperation in a new threat landscape [perspective]. *Frontiers in Bioengineering and Biotechnology, 7.* https://doi.org/10.3389/fbioe.2019.00099.

Ryan, J. R. (2016). *Biosecurity and bioterrorism: Containing and preventing biological threats* (2nd ed.). Elsevier.

Schabacker, D. S., Levy, L.-A., Evans, N. J., Fowler, J. M., & Dickey, E. A. (2019). Assessing cyberbiosecurity vulnerabilities and infrastructure resilience [perspective]. *Frontiers in Bioengineering and Biotechnology, 7.* https://doi.org/10.3389/fbioe.2019.00061

The Republic of Singapore. (2005). *Biological agents and toxins act, 2005.* Retrieved from https://sso.agc.gov.sg/Act/BATA2005

Wheelis, M., Rózsa, L., & Dando, M. (Eds.). (2006). *Deadly cultures: Biological weapons since 1945.* Harvard University Press.

WHO. (2006). *Biorisk management: Laboratory biosecurity guidance.* WHO.

WHO. (2020a). *Biosafety programme management.* WHO.

WHO. (2020b). *Laboratory biosafety manual* (4th ed.). WHO.

# 5 Performing Risk Assessments in Biological Laboratories

*Reynand Jay C. Canoy*
Scilore LLC, Muntinlupa City
Philippines

Institute of Biology, College of Science
University of the Philippines
Diliman, Quezon City, Philippines

*Rodel Jonathan S. Vitor II*
Department of Biology, College of Science
De La Salle University
Manila, Philippines
National Training Center for Biosafety and Biosecurity
National Institutes of Health
University of the Philippines
Manila, Philippines

## 5.1 INTRODUCTION

Working in any laboratory comes with dangers or risks. If these risks are not properly addressed and mitigated, then these can cause significant harm. However, these risks do not only apply to the people working inside the laboratory but also to those outside of the laboratory facility and to those in the general community. Risk assessment is an essential tool that, when used correctly, can effectively address and mitigate the ever-present risks in the laboratory.

Perhaps the greatest risk in biological laboratories is the Laboratory-Acquired Infection (LAI). As the term denotes, this is an infection or illness that is acquired due to exposure to activities performed inside the laboratory. As such, laboratory technicians and researchers are the ones most susceptible to LAI. However, infectious aerosols may remain suspended and can travel distances (Merhi et al., 2022), possibly resulting in secondary infections. Consequently, all individuals in the laboratory facility and in the surrounding community are at risk of LAIs.

The occurrence of LAI was reported as early as 1897 in the United Kingdom (Meyer & Eddie, 1941). A survey on reported cases of LAIs revealed that they are caused by bacterial pathogens, viruses, parasites, and fungi (Pike, 1976; Sewell, 1995;

 DOI: 10.1201/9781003426219-5

Siengsanan-Lamont & Blacksell, 2018; Singh et al., 2015; Sulkin, 1961; Sulkin & Pike, 1951; Wurtz et al., 2016). Among the reported cases, almost all were not linked to any specific incident or accident (Pike, 1976; Wurtz et al., 2016), suggesting that LAIs arise inconspicuously. In biosafety level 3 and level 4 laboratories, most of the reported LAIs arose from exposure to infectious aerosols, while the rest stemmed from inadequate personal protective equipment, animal bites and scratches, splashes, spills, and non-compliance with safety rules. Interestingly, of these underlying causes, human errors account for most cases of LAI (Wurtz et al., 2016).

In this chapter, risk assessment as part of the entire biorisk management system of a biological laboratory is focused on. This is to bring perspective that a robust risk assessment is essential to effectively allocate resources necessary for a safe laboratory environment. As risk assessment serves as a backbone of a biorisk management system, this can further lead to a sustainable laboratory that is able to prioritize risk management strategies vis-à-vis the available limited resources. Therefore, it is important to note that although the WHO Biosafety Manual and the US CDC Biosafety Manual for Biological Laboratories offer guidance and framework for biological risk management, the focus of this chapter is to provide the fundamental steps and workflow in performing a risk assessment in a biological laboratory.

## 5.2 WHAT IS A RISK?

Aside from LAI, there are other risks that are inherently present or can arise in a laboratory. Table 5.1 shows some of these risks, which are not necessarily exclusive to biological laboratories.

In layman's terms, 'risk' is synonymous with the word 'danger'. To a certain degree in casual conversation, they can be used interchangeably. Saying that there is a certain amount of 'danger' of getting hit by a car when we cross a street is

**TABLE 5.1**
**List of Common Risks, Hazards, and Threats in Biological Laboratories**

| Risk | Hazard | Threat |
|---|---|---|
| Laboratory Acquired Infection | Biological/clinical sample/organism/pathogens | Terrorists |
| Intentional/unintentional release of pathogens | Biological wastes | Disgruntled staff |
| Needle pricks | Laboratory animals and animal excreta | Staff with debts/problems |
| Failure of lab procedure | Animal products | State agents |
| Contamination | Insects | Departing employees |
| Slips, trips, falls | Sharps | Third-party vendors |
| Physical injuries | Glassware | Inside agents |
| Unauthorized access | Inexperienced staff | Institutional competitors |
| Damage of property/equipment | Irresponsible staff | |
| Loss/leakage/theft of materials/data/information | Untrained staff | |
| | Not certified/maintained equipment | |

essentially the same as saying that there is a certain amount of 'risk' of getting hit by a car when we cross a street.

In technical terms, risk is the likelihood of an undesirable event happening that involves a specific hazard or threat and has consequences. In other words, risk is the potential that a chosen action or inaction will lead to an undesirable outcome. It is important to note that inaction, or doing nothing, can also cause risks in the laboratory.

## 5.3 HOW DOES A RISK COME ABOUT?

To effectively address and manage these risks, it is essential to understand how they originate. Following its technical definition, a risk is the product of a hazard or a threat that arises in a particular situation. But first, let us examine what hazards and threats are. In casual conversations and some textbooks, the word 'hazard' is used interchangeably with the word 'risk'. However, in the biorisk management system, we make a clear distinction between hazards and risks. A hazard is the source or the specific entity that causes the risk. In other words, a hazard is the causative agent of a risk. In our first example, the identified danger or risk is that of you getting hit by the car. This includes the physical injuries that you would sustain if you were hit by the car. To identify the hazard in this case, we must ask what caused the physical injuries. And the answer to this is the car itself. Therefore, the hazard is the car. In the laboratory setting when you have LAI as the risk, your hazard might be the collected blood sample inside the tube that potentially contains infectious microorganisms.

In the context of biosafety, we use the term hazard, while in biosecurity we use the term threat. Hence, both hazards and threats are the causative agents of risks. Table 5.1 also shows some of the common laboratory hazards and threats. Please take note that a particular entity can become both a hazard and a threat, depending on the context. For example, a culture of *Bacillus anthracis* is a hazard that can cause LAI, and it can also be a threat due to unauthorized access and intentional release.

## 5.4 HOW TO PERFORM A RISK ASSESSMENT?

There are several approaches you can take when performing a risk assessment in a biological laboratory, made available by three internationally recognized references.

The Laboratory Biosafety Manual (4th Edition) of the World Health Organization (WHO) outlines a five-step risk assessment framework which is anchored on the Plan-Do-Check-Act cycle. The five steps are:

1) gathering information,
2) evaluating the risks,
3) developing a risk control strategy,
4) selecting and implementing risk control measures, and
5) reviewing risks and risk control measures.

Similarly, the Biosafety in Microbiological and Biomedical Laboratories (6th Edition) of the US Centers for Disease Control and Prevention (CDC) introduces a six-step management process:

1) identifying hazardous characteristics of the agent and assessing the inherent risks,
2) identifying laboratory procedure hazards,
3) determining the appropriate biosafety level and selecting additional precautions,
4) reviewing risk assessment and selected safeguards before implementation,
5) evaluating the proficiencies of staff, and
6) revisiting regularly and verifying risk management strategies.

On the other hand, the International Standard ISO 35001:2019 for the Biorisk Management of Laboratories and Other Related Organisations emphasizes on the various activities and supporting information when conducting risk assessments. This standard underscores that both likelihood and consequences are integral components of risk assessment and must always be included in such evaluation processes.

While both WHO Laboratory Biosafety Manual and BMBL provide a risk assessment framework that encompasses the entire biorisk management system, including the implementation and monitoring of control measures, this chapter focuses exclusively on risk assessment. Regardless of the specific approach you choose, performing risk assessments can be broken down into three essential steps:

1) identification of the risk, which stands as paramount,
2) characterization of the identified risk in terms of how likely it will occur and how severe the potential damage could be if it materializes, and
3) evaluation of the identified and characterized risk based on its significance to the laboratory or institution, as this will directly shape or influence your overall biorisk management system.

In this book chapter, we will delve deeper into these three important aspects in performing a risk assessment.

### 5.4.1 Risk Identification

The identification of risks can reasonably be considered the most important part of performing a risk assessment. Without the risk being properly identified, your risk assessment will lead nowhere. In risk identification, it is essential that we always go back to its technical definition. A risk is the product of a hazard or a threat given in a particular situation. As such, there are two components of risk: the hazard/threat and the situation surrounding the particular hazard/threat.

In our example, the risk is of you getting hit by the car. In this case, we have already identified the hazard, which is the car. But to properly identify the risk, we also need to identify the situation. In this case, the situation is you crossing the street.

But will the risk change if you are not crossing the street? If you are simply standing by the traffic light on a pedestrian lane while waiting for the green signal, will it be different? The two scenarios of you actually crossing the street and you simply standing by the traffic light on a pedestrian lane are two situations that can modify the risk of you getting hit by the car. Your hazard is still the same, which is the car, and the situation can either be you crossing the street or you standing still by the traffic light. Therefore, to properly identify the risk, you can either say 'the risk of you getting hit by a car while crossing the street' or 'the risk of you getting hit by a car while standing still by the traffic light on a pedestrian lane waiting for the green traffic signal'.

In the risk of LAI in the laboratory, the hazard can be your extracted blood inside the collection tube with potentially unknown infectious microorganisms. The situation can either be you opening the tube inside a certified biosafety cabinet or you opening the tube on the benchtop and outside a biosafety cabinet. Therefore, to properly identify the risk in this case, you can say 'the risk of LAI while opening a blood tube with potentially unknown infectious microorganisms inside a certified biosafety cabinet' or 'the risk of LAI while opening a blood tube on the benchtop and outside a certified biosafety cabinet'.

In the properly identified risks in the cited examples, there are always two components: the hazard and the situation.

### 5.4.2 Risk Characterization

Your risk assessment will either rise or fall based on your risk characterization. It is a critical step because it assigns value to the identified risk. This value depends on how likely the risk is to happen, as described by the situation, and how severe the potential damage is if it were to happen.

After identifying the risk with a specific hazard and a qualifying situation, the next step is to characterize the risk in terms of its likelihood and consequence. The likelihood of a risk is the probability that the risk will happen in a particular situation. In the laboratory, the likelihood of a risk is determined by considering all the control measures currently being implemented. That is why it is important to qualify the situation of the identified risk. On the other hand, the consequence of a risk is dictated by the severity of the damage the risk will cause if it occurs. Therefore, the consequence of a risk does not take into account the mitigation measures currently being implemented. It simply assumes that the risk will occur and estimates how severe the damage will be, regardless of whether there are mitigation measures in place or not.

In our pedestrian example, the likelihood of the risk of you getting hit by a car can be different depending on the situation. When you are crossing the street, the likelihood is higher compared to if you are simply standing by the traffic light on a pedestrian lane. As is evident, the likelihood changes as the situation changes. On the other hand, the consequence of the risk of you getting hit by a car does not change whether you are crossing the street or standing still by the traffic light on a pedestrian lane. Regardless of the situation, the damage it will inflict on you will be roughly the same. Applying this logic to our risk of LAI in the laboratory, the likelihood is also different if you are opening the blood tube in a certified biosafety cabinet compared to if you are opening the blood tube outside a biosafety cabinet. But the consequence of LAI will always be the same wherever you open the blood tube.

As risk characterization gives value to an identified risk in terms of its likelihood and consequence, you can assign either qualitative or quantitative values (Table 5.2 and Table 5.3). Having this scale will help you characterize your risk better. It is also possible to modify the scale yourself to suit your needs.

One common mistake in risk assessment is changing the consequence value of the risk after you implement the necessary control measures. This is somewhat understandable as implementing mitigation measures lowers the likelihood of the risk, so the consequence is mistakenly lowered along with likelihood as well. However, according to our definition, the consequence of a risk is not affected by the mitigation measure being implemented, which makes lowering the consequence of risk after risk mitigation erroneous. This is true in most cases. However, there are a few

**TABLE 5.2**
**Descriptions of Likelihood and Consequences**

| Likelihood | | | Consequence | | |
|---|---|---|---|---|---|
| **Level** | **Likelihood Descriptors** | **Description** | **Level** | **Consequence Descriptors** | **Consequence Description** |
| 1 | Rare | May occur only in exceptional circumstances | 1 | Insignificant | No injuries, low financial loss |
| 2 | Unlikely | Could occur at some time | 2 | Minor | First aid treatment, on-site release immediately contained |
| 3 | Possible | Might occur at some time | 3 | Moderate | Medical treatment required, on-site release contained with outside assistance, high financial loss |
| 4 | Likely | Will probably occur in most circumstances | 4 | Major | Extensive injuries, loss of production capability, off-site release with detrimental effects, major financial loss |
| 5 | Almost certain | Is expected to occur in most circumstances | 5 | Catastrophic | Death, toxic release off-site with detrimental effect, huge financial loss |

**TABLE 5.3**
**Sample Biosafety Risk Assessment Table**

| | Identification | | | | | Characterization | | Evaluation | | |
|---|---|---|---|---|---|---|---|---|---|---|
| Stage | Risk | Hazard/ Threat | Risk Group Level | Route of Exposure | Situation and Existing Control Measures | Likelihood | Consequence | Score | Acceptance | Decision |
| Receipt of Samples | LAI | Blood tubes with potential pathogens | 2 | Aerosol by inhalation | Blood tubes are triple-packed and compliant to IATA guidelines on sample transport | 1 | 4 | Low | Accept | Allow activity to proceed |
| Temporary Storage of Samples | LAI | Blood tubes with potential pathogens | 2 | Aerosol by inhalation | Blood tubes are put in a stable rack inside a biomedical refrigerator | 1 | 4 | Low | Accept | Allow activity to proceed |

*(Continued)*

**TABLE 5.3 (CONTINUED)**
**Sample Biosafety Risk Assessment Table**

| | Identification | | | | | Characterization | | Evaluation | | |
|---|---|---|---|---|---|---|---|---|---|---|
| Stage | Risk | Hazard/ Threat | Risk Group Level | Route of Exposure | Situation and Existing Control Measures | Likelihood | Consequence | Score | Acceptance | Decision |
| Processing of Samples | LAI | Blood tubes with potential pathogens | 2 | Aerosol by inhalation | Blood tubes are opened and samples are processed inside a biosafety cabinet that was certified more than one year ago | 3 | 4 | High | Reject | Do not allow this activity to proceed until the biosafety cabinet is certified by competent and duly-certified personnel |
| Analysis of Sample Results | Incorrect pathogen diagnosis | Sample raw results | n/a | n/a | Personnel doing the analysis is highly competent and well-trained | 1 | 4 | Low | Accept | Allow activity to proceed |

*(Continued)*

**TABLE 5.3 (CONTINUED)**
**Sample Biosafety Risk Assessment Table**

| | Identification | | | | | Characterization | | Evaluation | | |
|---|---|---|---|---|---|---|---|---|---|---|
| **Stage** | **Risk** | **Hazard/ Threat** | **Risk Group Level** | **Route of Exposure** | **Situation and Existing Control Measures** | **Likelihood** | **Consequence** | **Score** | **Acceptance** | **Decision** |
| Release of Sample Results | Mistakenly released results | Sample analyzed results | n/a | n/a | Personnel in-charge of the release of results is highly competent and well-trained | 1 | 4 | Low | Accept | Allow activity to proceed |
| Long-Term Storage of Samples, Processed Samples, or Derivatives | Sample spoilage | Sample DNA | 2 | Aerosol by inhalation | DNA is stored in a five-year-old freezer that has never been subjected for a preventive maintenance check and the temperature is manually monitored by a laboratory personnel who is frequently absent from work | 2 | 4 | Medium | Accept with caution | Allow activity to proceed but at least have the temperature monitoring done by a reliable laboratory personnel |

*(Continued)*

**TABLE 5.3 (CONTINUED)**
**Sample Biosafety Risk Assessment Table**

| | Identification | | | | | Characterization | | Evaluation | | |
|---|---|---|---|---|---|---|---|---|---|---|
| **Stage** | **Risk** | **Hazard/ Threat** | **Risk Group Level** | **Route of Exposure** | **Situation and Existing Control Measures** | **Likelihood** | **Consequence** | **Score** | **Acceptance** | **Decision** |
| Decontamination of Samples, Processed Samples, or Derivatives | LAI | Sample waste products | 2 | Aerosol by inhalation | Two-year-old autoclave that has never been subjected for a preventive maintenance check, calibration, and validation | 3 | 4 | Medium | Reject | Do not allow this activity to proceed until the autoclave is properly checked, calibrated, and validated |
| Waste Disposal of Samples, Processed Samples, or Derivatives | Foul smell | Sample waste products | 2 | Aerosol by inhalation | Decontaiminated wastes are stored in a waste holding room and collected by third-party waste collector every day | 1 | 2 | Low | Accept | Allow activity to proceed |

instances in which a mitigation measure can actually lower the consequence value of a given risk. In our LAI example, if your mitigation measure is to inoculate the personnel and the people in the surrounding community with an effective vaccine, or if you make an effective medication accessible to the general public, then you can actually lower the consequence of the risk. Hence, utmost diligence must be exercised in doing risk characterization.

### 5.4.3 Risk Evaluation

The values you obtain in a risk characterization are useless if they are not put into their proper context. In risk evaluation, you give meaning to your identified and characterized risks. It may seem to be the easiest part of risk assessment; however, it is the most difficult and the most crucial step. This step is crucial because the mitigation of risks will always depend on how the risks are evaluated. One simple way to evaluate a risk is by determining whether you accept it or not. While this may seem simple, accepting this risk also means that you assume the responsibilities and liabilities should the risk materialize.

In our pedestrian example, whether you are crossing the street or simply standing still by the traffic light, it means that you accept the possibility that you will be hit by a car, although the likelihood of that happening is different based on the situation. In this example, accepting or rejecting the risk of you getting hit by a car might seem menial. However, in the laboratory as in our LAI example, accepting the risk means that you take on responsibility and liabilities for any damages that might incur should laboratory personnel contract LAI. Damages might include hospital and recuperation expenses for the personnel, as well as the tarnished reputation of the laboratory facility, as perceived by the surrounding community. Therefore, the laboratory top management must weigh their options: either pay for any monetary damages resulting from LAI or invest in preventive measures to avoid LAI altogether. As this involves resource allocation, the decision can be challenging.

Rejecting a risk may seem to be the straightforward choice, as you would avoid the risk altogether. However, this is not always feasible, especially if the activity is essential. Using our pedestrian example, what if you need to cross the street to reach your destination? Simply avoiding the action is not an option since it prevents you from arriving at your desired destination. Likewise, in our LAI example, opening the blood tubes is the first step for you to conduct molecular diagnostic testing. You cannot avoid opening the blood tubes, as this is your primary task or job.

In many cases in the laboratory, you must decide under which conditions an activity will proceed. Plotting the risk on a graph in terms of likelihood and consequence and assigning a value score like 'low', 'medium', or 'high' can aid this decision (Figure 5.1). Using this scale, the risk of getting hit by a car can be evaluated as low if you are simply standing by the traffic light of a pedestrian lane, while it is high if you are crossing the street. In our LAI example, the risk is low if you open the tubes inside a certified biosafety cabinet while it is high if you open them outside a biosafety cabinet. With these values, you can set up a policy or rule that you will only be accepting low risks, and thus in the LAI example, you will only open blood

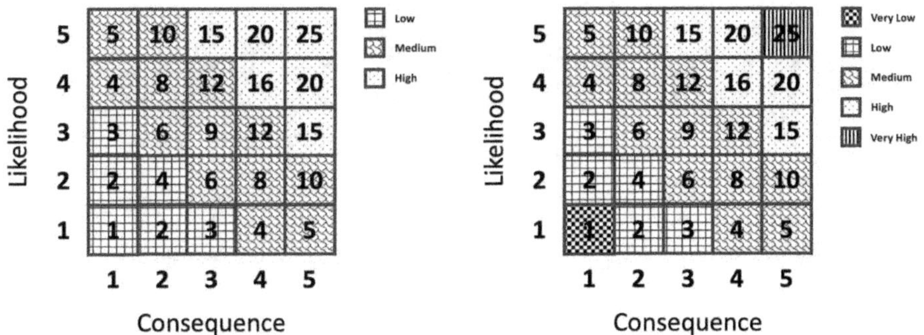


**FIGURE 5.1** Risk evaluation using a risk impact matrix. A three-level (left) and a five-level (right) evaluation of the risk is shown. The overall risk level increases as the scores of likelihood and consequence increase. The top management of an institution or a laboratory ultimately determines the corresponding mitigation measures that are appropriate for each evaluated risk. Having more risk levels (five instead of three) can aid the top management in better prioritizing and allocating resources within the overall biorisk management system.

tubes inside a certified biosafety cabinet. Depending on the circumstances, further dividing 'low' and 'high' into more precise tiers can enhance your risk evaluation (Figure 5.1).

Risk evaluation is inherently subjective and thus should involve group input. While trained biorisk officers or professionals can offer recommendations, top management typically makes the final decision. This ensures that the risk assessment process receives the appropriate attention and significance. Also, this makes the stakeholders, including the trained biorisk officers or professionals, not singularly liable for any outcomes.

Risk evaluation can be influenced by the perceptions of laboratory personnel, the institution, and people in the surrounding community. Several factors can shape these perceptions, including education, professional and financial goals, reputation, legal considerations, compliance with rules and regulations, community health, and opportunities for education or professional advancement. Since the institution's top management primarily decides risk evaluation, if an activity presents unacceptable risks, the institution will either cease the activity or find ways to reduce the risks to acceptable levels.

There is also this concept of being risk-averse and risk-tolerant. For the same identified and characterized risks (risk score of 9), a risk that is evaluated as high and is not accepted by risk-averse institutions would be evaluated as low and is accepted by risk-tolerant institutions (Figure 5.2). If risk-averse institutions need to proceed with activities that have high risks, they must find ways and invest in resources to control and mitigate the risks to bring them down to acceptable levels. On the other hand, risk-tolerant institutions would proceed with the concerned activities with low risks and would only spend minimal resources, if any, to maintain them at acceptable levels. Indeed, this clearly demonstrates that risk evaluation drives investment decisions.

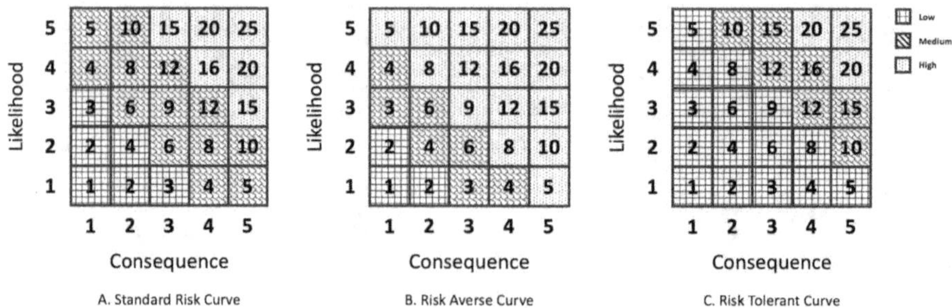


**FIGURE 5.2** Risk impact matrices depicting the standard risk curve (A), risk-averse curve (B), and risk-tolerant curve (C). For a risk characterized similarly, it can be evaluated as high by those who are risk-averse or as low by those who are risk-tolerant. The risk perception and appetite of the people involved in the risk assessment drive the tolerance or aversion toward risk. Consequently, this significantly affects the development of risk mitigation strategies and the allocation of resources for them.

## 5.5 HOW TO INTERPRET RISK ASSESSMENTS?

Risk assessment within the context of biorisk management is a systematic process designed to identify, characterize, and evaluate potential dangers associated with working with biological materials in the laboratory. It comprehensively examines the inherent and acquired hazards and threats in the laboratory, along with the likelihood and consequences of their eventual occurrence. By understanding these elements, risk assessment provides a clear understanding of how the possible dangers in the laboratory can happen and the magnitude of the damage should the dangers occur. In this manner, top management can make informed decisions in dealing with the risks in the laboratory.

One of the most crucial roles of risk assessment in biorisk management is to enable the crafting of strategies to eliminate or mitigate the risks in the laboratory. It serves as a point of measurement as to how much the top management can tolerate the risks involved. Without it, there is no scale that the top management can use in deciding which approach to take in mitigating the risks. For example, when a laboratory is involved in research on highly contagious pathogens, a risk assessment may reveal that the risk of accidental release is higher than anticipated. In response to this finding, the laboratory might implement stricter containment protocols, upgrade its safety equipment, and enhance staff training. This realignment of safety measures is directly driven by the information obtained from risk assessment. Fundamentally, risk assessment guides organizations and institutions on how to allocate resources and develop effective safety protocols. By tailoring these measures to the specific risks identified, biorisk management becomes a proactive endeavor rather than a reactive one. Such proactive management is essential in preventing accidents and avoiding the potential consequences of mishandling dangerous biological agents.

Beyond guiding risk mitigation strategies, risk assessment is an essential tool for ensuring that risks remain at acceptable levels. The term 'acceptable risk' is crucial in biorisk management, as it defines the threshold beyond which the potential consequences of a biological incident become unacceptable. In the context of a high-security biological research facility, risk assessment plays an integral role in determining the maximum allowable risk level for a particular activity. If the assessed risk exceeds this set acceptable threshold, then corrective actions must be taken. These actions may include additional safety measures, containment improvements, or even reconsidering the necessity of stopping the activity itself. Furthermore, risk assessment is not a static process. It is an ongoing, dynamic endeavor. As new information becomes available or as circumstances change, risk assessments should be revisited and updated. This adaptability ensures that as our knowledge of biological risks evolves, our risk mitigation strategies and safety protocols evolve alongside, thus maintaining risks at acceptable levels.

## 5.6 WHO SHOULD PERFORM RISK ASSESSMENTS?

As discussed in this book chapter, performing risk assessments involves the clear identification, comprehensive characterization, and informed evaluation of risks. In this context, trained biorisk officers or professionals do not perform this task alone. Instead, they lead a team of stakeholders in performing risk assessments. The team can encompass all relevant individuals, laboratory personnel, and experts knowledgeable about the specific hazards or threats in the laboratory. Together with the lead biorisk officer or professional, they can effectively identify and characterize risks. For example, if an assessment in a microbiology laboratory is underway, then stakeholders from the microbiology laboratory, including personnel and experts in the field, should be integrally involved. This will ensure that all inherent risks are thoroughly identified and characterized as activity- and protocol-specific risk assessments are conducted. After that, top management needs to engage in the evaluation of risks based on how they were identified and characterized. The participation of the top management is crucial as the risk evaluation significantly influences the investment decisions of the institution. Likewise, there may be cases where representatives of the community, wherein the laboratory facility is located, also need to be included in the risk assessment, especially if the risks being assessed can directly affect the community. Therefore, performing risk assessments is a group endeavor, and the trained biorisk officer or professional primarily provides support or guidance for its proper implementation.

## 5.7 WHEN SHOULD RISK ASSESSMENTS BE PERFORMED?

Frequency of risk assessments is largely dependent on the policy set by the top management of the institution. This can be influenced by the types of activities being performed and the specific biological materials being handled within the laboratory facility. If the institution also adheres to other international management standards, such as ISO 9001, then the frequency of risk assessments usually aligns with those standards. As such, risk assessments are to be performed at least once a year. Depending on the risks that the laboratory facility manages, it may need to conduct more frequent risk assessments to monitor the adequacy of the mitigation strategies being implemented. This was evident during the COVID-19 global pandemic, where regular risk assessments were performed as soon as additional information about the virus became available, facilitating the revisions in how the disease was diagnosed and managed.

## 5.8 BIOSAFETY RISK ASSESSMENT

One approach you can consider in performing risk assessments in biological laboratories is to follow the life cycle of biological samples. You can start by tracing the associated risks from the receipt of samples; temporary storage of samples prior to processing; actual processing of samples; analysis and release of sample results; long-term storage of samples, processed samples, or derivatives; and decontamination and waste disposal of samples, processed samples, or derivatives. Each stage of this life cycle has biosafety and/or biosecurity risks which need to be properly assessed to be effectively mitigated and monitored.

Table 5.3 shows an example of a risk assessment conducted by a laboratory that receives and processes blood samples for the molecular detection of human pathogens.

## 5.9 BIOSECURITY RISK ASSESSMENT

Biosecurity risk assessment is a vital practice in ensuring the safety and security of biological materials and facilities. While it shares similarities with biosafety risk assessment, its scope is broader as it considers vulnerabilities in the life cycle of biological samples. Such a comprehensive approach is vital essential for safeguarding against intentional threats, theft, or misuse of sensitive materials within the laboratory.

In conducting biosecurity risk assessments, it is important to begin by examining the various stages in the life cycle of biological samples. One key vulnerability to assess is at the point of sample receipt within the laboratory. Here, the focus should be on identifying potential weaknesses or security gaps that malicious actors might exploit to steal or tamper with the sample. By thoroughly analyzing each stage, from sample acquisition to disposal, institutions can better understand their vulnerability to biosecurity threats.

In addition, institutions can utilize specialized tools and frameworks to enhance the effectiveness of their biosecurity risk assessments. Two valuable resources for this purpose are the Novel Dutch Self-Assessment Biosecurity Toolkit (https://www.biosecurityselfscan.nl/home) and the National Laboratory Biosecurity Assessment and Monitoring Checklist (Brizee et al., 2019).

The Novel Dutch Self-Assessment Biosecurity Toolkit offers a unique approach to biosecurity risk assessment. With this toolkit, users can identify the specific biosecurity pillars or aspects where they are most vulnerable. Such targeted assessment allows institutions to focus their efforts on areas that require the most attention and improvement. So, this toolkit is particularly useful in tailoring biosecurity measures to address an institution's specific weaknesses effectively.

On the other hand, the National Laboratory Biosecurity Assessment and Monitoring Checklist provides a structured set of questions and guidelines for evaluating various biosecurity components within a laboratory or research facility. These questions cover a wide range of aspects, from access control to personnel training and from sample handling to emergency response protocols. The checklist can be customized to suit the unique circumstances and priorities of each country or institution. All in all, it serves as a comprehensive tool for identifying vulnerabilities and assessing biosecurity readiness.

The primary goal of both the Novel Dutch Self-Assessment Biosecurity Toolkit and the National Laboratory Biosecurity Assessment and Monitoring Checklist is to help institutions in conducting comprehensive vulnerability assessments. Doing this is a critical step in enhancing biosecurity measures and mitigating potential threats. By identifying weaknesses and areas of concern, institutions can develop targeted strategies to address these vulnerabilities and improve their overall biosecurity posture.

## REFERENCES

Brizee, S., Passel, M. W. J. van, Berg, L. M. van den, Feakes, D., Izar, A., Lin, K. T. B., Podin, Y., Yunus, Z., & Bleijs, D. A. (2019). Development of a biosecurity checklist for laboratory assessment and monitoring. *Applied Biosafety*, *24*(2), 83–89. https://doi.org/10.1177/1535676019838077

Merhi, T., Atasi, O., Coetsier, C., Lalanne, B., & Roger, K. (2022). Assessing suspension and infectivity times of virus-loaded aerosols involved in airborne transmission. *Proceedings of the National Academy of Sciences*, *119*(32), e2204593119. https://doi.org/10.1073/pnas.2204593119

Meyer, K. F., & Eddie, B. (1941). Laboratory infections due to Brucella. *The Journal of Infectious Diseases*, *68*(1), 24–32.

Pike, R. M. (1976). Laboratory-associated infections: Summary and analysis of 3921 cases. *Health Laboratory Science*, *13*(2), 105–114.

Sewell, D. L. (1995). Laboratory-associated infections and biosafety. *Clinical Microbiology Reviews*, *8*(3), 389–405. https://doi.org/10.1128/cmr.8.3.389

Siengsanan-Lamont, J., & Blacksell, S. D. (2018). A review of laboratory-acquired infections in the Asia-Pacific: Understanding risk and the need for improved biosafety for veterinary and zoonotic diseases. *Tropical Medicine and Infectious Disease*, *3*(2), Article 2. https://doi.org/10.3390/tropicalmed3020036

Singh, L., Pushker, N., Saini, N., Sen, S., Sharma, A., Bakhshi, S., Chawla, B., & Kashyap, S. (2015). Expression of pro-apoptotic Bax and anti-apoptotic Bcl-2 proteins in human retinoblastoma. *Clinical and Experimental Ophthalmology*, *43*(3), 259–267. https://doi.org/10.1111/ceo.12397

Sulkin, S. E. (1961). Laboratory-acquired infections. *Bacteriological Reviews*, *25*(3), 203–209. https://doi.org/10.1128/br.25.3.203-209.1961

Sulkin, S. E., & Pike, R. M. (1951). Survey of laboratory-acquired infections. *American Journal of Public Health and the Nation's Health*, *41*(7), 769–781. https://doi.org/10.2105/AJPH.41.7.769

Wurtz, N., Papa, A., Hukic, M., Di Caro, A., Leparc-Goffart, I., Leroy, E., Landini, M. P., Sekeyova, Z., Dumler, J. S., Bădescu, D., Busquets, N., Calistri, A., Parolin, C., Palù, G., Christova, I., Maurin, M., La Scola, B., & Raoult, D. (2016). Survey of laboratory-acquired infections around the world in biosafety level 3 and 4 laboratories. *European Journal of Clinical Microbiology and Infectious Diseases*, *35*(8), 1247–1258. https://doi.org/10.1007/s10096-016-2657-1

# 6 Minimizing Environmental Footprints of Biological Wastes through Biosafety and Biosecurity Measures

*Jonathan Jaime G. Guerrero*
University of the Philippines, Manila
College of Medicine and College of Public Health
Manila, Philippines

*Angelo R. Agduma*
University of Southern Mindanao Kabacan
College of Science and Mathematics
Department of Biological Sciences
Ecology and Conservation Research Laboratory
Kabacan, Cotabato, Philippines and Guangxi University
College of Forestry, Nanning City
Guangxi, P.R. China

*Krizler C. Tanalgo*
University of Southern Mindanao Kabacan
College of Science and Mathematics
Department of Biological Sciences
Ecology and Conservation Research Laboratory
Kabacan, Cotabato
Philippines

*Rhea G. Abisado-Duque*
Ateneo de Manila University
Department of Biology
Quezon City

*Frederick John B. Navarro*
Scilore, LLC, Ayala Alabang
Muntinlupa City

DOI: 10.1201/9781003426219-6

## 6.1 INTRODUCTION

Human activities that drive development generate waste that eventually find their way into the environment. Larger cities where commerce and industrial innovations are centralized, and have larger populations, are likely to contribute more to waste generation. In the Philippines, for example, 25% of the country's waste is generated in the component cities of Metro Manila, the country's capital (Magalang, 2014). Cities across the Philippines which are key destinations for tourism, preferred manufacturing sites of corporations, and prime metropolitan gateways are likewise contributory to the national solid waste generation (Ancog et al., 2012; Lunag et al., 2019; Olalo et al., 2022).

Managing wastes takes several forms including composting, landfills, and incineration, to name a few (Karak et al., 2012), and depends highly on where the country is on the global economic stage, as in the case of Asian countries (Shekdar, 2009). In many parts of the world, including the Philippines, the management of common household or solid waste is legally prescribed (Paul et al., 2012). The Philippine Ecological Waste Management Act of 2000, or RA 9003, was signed into law in 2001 mandating local government units (LGUs) to implement policies and programs to manage solid waste within their jurisdiction (Acosta et al., 2012). Household and business establishments are major players in this implementation as seen in many local studies (Bernardo, 2008; Holmer & Guanzon, 2003; Reyes & Furto, 2013; Vivar et al., 2015; Wynne et al., 2018). Essential to this are university-level initiatives to promote awareness among faculty and students, support local government agenda along solid waste management, and establish community and private partnerships in advancing environmental governance (Anacio, 2017; Cuaresma, 2019; Era, 2019; Gequinto, 2017).

Part of this solid waste are those generated from medical and diagnostic laboratories, and other biological wastes from research and academic institutions, and agriculture facilities. These wastes are unique in the risks they pose to handlers and to the environment and thus require careful and deliberate management practices. Some countries have separate medical waste legislation but may lack either clarity or implementation (Coker et al., 2009; Windfeld & Brooks, 2015). Universities who conduct research on biological materials have management systems in place, such as in the United States (Mecklem, n.d.), but may not be true in low-income countries due to multiple barriers (lack of resources, personnel, incentives, and motivation) (Yekkalar et al., 2015).

The challenge of solid waste management, especially of infectious wastes from hospitals, diagnostic and research laboratories, and even academic institutions working on biological materials, is a biosafety and biosecurity concern. For instance, dumping of wastes in landfills pose the risk of infection among garbage collectors and the informal sector who rely heavily on recoverable materials. Developing countries have integrated the informal sector in waste management as a boost to their livelihood (Wilson et al., 2006) and are taking an active role especially in the collection and recycling processes. For example, at least 13% of wastes generated in Bandung, Indonesia, were collected by the informal sector, diverting these into recyclable materials (Sembiring & Nitivattananon, 2010). In the Mekong Delta region, Vietnam, the informal sector forms an intricate network with the formal sector in the waste management system (Tong et al., 2021). In the Philippines, a waste worker association is established in Iloilo City as one of the means to integrate the informal sector in the municipal solid waste management program (Paul et al., 2012). The informal sector in Thailand also reduces the burden of waste disposal through material recovery and recycling (Chiemchaisri et al., 2007).

Furthermore, contamination of soil and water ecosystems is an age-old problem, especially in areas with no sanitation facilities and nearby major landfills. Microbiological contamination is a serious risk to public health (Pedley & Howard, 1997) by rendering groundwater unreliable for domestic use (Mor et al., 2006). Toxic metals, while non-biological in nature, essentially pose the same risk by contamination of groundwater or bioaccumulation into the human food supply (Parvin & Tareq, 2021). Unmanaged biomedical wastes are implicated in the death of an estimated 5.2 million people, 4 million of which are children (Rahman et al., 2020a). Inefficient treatment procedures, limited resources, and improper disposal of medical wastes create serious risks, not only for healthcare workers but to society in general (El-Ramady et al., 2021). In the current COVID-19 pandemic, healthcare waste management in the Philippines is also critical in the control of the disease (Atienza et al., 2022), a challenge shared among its ASEAN neighbors (Agamuthu et al., 2022; Budihardjo et al., 2022; Nguyen et al., 2022; Srikanth et al., 2022).

The risk posed to biodiversity, from microbes to flora and fauna, is equally important. Pharmaceutical wastes can impact aquatic organisms, enter the food chain, and disrupt physiology, which can cause trophic cascades (Néstor & Mariana, 2019; Sharma & Kaushik, 2021). Antimicrobial resistance can also arise from environmental contamination (White & Hughes, 2019) and the uncontrolled use of antibiotics in agriculture (Samreen et al., 2021). Soil pollution brought by medical wastes can lead to loss of ecosystem services (Raimi et al., 2022a) and by extension a peril to biodiversity.

The environmental aspect of a biorisk management system acknowledges that good sanitation, including good hygiene and safe water, is key component and contributory to overall health and socioeconomic development (Mara et al., 2010; Van Minh & Hung, 2011). Biosafety and biosecurity concerns arising from environmental science and management, therefore, require thorough investigations.

## 6.2 CURRENT SITUATION OF BIOLOGICAL WASTES AND SOURCES

The high production rate and unsustainable management of biological waste worldwide were exacerbated when the novel coronavirus (SARS-COV-2; COVID-19) pandemic hit the global population (Singh et al., 2022). Tons of biohazardous waste were generated throughout the diagnosis, transport, isolation, testing, immunization, and treatment of COVID-19 patients and research of the infectious agent. Examples are blood samples, cell lines, microbiological cultures and stocks, expired vaccines, medical and laboratory instruments, body parts and fluids, and contaminated sharps (Capoor & Parida, 2021; Das et al., 2021; Khan et al., 2019; Kuppusamy et al., 2022). Before the pandemic, hospitals in Wuhan City in China produced about 40 metric tons daily (mt/d) of biomedical waste. However, during the height of the pandemic, the city's healthcare facilities discharge grew to about 240 mt/d of waste (Hossain et al., 2020; Yu, 2020). On the other hand, India generated an average of 586 mt/d between 2017 and 2019. This increased to 850 mt/d during the pandemic (Chand et al., 2021; CSE India, 2021). Manila in the Philippines produced about 47 mt/d of biomedical waste in 2006 (Visvanathan, 2006) and had to deal with as much as 280 mt/d of additional biomedical waste during the pandemic period (ADB, 2020). A government hospital in Manila alone generated about 300 kg each day (Cabico, 2020), representing only a tiny fraction of the country's 975 mt/d discharge of biomedical waste (Fernandez, 2022). In 2020, the projected additional infectious medical waste due to COVID in other major cities in Asia were as follows: Jakarta, Indonesia: 212 mt/d, Kuala Lumpur, Malaysia: 154 mt/d, Bangkok, Thailand: 210 mt/d, and Ha Noi, Vietnam: 160 mt/d (ADB, 2020).

Medical and biological diagnostic, research, and teaching laboratories use human and animal structures and substances and microbiological cultures or specimens and carry out biological, immunological, biochemical, hematological, and other procedures for diagnosis and studies. Sharps such as razors, needles, scissors, and clippers, blood-soaked paper towels, bandages, and cut human hair from barbershops and beauty parlors are biohazardous. Like hospitals, retirement homes also generate a pile of waste, such as bandages, syringes, and adult diaper pants contaminated with blood or bodily fluids (US-Bio-Clean, 2014). Other probable sources of biological waste are wildlife parks, zoos, wildlife markets, animal farms, slaughterhouses, and wet markets (Al-Gheethi et al., 2021; Some et al., 2021). To protect the people and the environment, human actions, such as of administrators, workers, and the general public, in the generation and proper management of biological wastes are of prime importance.

## 6.3 EFFECTS OF BIOLOGICAL WASTES ON MICROBIAL DIVERSITY

Hospital waste have a higher diversity of potentially infective microbial species (Maina et al., 2018), than in residential waste (Kalnowski et al., 1983), but higher microbial load may be present in the former than in the latter (Egbenyah et al., 2021; Möse & Reinthaler, 1985). Although the majority of microorganisms are common to both sources, *Klebsiella pneumoniae*, *Acinetobacter* spp., and *Clostridium tetani* are exclusively found in the waste of healthcare institutions (Egbenyah et al., 2021). Gloves and gowns of nurses and patients, which would probably go into waste later on, can be contaminated with methicillin-resistant *Staphylococcus aureus* (MRSA) from wounds and urine of patients during morning care routines (Boyce et al., 1997). Besides *S. aureus* and *K. pneumoniae*, hospital waste can also contain *Bacillus subtilis* and *E.coli* (Anitha & Jayraaj, 2012).

The possibility that wastewater and soil from various sources can contain pathogens cannot be disregarded (Capoor & Parida, 2021; Manzoor & Sharma, 2019). They play a significant role in the growth and spread of a vast diversity of microorganisms and in developing antibiotic-resistant strains (Maria et al., 2019). Bacterial pathogens that are known to cause nosocomial infections, such as *Pseudomonas* spp., *Escherichia coli.*, *Klebsiella* spp., *Serratia marcescens*, *Proteus vulgaris*, and *S. aureus*, among others, were isolated by Maina et al. (2018) from surfaces, wastes, and dumpsites of hospitals. Also, since they grow well in decomposing waste in soils, they are held responsible for the emergence and spread of community-acquired infections (Maina et al., 2018). Ubiquitous major bacterial groups in wastewaters of different hospitals and sewage disposal ponds include Proteobacteria, Bacteriodetes, Firmicutes, and Actinobacteria (Cruz et al., 2021; Kaur et al., 2020). Specific bacterial and fungal isolates were *Aeromonas, Kluyvera, E. coli, Salmonella, Shigella, S. aureus, Faecalibacterium, Prevotella, Streptococcus, Acromobacter, Klebsiella, Enterobacter, Pseudomonas, Candida, Aspergillus, Terreus, Fusarium, Mucor, Paecilomyces, Cryptococcus, Penicillium, Trichophyton*, and *Rhizopus* (Katleho et al., 2022; Kaur et al., 2020).

Due to the widespread use of antimicrobials in wastewater treatments, antibiotic resistance among bacteria develops (Perry et al., 2021). In addition, antibiotics and antibiotic-resistant bacteria (ARB) enter the hospital waste through excretion from patients. In developing countries, the treatment system of wastes may not always be available in hospitals. This creates a favorable condition not only for microbial growth but also for the sharing of resistance genes among them (Riaz et al., 2020). Thus, healthcare facilities are emergence and spread hotspots for ARB and other potentially infectious microorganisms compared to other sources (Cruz et al., 2021; Hocquet et al., 2016) and therefore play a major role in their downstream route to soil, drinking water sources, surface water, and marine environments where they may alter the ecosystem processes and services (Hocquet et al., 2016; Riaz et al., 2020; Zhang et al., 2020).

## 6.4 IMPACTS OF BIOLOGICAL WASTES ON BIOLOGICAL DIVERSITY AND ECOSYSTEMS

When biological waste is not adequately disposed of, it can accumulate, displace, and negatively affect natural habitats, harming wildlife species of plants and animals that widely depend on pristine environments for survival (Evode et al., 2021). For example, the increasing use and accumulation of waste generated from personal protective equipment (PPE) during the COVID-19 pandemic concerns various health and environmental sectors (Kutralam-Muniasamy et al., 2022). These items often end up in inappropriate locations, such as beaches, rivers, and cities, resulting in littering (Canning-Clode et al., 2020; De-la-Torre & Aragaw, 2021).

When biological waste accumulates and enters the ecosystem, one of its significant impacts on biodiversity and the natural environment is its contribution to the increase in soil and water pollution. Soil and water pollution poses numerous risks that extend beyond the environment as it can negatively impact human and ecological health (Raimi et al., 2022b). Material waste from biological facilities has a high potential for soil and water contaminants, leading to environmental pollution. Biological pollutants released from healthcare and veterinary facilities can contaminate water bodies and are hazardous to aquatic organisms, and the effect can be magnified throughout the natural food chain (Khan et al., 2021; Mohan et al., 2021). Similarly, biological pollutants in the soil can harm plants and crops that serve as a refuge and habitat for various wildlife populations (Kemper, 2008; Sayadi & Pathak, 2010). Consequently, the impacts of biological waste on soil and water environments can lead to significant declines in a diverse population of sensitive and critical species of wildlife. Furthermore, threats from pollution, including biological waste, particularly to aquatic ecosystems alone, can affect at least 1.1 to 2.5 billion people between 2000 and 2050, mainly from poorer and developing economies (Wen et al., 2017).

Furthermore, exposure to biological waste pollutants in various ecosystems can have adverse effects on the physiology and ecological interactions of wildlife, whether it is short-term or long-term exposure (Arnold et al., 2014; Rhind, 2009). The effect of exposure to biological wastes can vary depending on organisms or ecosystems. Freshwater protozoans and metazoan parasites, for example, are dominant stressors for aquatic biota and affect the physiological process and population structure of their hosts. Still, when exposed to various wastes, including biological hazards, their effects on the host are also altered (Morley, 2009). Exposure of parasites to different biological waste products can increase the risk of developing parasite-resistant strains (Lebarbenchon et al., 2008). This disruption to ecological interactions is expected to continue and will be exacerbated by the continuous growth in the biological and pharmaceutical industries in modern times.

Finding solutions to minimize the accumulation of biological wastes and their associated risks is one of the priorities of many facilities. The open burning and incineration of biological waste are examples of management approaches to control waste production. However, it could emit harmful compounds, including dioxins, furans, and particulate matter, further creating another negative environmental cost (Velis & Cook, 2021). Efforts to address the impacts of biological waste pollution on biodiversity and the environment should prioritize the development of a sustainable and safe management of biological waste and its sources. This involves conducting appropriate environmental impact assessments, avoiding areas with high biodiversity and wildlife populations, and increasing public awareness of the adverse effects of biological waste. This initiative is crucial, particularly during the COVID-19 pandemic, where an increase in biological waste was observed as a trade-off for safety (Kanwar et al., 2022).

Addressing the negative impact of biological wastes requires addressing pollution and its sources, as they are interconnected and cannot be effectively tackled in isolation. Protecting biodiversity from pollution and biological waste involves reducing pollution at the source, promoting effective waste management systems, using effective decontamination of biological hazards, supporting conservation efforts, and raising awareness about the importance of protecting biodiversity. Furthermore, statutory policies and protocols should be established and developed to ensure that governments have the necessary regulatory frameworks in place to assess and manage the risks associated with biological waste and pollution toward biodiversity (Tickner et al., 2020). Identifying areas with high vulnerability to biological waste pollution and quantifying and understanding their negative impacts (e.g., socio-economic costs) are also other steps in reducing the negative impacts of biological wastes on biodiversity. These measures can help preserve ecosystem health and ensure species survival.

## 6.5 EFFECTS OF BIOLOGICAL WASTES ON HUMAN POPULATIONS

Human exposure to biological wastes poses a serious public health concern, especially in developing countries where waste management systems are poorly developed, inadequately funded, and badly implemented if not completely neglected (Apostol et al., 2022; Dihan et al., 2023). The impact of biological waste on human health depends on the magnitude and type of waste, as well as the route by which the biological wastes are introduced in the community. For example, inhalation of airborne infectious agents found in the biological wastes can cause respiratory diseases such as COVID-19 and exacerbate other diseases such as asthma and allergies (Agache et al., 2023; Barcelo, 2020; Sun et al., 2022). Infectious agents can also get into the food supply, resulting in the spread of foodborne illnesses (Osafo et al., 2022).

Untreated or improperly treated wastewater containing biological agents could also significantly contribute to the transmission of waterborne diseases such as cholera and typhoid fever, and even COVID-19 potentially (Gwenzi & Sanganyado, 2019; Hu et al., 2022; Kilaru et al., 2023). In developing countries where wastewater is generally poorly managed and mishandled, the potential of wastewater in spreading infection transmitted through ingestion, water vapor inhalation, and skin contact is significantly high (Barcelo, 2020; Dihan et al., 2023; Ketema et al., 2023). Reckless disposal of biological waste is not only a threat to the surface water but can also make its way to the groundwater, which is a very important water resource (Behera et al., 2022; Shayo et al., 2023; Sia Su, 2005). Furthermore, the spread of waterborne diseases is aggravated during typhoon seasons in flood prone areas as observed in the Philippines (Rocha et al., 2022; Sethi et al., 2023).

Biological wastes can also serve as a vehicle for multidrug-resistant bacterial infections which are complicated to treat and thus associated with increased mortality and morbidity (Ikhimiukor et al., 2022; Rosenberg Goldstein et al., 2012). Improper biological waste disposal does not only increase the chances of human exposure to antibiotic resistant microorganisms but also contributes to the development of antimicrobial resistance because the antibiotic resistance genes can be acquired by bacteria through horizontal gene transfer (Alam et al., 2019; Madigan et al., 2021). With the spread of antibiotic-resistant pathogens, the cases of nosocomial and opportunistic infections could also shoot up (Gidey et al., 2023).

Biological wastes have not only magnified health burdens on humans but also economic and social implications (Chokhandre et al., 2017; Sawant et al., 2021). Establishing and operating a waste treatment facility and treatment of infectious diseases are both costly. In one way or another, biological wastes affect the livelihood and daily lives of the community. Proper disposal of biological waste is essential to mitigate issues arising from the accumulation of biological waste. Thus, guidelines on proper management and disposal of biological waste in every barangay health center, clinics, hospitals, industries, and academic and research institutions must be developed and implemented to minimize biosafety-related concerns, disease transmission, and emergence and re-emergence of infectious diseases.

## 6.6 HAZARDOUS WASTE CLASSIFICATION AND SEGREGATION

From an institutional perspective, the amount and number of waste types can challenge its waste management system (WMS) (Ashbrook & Reinhardt, 2002; Rau et al., 2000). Thus, proper waste classification and segregation is the first crucial step for an effective WMS to protect human health and the environment and reduce significant amounts of hazardous waste and disposal costs.

When developing WMS, finding relevant waste codes ensures that local and national regulations are met for waste classification. Most countries have hazardous waste (HW) classification regulations that are based on international treaties and guidelines. Waste classification methods in the USA, China, Japan, and EU-member states combine source-oriented, product-oriented, process-oriented, and hazard characteristic systems. The process-oriented system is widely used for declaration and registration. It is simple and feasible for generators to manage the process based on waste properties (Wen et al., 2014). Product and industry-oriented systems are simple and efficient for waste generators with a regulatory understanding of the production. In addition, these systems generate data on the total environmental load for policy development on environmental management in companies and authorities (Buenrostro et al., 2001; Rønning et al., 2000). However, these systems may be impractical for facilities with unidentified, new, or uncharacterized wastes from educational or research activities whose effects on health and the environment are not known. Thus, a classification system using diversified hazard characteristics would be appropriate for hazard-specific wastes.

Hazardous waste can be described based on its physical, chemical, biological, and radioactive properties, generally categorized into chemical, biological, radioactive, or multi-hazardous wastes (Rau et al., 2000). Most universities in the US treat non-infectious human-cell-culture waste as biohazardous, conforming to the regulatory definition of waste. Sharp items like hypodermic needles, syringes with needles, and scalpel blades are classified as biohazardous sharps regardless of contamination status (Mecklem & Neumann, 2003). Integrating the ‘reduce, reuse, recycle’ concepts in institutional medical WMS reduces the regulated waste volume by 40% (Bai et al., 2013). In countries without legislation or non-inclusion of chemical waste in the national lists, chemical wastes are classified using the globally harmonized systems (GHS) scheme (Aja et al., 2016; Giovanni et al., 2021). In some cases, the chemical properties of reaction products are unknown. Thus, ignitability, corrosivity, reactivity, and toxicity testing are warranted before classification (Jha et al., 2022). These efforts reflect that countries are motivated to adopt the GHS scheme to reduce trade and transport barriers.

Facilities categorize radioactive waste based on the purpose of its derivation. The qualitative approach classifies waste according to its origin and physical state, while the quantitative approach classifies liquid waste based on concentration (Hosan, 2017). Mixed wastes are a subclass of multi-hazardous waste, which contain low-level radioactive waste in combination with chemical or biological waste. The concentration of the radioactive material is low to be regulated but requires special consideration to mitigate radiological hazards (Rau et al., 2000). Thus, properly characterizing hazardous waste helps identify the associated risk. Moreover, risk evaluation on the degree of consequence and the likelihood of exposure is necessary for assessing appropriate segregation strategies.

Most of the time, waste segregation is best done at the point of generation. Improvement of segregation practice includes engineering and administrative controls, personal knowledge, and attitude. In Ethiopia, color-coded bins at generation points motivates healthcare workers (HCWs) to practice segregation, which is 9.5 times more likely to perform compared to HCWs with no access to bins (Ibrahim et al., 2023; Sahiledengle, 2019). The majority of HCWs perform incorrect segregation if bins are not color-coded (82%) or lack a labeled biohazard symbol (63%) (Mesfin et al., 2014). Bai et al. (2013) proposed a color-coding system for eight classes of medical wastes to address deficient elements in the existing color-coding system (Table 6.1), which improves segregation practices by 80%, highlighting that colored bins were placed in the point of generation with international symbols and appropriate wording.

Administrative oversight improves WMS, which can be done by assigning a focal person for each area in a month and on a rotation basis. Reviewing and revising WMS, SOPs, and instructive posters motivates HCW. HCWs with WMS training will most likely have good segregation practices compared to staff lacking knowledge and awareness (Ibrahim et al., 2023; Johnson et al., 2013; Sapkota et al., 2014).

In Ethiopia, only 54% of the HCWs had good segregation practices in a self-assessment (Sahiledengle, 2019) and 46% performed correctly in actual settings (Mesfin et al., 2014). Aside from the mentioned factors, compliance and favorable attitude of HCWs were correlated to their demographic profile (sex, age, profession, length of service, and working hours) (Ibrahim et al., 2023; Mesfin et al., 2014). Other programs that positively motivate HCWs and waste handlers are the provision of a complete set of personal protective equipment (PPE) and immunization against hepatitis B and tetanus, which significantly improve waste management practices by 60% (Sapkota et al., 2014). Hence, enhancing classification and segregation practices in combination with other mitigation measures minimize the burden of infectious and hazardous waste, especially to waste handlers that ensure treatment before disposal is performed.

## 6.7 DECONTAMINATION PROCEDURES

Decontamination, the process of treating biological waste to make it safe for handling and disposal, is integral to the success and safety of laboratory operations (Khan et al., 2019; WHO, 2020a; Zaki & Campbell, 1997). It involves killing of

**TABLE 6.1**
**Suggested Color Codes for Different Types of Medical Waste (Bai et al., 2013)**

| Color Code | Medical Waste | Description of Wastes |
|---|---|---|
| Red | Microbiological wastes | Wastes that have potentially high concentrations of pathogenic organisms. These include microbial cultures (fungi, bacteria, viruses, and algae) or specimens in culture media (broth or agar), and stocks of infectious agents. |
| Yellow | Pathological wastes | These are infectious cadavers and carcasses, body parts, organs, or tissues, which are considered biohazardous because they are potential sources of pathogens and zoonotic diseases. |
| Blue | Sharps waste | These are discarded materials that could cause cuts or puncture wounds that can induce subdermal inoculation of infectious agents. |
| Brown | Pharmaceutical waste | These are discarded and unused pharmaceutical products, or materials associated with the production, distribution, and administration of drugs and medications. |
| Orange | Chemical waste | It consists of discarded, solid, liquid, and gaseous chemicals from experimental work, cleaning, housekeeping, and disinfecting procedures. |
| Silver | Radioactive waste | Waste that comes in radiotherapy and nuclear medicine procedures. It consists of pathological wastes, medical equipment, and PPE that has been contaminated with radionuclide. |
| Black | Non-recyclable waste | Waste that cannot be effectively or economically recycled through standard recycling processes. |
| Green | Recyclable waste | Waste that can be collected, processed, and reused to create new products or materials. |

all microorganisms, including the endospores, viruses, and prions (Madigan et al., 2021). In decontamination, waste segregation must be observed (Attrah et al., 2022; WHO, 2020b). The biological waste must be stored under cold conditions and properly transported if prompt and onsite decontamination is impossible (Agamuthu & Barasarathi, 2021; WHO, 2020a, 2020b). The most commonly used decontamination procedures in developing countries include the use of the autoclave, pressure cooker, incinerator, and chemical disinfectants.

An autoclave uses moist heat (steam under pressure) to eliminate all microorganisms (Madigan et al., 2021; WHO, 2020a; Willey et al., 2023). The application of pressure (15 psi) is necessary to get to the temperature (121 °C) required to eliminate the endospores (Tortora et al., 2019). Exposure to moist heat results in protein denaturation, cell membrane damages, and nucleic acid degradation (Tortora et al., 2019; WHO, 2020a; Willey et al., 2023). Compared to dry heat, moist heat is a physical method that has a better penetrating power, thus resulting in faster killing of microbes (Madigan et al., 2021). Autoclaving is commonly used because it is efficient, safe, and cost-effective (Ahmed et al., 2015; Chisholm et al., 2021; Ciplak & Kaskun, 2015; Diaz et al., 2005; Khan et al., 2019; Pathak et al., 2021; Shinee et al., 2008). It can be used to decontaminate solid wastes such as agar culture plates, pipette tips, sharps, and glassware, but not appropriate for large animal carcasses, bulky solid waste, and completely dry solid wastes such as laboratory gowns and gloves (Diaz et al., 2005; WHO, 2020b). Any liquid wastes except those contaminated with hazardous and flammable chemicals (e.g., bleach and ethanol) can also be autoclaved (WHO, 2020a, 2020b).

The effectiveness of an autoclave for decontamination relies on its proper operation so its validation, preventive maintenance, and usage must be closely monitored (Cascardo et al., 2020; WHO, 2020a, 2020b). For example, the autoclave default settings are found to be ineffective for some medical wastes (Cascardo et al., 2020; Garibaldi et al., 2017). Although holding time of 3 min at 134 °C, 10 min at 126 °C, 15 min at 121 °C, or 25 min at 115 °C generally works, the holding time must be extended if large volumes of waste are involved (Madigan et al., 2021; WHO, 2020b). It is also important that the waste packaging material is steam permeable, can be fully air evacuated, and not tightly sealed (WHO, 2020b). Moreover, overloading of autoclave and blocking of valves and drains must be avoided to allow even penetration of steam (WHO, 2020a, 2020b). Studies also showed that the addition of water in waste bags especially for solid wastes ensures effective decontamination (Lauer et al., 1982; Pienpatanakij et al., 2016). In validating the effectiveness of an autoclave cycle, chemical (e.g., autoclave tapes) and the biological indicator, *Geobacillus stearothermophilus*, can be used (Cascardo et al., 2020; Panta et al., 2019; WHO, 2020b). To ensure safety, users must wear proper PPE, provided with SOP, and be trained. Precleaning must be avoided and secondary containers must be used at all times (WHO, 2020a, 2020b).

A pressure cooker, which also uses moist heat, can be a good alternative for an autoclave (WHO, 2020b). It is a common equipment used in academic laboratories in developing countries such as the Philippines because of its lower cost. The principles followed in the validation, operation, and maintenance of an autoclave are also applicable for pressure cookers (Swenson et al., 2018; Tortora et al., 2019). Domestic pressure cookers may also be used for decontamination. However, they must undergo validation, safety, and efficiency tests first (Swenson et al., 2018; Zaki & Campbell, 1997).

Incineration, another physical method, employs dry heat in killing all microorganisms (Awodele et al., 2016; Bendjoudi et al., 2009; Chisholm et al., 2021; Diaz et al., 2005; Kühling & Pieper, 2012; Tadesse & Kumie, 2014). It involves burning (full combustion), an appropriate decontamination method for solid and bulky wastes such as large animal carcasses, biomedical waste, plastics, and sharps (Agamuthu & Barasarathi, 2021; Ahmed et al., 2015; Attrah et al., 2022; Henneman et al., 2020). Compared to autoclaves, incineration requires a longer decontamination time and a higher temperature (Olanrewaju & Fasinmirin, 2019; WHO, 2020b). An incinerator burns the waste in the primary chamber (≥800 °C) first, and then in the secondary chamber (≥1000 °C) (WHO, 2020b; Zaki & Campbell, 1997). Furnaces and burn pits, although common in developing countries due to low cost, are not recommended as alternatives for biosafety reasons (Attrah et al., 2022; Diaz et al., 2005; WHO, 2020b). During incineration, the solid laboratory wastes are converted to ash, carbon dioxide, and water (Olanrewaju & Fasinmirin, 2019; WHO, 2020b). However, it could also lead to the concentration of toxic chemicals (e.g., heavy metals) (Adama et al., 2016). Hence, ashes from the incinerator must be properly discarded (WHO, 2020b). Incinerators must be monitored and maintained regularly. Its effectiveness can be indicated by the absence of viable microbes in the exhaust gases (Zaki & Campbell, 1997).

Biological waste can also be decontaminated using chemical sterilants or disinfectants (Madigan et al., 2021; Munson et al., 2018; WHO, 2020a). Although chemical disinfectants are typically used to decontaminate surfaces after work and to clean a biological spill, they have become a common decontamination procedure for broth cultures, sharps, pipette tips, PPE (e.g., gloves and gowns), and materials contaminated with human and animal specimens (Alfy et al., 2022; Bendjoudi et al., 2009; Chisholm et al., 2021; Diaz et al., 2005; WHO, 2020b, 2020a). However, chemical disinfectants are not always preferred because of their unreliable efficacy, contact dependence, toxicity, high cost, incompatibility with waste components (e.g. radioactive substances), and environmental hazards (Attrah et al., 2022; Diaz et al., 2005; WHO, 2020b). The most commonly used chemical disinfectants in academic and research institutions are sodium hypochlorite, quaternary ammonium compounds (QACs), and alcohol. The selection of the type of chemical disinfectant is dependent on several factors such as the waste composition (Diaz et al., 2005; WHO, 2020b; Willey et al., 2023).

Sodium hypochlorite is a readily accessible and cheap oxidizing agent that is active against microorganisms (Madigan et al., 2021; WHO, 2020b; Willey et al., 2023). However, it works only if freshly diluted and in low organic matter containing waste. Its most common form is 10% bleach with a required contact time of 30 min (Alfy et al., 2022; WHO, 2020b; Willey et al., 2023). It must be handled carefully because it is skin irritating and corrosive (Diaz et al., 2005; Kohn et al., 2017; WHO, 2020b). QACs are cationic detergents which are effective against non-spore forming bacteria (except *Mycobacterium tuberculosis*) and enveloped viruses (Madigan et al., 2021; WHO, 2020b; Willey et al., 2023). They can kill microorganisms through cytoplasmic membrane disruption and protein denaturation (Willey et al., 2023). QACs, such as benzalkonium chloride, are more stable, safer, and less irritating than hypochlorites but are more expensive (Willey et al., 2023). They are also non-toxic and non-corrosive. However, they can also be inactivated by organic material, hard water, and soap (WHO, 2020b; Willey et al., 2023). Alcohol (e.g., 60 to 85% ethanol, isopropanol, or propanol), on the other hand, is a widely used bactericidal, fungicidal, and virucidal for surfaces and small instruments (Willey et al., 2023). It requires a contact time of 10–15 min and works against non-spore forming bacteria, enveloped viruses, and some non-enveloped viruses by dissolving lipids and denaturing proteins (Madigan et al., 2021; WHO, 2020b). Unlike hypochlorite and QACs, alcohol is resistant to organic matter inactivation but is highly volatile and can be inhibited by coagulated proteins (WHO, 2020b; Willey et al., 2023).

## 6.8 STERILIZATION PROCEDURES

To manipulate microorganisms in the laboratory successfully, all the materials and tools to be used must be sterilized prior to use (Swenson et al., 2018). Similar to decontamination, sterilization also involves killing or removal of all the microorganisms, including the endospores and viruses using physical and chemical methods (Madigan et al., 2021; Willey et al., 2023). The sterilization procedures commonly used in developing countries such as steam and dry heat sterilization, irradiation, and chemosterilants are given emphasis here.

Steam sterilization, the most widely used physical sterilization procedure, is carried out using an autoclave or a pressure cooker (Fast et al., 2017; Qamar et al., 2020; Robertson et al., 2021). It can be used to sterilize any materials that are not heat-labile such as culture media, empty glassware, and pipette tips. When sterilizing dry materials, it is recommended to use paper instead of aluminum foil as a wrapper because aluminum foil is impermeable to steam (Tortora et al., 2019). The autoclave cycle for sterilization of culture media is typically set at 15 psi, 121 °C for 15 min. The holding time can be longer if dealing with large volumes, but it must be carefully considered to avoid denaturation of nutrients (Tortora et al., 2019).

Sterilization can also be done using dry heat, which kills by protein denaturation and cell constituent oxidation (Fast et al., 2017; Jildeh et al., 2021; Tortora et al., 2019). An example of this is the direct flaming of a wire loop until it glows red using an alcohol lamp or a bunsen burner (Willey et al., 2023). The use of an oven with a temperature of 160–170 °C for 2–3 h is another example (Tortora et al., 2019). Although time consuming and not appropriate for rubber or plastics, oven sterilization is preferred for oil, powder, soil, or if the material is prone to corrosion (Aggangan et al., 2019; Willey et al., 2023). The effectiveness of oven sterilization can be validated using the biological indicators, *Bacillus atrophaeus* and *Bacillus subtilis* (WHO, 2020b).

Irradiation is also a widely used sterilization method due to its simplicity, cost efficiency, and relatively short processing time (Jildeh et al., 2021). The effectiveness of irradiation can be validated using *Bacillus pumilus* (WHO, 2020b). Radiation sterilization can be achieved using ultraviolet radiation (UV) and ionizing radiation (Jildeh et al., 2021). UV, at about 260 nm, is a non-ionizing radiation that can be used to sterilize work surfaces, circulating air, plastics, biosafety cabinet, laminar flow hoods, and even medical supplies (e.g., surgical supplies) (Madigan et al., 2021; Mamahlodi, 2019; Pawar et al., 2021; WHO, 2020b). UV exposure results in pyrimidine dimer formation which eventually leads to microbial killing (Madigan et al., 2021; Willey et al., 2023). The effectiveness of UV is dependent on the type of microorganism (effective against vegetative cells and viruses only), direct contact, dose, and relative humidity (Mamahlodi, 2019; Willey et al., 2023). Users must be properly trained because UV can cause skin cancer, burns, and eye damage (Mamahlodi, 2019; Tortora et al., 2019). To ensure effectiveness, UV lamps must be cleaned and replaced regularly (Mamahlodi, 2019; Pawar et al., 2021). For bulky materials and wrapped packages, ionizing radiation is a more appropriate method because of its better penetrating power and higher energy content (Madigan et al.,

2021; WHO, 2020b; Willey et al., 2023). Ionizing radiation, such as high-energy electron beams, gamma rays, and X rays, can be used to sterilize gloves, plastic syringes, antibiotics, hormones, and other medical supplies (IAEA, 2008; Tortora et al., 2019; Willey et al., 2023). It can kill microbial cells, endospores, and some viruses through production of reactive hydroxyl radicals which can damage the DNA and other cell components (Willey et al., 2023).

Filtration is a physical method of sterilization that is used for heat-sensitive liquids (Othman et al., 2021; Willey et al., 2023). The most commonly used membrane filter is a high-tensile-strength polymer typically made from polyvinylidene fluoride, polycarbonate, cellulose acetate, and cellulose nitrate (Madigan et al., 2021; Tortora et al., 2019; Willey et al., 2023). A pore size of 0.22 and 0.45 μm can be used to remove bacterial and fungal contaminants while 0.01 μm for viruses (Madigan et al., 2021; Tortora et al., 2019). The liquid (e.g., antibiotic, oil, culture media, sugar, and pharmaceutical solutions) to be sterilized is forced through the filtration apparatus using a syringe or a pump into a sterile container (Madigan et al., 2021; Willey et al., 2023).

There are also chemicals (chemosterilants) that can be used for sterilization (cold sterilization) (Madigan et al., 2021; Tortora et al., 2019). One example is ethylene oxide (EO), an alkylating agent, that can kill both vegetative cells and endospores by cross-linking proteins and nucleic acids which in turn interferes with important cellular processes (Jildeh et al., 2021; Robertson et al., 2021). This method, which requires several hours of exposure, is used for heat-labile materials such as disposable Petri dishes (Willey et al., 2023). Pure EO is toxic and explosive, so it is typically used in combination with carbon dioxide or other non-flammable gas (Tortora et al., 2019).

## 6.9 PROPER WASTE DISPOSAL AND BIOLOGICAL TREATMENT

Microbiological cultures and other infectious liquid wastes, including human or animal blood, blood products, nasal secretions, sputum, saliva, urine, feces, sweat, vomitus, and other biological wastes should be securely placed in autoclavable bags and stored in designated containers labeled with 'Infectious/Biohazardous Waste'. These wastes should be promptly decontaminated through steam sterilization (autoclaving) at 121 °C for 15 minutes (15 parts per square inch) or longer if the waste has a larger bulk or volume. Sharps, such as needles, must be deposited in impervious, rigid, puncture-resistant containers to prevent injuries. These sharps containers should be positioned as close to the usage site as possible. Activities like clipping, recapping, bending, manipulating, or breaking needles should not be done due to the potential generation of infectious aerosols and the risk of personal injury. Containers must be appropriately labeled as 'Infectious Biological Sharps'. They must undergo steam sterilization to eliminate their biohazardous potential. Alternatively, overnight chemical decontamination using 10% bleach is recommended. Decontaminated biological sharps must be buried along with the container. Non-sharp biological wastes, on the other hand, such as contaminated cotton balls or plugs, gauze, bandages or dressing, soiled linen, spill towels or paper, disposable culture plates and tubes, and gloves can undergo chemical decontamination by soaking them in 10% bleach overnight for small volumes. However, for large volumes of these products, autoclaving is recommended. After treatment or decontamination, they can be treated as regular solid wastes. Moreover, cadavers, carcasses, organs, and plant parts should be carefully placed in bins or bags labeled with 'biohazardous material'. Immediate disposal is mandatory after study or use to prevent any potential spread of infectious agents. They must be incinerated or autoclaved for proper disposal. Following treatment, cadavers, carcasses, and large volumes of human and animal cells and tissues must be buried at least six feet underground.

## 6.10 WAY FORWARD

Keeping the environment free from the biological agents from laboratories is a key element of a sound biorisk management system. Biological wastes, as was outlined in this chapter, are the carriers of these agents to the environment posing potential harm to humans, animals, and the ecosystem. The effects are magnified through time, and are exacerbated by socioeconomic factors. Low-resourced laboratories are challenged even more by the inadequate resources. Thus, other important mitigation measures are needed and should be strengthened, to prevent unnecessary exposure and spillage beyond the laboratory. Good laboratory practices and procedures, as well as a responsive and tailor-fit policy, are cornerstones to ensuring environmental biosafety and biosecurity.

## ACKNOWLEDGMENT

Authors acknowledge the contribution of researchers in the field of environmental biosafety and biosecurity.

## REFERENCES

Acosta, V., Paul, J., Lao, C., Aguinaldo, E., & Valdez, M. D. C. (2012). Development of the Philippines national solid waste management strategy 2012–2016. *Procedia Environmental Sciences*, *16*, 9–16. https://doi.org/10.1016/j.proenv.2012.10.003

Adama, M., Esena, R., Fosu-Mensah, B., & Yirenya-Tawiah, D. (2016). Heavy metal contamination of soils around a hospital waste incinerator bottom ash dumps site. *Journal of Environmental and Public Health*, *2016*, 1–6. https://doi.org/10.1155/2016/8926453

ADB. (2020). *Managing infectious medical waste during the COVID-19 pandemic*. Asian Development Bank. https://www.adb.org/publications/managing-medical-waste-covid19

Agache, I., Laculiceanu, A., Spanu, D., & Grigorescu, D. (2023). The concept of one health for allergic diseases and asthma. *Allergy, Asthma and Immunology Research*, *15*(3), 290–302. https://doi.org/10.4168/aair.2023.15.3.290

Agamuthu, P., & Barasarathi, J. (2021). *Clinical waste management under COVID-19 scenario in Malaysia*, *39*(1). https://doi.org/10.1177/0734242X20959701

Agamuthu, P., Mehran, S. B., & Jayanthi, B. (2022). Clinical waste management in Malaysia and COVID-19 waste management case study. In S. K. Ghosh & P. Agamuthu (Eds.), *Health Care Waste Management and COVID 19 Pandemic: Policy, Implementation Status and Vaccine Management* (pp. 23–41). Springer Nature. https://doi.org/10.1007/978-981-16-9336-6_2

Aggangan, N. S., Cortes, A. D., & Reaño, C. E. (2019). Growth response of cacao (Theobroma cacao L.) plant as affected by bamboo biochar and arbuscular mycorrhizal fungi in sterilized and unsterilized soil. *Biocatalysis and Agricultural Biotechnology*, *22*, 101347. https://doi.org/10.1016/j.bcab.2019.101347

Ahmed, S. S., Alp, E., Ulu-Kilic, A., & Doganay, M. (2015). Establishing molecular microbiology facilities in developing countries. *Journal of Infection and Public Health*, *8*(6), 513–525. https://doi.org/10.1016/j.jiph.2015.04.029

Aja, O. C., Al-Kayiem, H. H., Zewge, M. G., & Joo, M. S. (2016). Overview of hazardous waste management status in Malaysia. In H. E.-D. M. Saleh & R. O. Abdel Rahman (Eds.), *Management of Hazardous Wastes*. InTech. https://doi.org/10.5772/63682

Alam, M.-U., Rahman, M., Abdullah-Al-Masud, Islam, M. A., Asaduzzaman, M., Sarker, S., Rousham, E., & Unicomb, L. (2019). Human exposure to antimicrobial resistance from poultry production: Assessing hygiene and waste-disposal practices in Bangladesh. *International Journal of Hygiene and Environmental Health*, *222*(8), 1068–1076. https://doi.org/10.1016/j.ijheh.2019.07.007

Alfy, M. M., El Sayed, S. B., & El-Shokry, M. (2022). Assessing decontamination practices at a medical microbiology research laboratory. *Journal of Biosafety and Biosecurity*, *4*(2), 124–129. https://doi.org/10.1016/j.jobb.2022.08.002

Al-Gheethi, A., Ma, N. L., Rupani, P. F., Sultana, N., Yaakob, M. A., Mohamed, R. M. S. R., & Soon, C. F. (2021). Biowastes of slaughterhouses and wet markets: An overview of waste management for disease prevention. *Environmental Science and Pollution Research International*, 1–14. https://doi.org/10.1007/s11356-021-16629-w

Anacio, D. B. (2017). Designing sustainable consumption and production systems in higher education institutions: The case of solid waste management. In W. Leal Filho, M. Mifsud, C. Shiel, & R. Pretorius (Eds.), *Handbook of Theory and Practice of Sustainable Development in Higher Education: Volume 3* (pp. 3–25). Springer International Publishing. https://doi.org/10.1007/978-3-319-47895-1_1

Ancog, R.. Archival, N., & Rebancos, C. (2012). Institutional Arrangements for Solid Waste Management in Cebu City, Philippines. *Journal of Environmental Science and Management*, *15*, 74–82

Anitha, J., & Jayraaj, I. A. (2012). (7) (PDF) Isolation and identification of bacteria from biomedical waste (BMW). *International Journal of Pharmacy and Pharmaceutical Sciences*, *4*(5), 286–388.

Apostol, G. L. C., Acolola, A. G. A., Edillon, M. A., & Valenzuela, S. (2022). How comprehensive and effective are waste management policies during the COVID-19 pandemic? Perspectives from the Philippines. *Frontiers in Public Health*, *10*. https://www.frontiersin.org/articles/10.3389/fpubh.2022.958241

Arnold, K. E., Brown, A. R., Ankley, G. T., & Sumpter, J. P. (2014). Medicating the environment: Assessing risks of pharmaceuticals to wildlife and ecosystems. *Philosophical Transactions of the Royal Society B: Biological Sciences*, *369*(1656), 20130569. https://doi.org/10.1098/rstb.2013.0569

Ashbrook, P. C., & Reinhardt, P. A. (2002, May 1). *Hazardous wastes in academia* (world) [Research-article]. ACS Publications; American Chemical Society. https://doi.org/10.1021/es00142a002

Atienza, V., Pintor, L., & Ancheta, A. (2022). Healthcare waste management and post-pandemic countermeasures: The case of the Philippines. In S. K. Ghosh & P. Agamuthu (Eds.), *Health Care Waste Management and COVID 19 Pandemic: Policy, Implementation Status and Vaccine Management* (pp. 71–102). Springer Nature. https://doi.org/10.1007/978-981-16-9336-6_4

Attrah, M., Elmanadely, A., Akter, D., & Rene, E. R. (2022). A review on medical waste management: Treatment, recycling, and disposal options. *Environments*, *9*(11), 146. https://doi.org/10.3390/environments9110146

Awodele, O., Adewoye, A. A., & Oparah, A. C. (2016). Assessment of medical waste management in seven hospitals in Lagos, Nigeria. *BMC Public Health*, *16*(1), 269. https://doi.org/10.1186/s12889-016-2916-1

Bai, V. R., Vanitha, G., & Ariff, A. R. Z. (2013). Effective hospital waste classification to overcome Occupational Health issues and reduce waste disposal cost. *Infection Control and Hospital Epidemiology*, *34*(11), 1234–1235. https://doi.org/10.1086/673461

Barcelo, D. (2020). An environmental and health perspective for COVID-19 outbreak: Meteorology and air quality influence, sewage epidemiology indicator, hospitals disinfection, drug therapies and recommendations. *Journal of Environmental Chemical Engineering*, *8*(4), 104006. https://doi.org/10.1016/j.jece.2020.104006

Behera, J. K., Mishra, P., Jena, A. K., Bhattacharya, M., & Behera, B. (2022). Understanding of environmental pollution and its anthropogenic impacts on biological resources during the COVID-19 period. *Environmental Science and Pollution Research*. https://doi.org/10.1007/s11356-022-24789-6

Bendjoudi, Z., Taleb, F., Abdelmalek, F., & Addou, A. (2009). Healthcare waste management in Algeria and Mostaganem department. *Waste Management*, *29*(4), 1383–1387. https://doi.org/10.1016/j.wasman.2008.10.008

Bernardo, E. C. (2008). Solid-waste management practices of households in Manila, Philippines. *Annals of the New York Academy of Sciences*, *1140*(1), 420–424. https://doi.org/10.1196/annals.1454.016

Boyce, J. M., Potter-Bynoe, G., Chenevert, C., & King, T. (1997). Environmental contamination due to methicillin-resistant Staphylococcus aureus: Possible infection control implications. *Infection Control and Hospital Epidemiology*, *18*(9), 622–627

Budihardjo, M. A., Humaira, N. G., Putri, S. A., Syafrudin, Yohana, E., Ramadan, B. S., Zaman, B., & Sutrisno, E. (2022). Indonesian efforts to overcome Covid-19's effects on its municipal solid waste management: A review. *Cogent Engineering*, *9*(1), 2143055. https://doi.org/10.1080/23311916.2022.2143055

Buenrostro, O., Bocco, G., & Cram, S. (2001). Classification of sources of municipal solid wastes in developing countries. *Resources, Conservation and Recycling*, *32*(1), 29–41. https://doi.org/10.1016/S0921-3449(00)00094-X

Cabico, G. K. (2020, August 27). In the Philippines, medical waste piles up as COVID-19 cases rise. *Earth Journalism Network*. https://earthjournalism.net/stories/in-the-philippines-medical-waste-piles-up-as-covid-19-cases-rise

Canning-Clode, J., Sepúlveda, P., Almeida, S., & Monteiro, J. (2020). Will COVID-19 containment and treatment measures drive shifts in marine litter pollution? *Frontiers in Marine Science*, *7*. https://www.frontiersin.org/articles/10.3389/fmars.2020.00691

Capoor, M. R., & Parida, A. (2021). Biomedical waste and solid waste management in the time of COVID-19: A comprehensive review of the national and international scenario and guidelines. *Journal of Laboratory Physicians*, *13*(2), 175–182. https://doi.org/10.1055/s-0041-1729132

Cascardo, E., Goenaga, S., Fossa, S., Bottale, A., Levis, S., & Riera, L. (2020). Development and validation of waste decontamination cycle in a biosafety Level 3 laboratory. *Applied Biosafety*, *25*(4), 225–231. https://doi.org/10.1177/1535676020933714

Chand, S., Shastry, C. S., Hiremath, S., Joel, J. J., Krishnabhat, C. H., & Mateti, U. V. (2021). Updates on biomedical waste management during COVID-19: The Indian scenario. *Clinical Epidemiology and Global Health*, *11*. https://doi.org/10.1016/j.cegh.2021.100715

Chiemchaisri, C., Juanga, J. P., & Visvanathan, C. (2007). Municipal solid waste management in Thailand and disposal emission inventory. *Environmental Monitoring and Assessment*, *135*(1), 13–20. https://doi.org/10.1007/s10661-007-9707-1

Chisholm, J. M., Zamani, R., Negm, A. M., Said, N., Abdel daiem, M. M., Dibaj, M., & Akrami, M. (2021). Sustainable waste management of medical waste in African developing countries: A narrative review. *Waste Management & Research: The Journal for a Sustainable Circular Economy*, *39*(9), 1149–1163. https://doi.org/10.1177/0734242X211029175

Chokhandre, P., Singh, S., & Kashyap, G. C. (2017). Prevalence, predictors and economic burden of morbidities among waste-pickers of Mumbai, India: A cross-sectional study. *Journal of Occupational Medicine and Toxicology*, *12*(1), 30. https://doi.org/10.1186/s12995-017-0176-3

Ciplak, N., & Kaskun, S. (2015). Healthcare waste management practice in the west Black Sea Region, Turkey: A comparative analysis with the developed and developing countries.*Journal of the Air & Waste Management Association*, *65*(12), 1387–1394.

Coker, A., Sangodoyin, A., Sridhar, M., Booth, C., Olomolaiye, P., & Hammond, F. (2009). Medical waste management in Ibadan, Nigeria: Obstacles and prospects. *Waste Management*, *29*(2), 804–811. https://doi.org/10.1016/j.wasman.2008.06.040

Cruz, A. F., Wijesekara, R. G. S., Jinadasa, K. B. S. N., Gonzales, B. J., Ohura, T., & Guruge, K. S. (2021). Preliminary investigation of microbial community in wastewater and surface waters in Sri Lanka and the Philippines. *Frontiers in Water*, *3*. https://www.frontiersin.org/articles/10.3389/frwa.2021.730124

CSE India. (2021). *46 per cent increase in COVID-19 biomedical waste generation in India in April–May 2021, says CSE's new statistical analysis*. Centre for Science and Environment. https://www.cseindia.org/46-per-cent-increase-in-covid-19-biomedical-waste-generation-in-india-in-april-may-2021-10849

Cuaresma, J. C. (2019). How green can you go? Initiatives of dark green universities in the Philippines. In W. Leal Filho & U. Bardi (Eds.), *Sustainability on University Campuses: Learning, Skills Building and Best Practices* (pp. 165–189). Springer International Publishing. https://doi.org/10.1007/978-3-030-15864-4_11

Das, A. K., Islam, Md. N., Billah Md, M., & Sarker, A. (2021). COVID-19 pandemic and healthcare solid waste management strategy – A mini-review. *Science of the Total Environment*, *778*, 146220. https://doi.org/10.1016/j.scitotenv.2021.146220

Daszak, P., Cunningham, A. A., & Hyatt, A. D. (2000). Emerging infectious diseases of wildlife—Threats to biodiversity and human health. *Science*, *287*(5452), 443–449. https://doi.org/10.1126/science.287.5452.443

De-la-Torre, G. E., & Aragaw, T. A. (2021). What we need to know about PPE associated with the COVID-19 pandemic in the marine environment. *Marine Pollution Bulletin*, *163*, 111879. https://doi.org/10.1016/j.marpolbul.2020.111879

Diaz, L. F., Savage, G. M., & Eggerth, L. L. (2005). Alternatives for the treatment and disposal of healthcare wastes in developing countries. *Waste Management*, *25*(6), 626–637. https://doi.org/10.1016/j.wasman.2005.01.005

Dihan, M. R., Abu Nayeem, S. M., Roy, H., Islam, Md. S., Islam, A., Alsukaibi, A. K. D., & Awual, Md. R. (2023). Healthcare waste in Bangladesh: Current status, the impact of Covid-19 and sustainable management with life cycle and circular economy framework. *Science of the Total Environment*, *871*, 162083. https://doi.org/10.1016/j.scitotenv.2023.162083

Egbenyah, F., Udofia, E. A., Ayivor, J., Osei, M.-M., Tetteh, J., Tetteh-Quarcoo, P. B., & Sampane-Donkor, E. (2021). Disposal habits and microbial load of solid medical waste in sub-district healthcare facilities and households in Yilo-Krobo municipality, Ghana. *PLOS One*, *16*(12), e0261211. https://doi.org/10.1371/journal.pone.0261211

El-Ramady, H., Brevik, E. C., Elbasiouny, H., Elbehiry, F., Amer, M., Elsakhawy, T., Omara, A. E.-D., Mosa, A. A., El-Ghamry, A. M., Abdalla, N., Rezes, S., Elboraey, M., Ezzat, A., & Eid, Y. (2021). Planning for disposal of COVID-19 pandemic wastes in developing countries: A review of current challenges. *Environmental Monitoring and Assessment, 193*(9), 592. https://doi.org/10.1007/s10661-021-09350-1

Era, M. de L. (2019). From response to responsibility: An academe–industry partnership on solid waste management in the Philippines. In B. Zutshi, A. Ahmad, & A. B. Srungarapati (Eds.), *Disaster Risk Reduction: Community Resilience and Responses* (pp. 173–188). Springer Nature. https://doi.org/10.1007/978-981-10-8845-2_11

Evode, N., Qamar, S. A., Bilal, M., Barceló, D., & Iqbal, H. M. N. (2021). Plastic waste and its management strategies for environmental sustainability. *Case Studies in Chemical and Environmental Engineering, 4*, 100142. https://doi.org/10.1016/j.cscee.2021.100142

Fast, O., Fast, C., Fast, D., Veltjens, S., Salami, Z., & White, M. C. (2017). Limited sterile processing capabilities for safe surgery in low-income and middle-income countries: Experience in the Republic of Congo, Madagascar and Benin. *BMJ Global Health, 2*(suppl 4), e000428. https://doi.org/10.1136/bmjgh-2017-000428

Fernandez, H. A. (2022, February 11). Piles of medical waste in Philippine hospital highlight country's Covid-19 trash problem. *Eco-Business.* https://www.eco-business.com/news/piles-of-medical-waste-in-philippine-hospital-highlight-countrys-covid-19-trash-problem/

Garibaldi, B. T., Reimers, M., Ernst, N., Bova, G., Nowakowski, E., Bukowski, J., Ellis, B. C., Smith, C., Sauer, L., Dionne, K., Carroll, K. C., Maragakis, L. L., & Parrish, N. M. (2017). Validation of autoclave protocols for successful decontamination of Category A medical waste generated from care of patients with serious communicable diseases. *Journal of Clinical Microbiology, 55*(2), 545–551. https://doi.org/10.1128/JCM.02161-16

Gequinto, A. C. (2017). Solid waste management practices of select state universities in CALABARZON, Philippines *Asia Pacific Journal of Multidisciplinary Research 5*(1), 1–8.

Gidey, K., Gidey, M. T., Hailu, B. Y., Gebreamlak, Z. B., & Niriayo, Y. L. (2023). Clinical and economic burden of healthcare-associated infections: A prospective cohort study. *PLOS ONE, 18*(2), e0282141. https://doi.org/10.1371/journal.pone.0282141

Giovanni, C., Marques, F. L. N., & Günther, W. M. R. (2021). Laboratory chemical waste: Hazard classification by GHS and transport risk. *Revista de Saúde Pública, 55*, 102. https://doi.org/10.11606/s1518-8787.2021055003259

Gwenzi, W., & Sanganyado, E. (2019). Recurrent cholera outbreaks in sub-Saharan Africa: Moving beyond epidemiology to understand the environmental reservoirs and drivers. *Challenges, 10*(1), 1. https://doi.org/10.3390/challe10010001

Henneman, J. R., Johnson, J. A., & Minihan, M. A. (2020). Challenges and solutions with agricultural animal high containment waste disposal. *ILAR Journal, 61*(1), 32–39. https://doi.org/10.1093/ilar/ilab015

Hocquet, D., Muller, A., & Bertrand, X. (2016). What happens in hospitals does not stay in hospitals: Antibiotic-resistant bacteria in hospital wastewater systems. *The Journal of Hospital Infection, 93*(4), 395–402. https://doi.org/10.1016/j.jhin.2016.01.010

Holmer, R., & Guanzon, Y. (2003). *Composting of Organic Wastes: A Main Component for Successful Integrated Solid Waste Management in Philippine Cities.* https://doi.org/10.13140/2.1.4224.8485

Hosan, M. (2017). Radioactive waste classification, management and environment. *Engineering International, 5*(2). https://doi.org/10.18034/ei.v5i2.1082

Hossain, I., Mullick, A., Bari, S., & Islam, M. (2020). Pandemic COVID-19 and biomedical waste handling: A review study. *J Med Sci Clin Res, 8*(5), 497–502.

Hu, B., Hou, P., Teng, L., Miao, S., Zhao, L., Ji, S., Li, T., Kehrenberg, C., Kang, D., & Yue, M. (2022). Genomic investigation reveals a community typhoid outbreak caused by contaminated drinking water in China, 2016. *Frontiers in Medicine, 9*. https://www.frontiersin.org/articles/10.3389/fmed.2022.753085

IAEA. (Ed.). (2008). *Trends in Radiation Sterilization of Health Care Products*. International Atomic Energy Agency.

Ibrahim, M., Kebede, M., & Mengiste, B. (2023). Healthcare waste segregation practice and associated factors among healthcare professionals working in public and private hospitals, Dire Dawa, Eastern Ethiopia. *Journal of Environmental and Public Health, 2023*, e8015856. https://doi.org/10.1155/2023/8015856

Ikhimiukor, O. O., Odih, E. E., Donado-Godoy, P., & Okeke, I. N. (2022). A bottom-up view of antimicrobial resistance transmission in developing countries. *Nature Microbiology, 7*(6), 757–765. https://doi.org/10.1038/s41564-022-01124-w

Jha, R., Dwivedi, S., & Modhera, B. (2022). *Measurement and Practices for Hazardous Waste Management* (pp. 89–115). https://doi.org/10.1016/B978-0-12-824344-2.00011-2

Jildeh, Z. B., Wagner, P. H., & Schöning, M. J. (2021). Sterilization of objects, products, and packaging surfaces and their characterization in different fields of industry: The status in 2020. *Physica Status Solidi (a), 218*(13), 2000732. https://doi.org/10.1002/pssa.202000732

Johnson, K., González, M., Dueñas, L., Gamero, M., Relyea, G., Luque de Johnson, L., & Caniza, M. (2013). Improving waste segregation while reducing costs in a tertiary-care hospital in a lower-middle-income country in Central America. *Waste Management and Research : The Journal of the International Solid Wastes and Public Cleansing Association, ISWA, 31*(7). https://doi.org/10.1177/0734242X13484192

Kalnowski, G., Wiegand, H., & Rüden, H. (1983). Microbial contamination of hospital waste. *Zentralblatt Fur Bakteriologie, Mikrobiologie Und Hygiene. 1. Abt. Originale B, Hygiene, 178*(4), 364–379

Kanwar, V. S., Sharma, A., Rinku Kanwar, M., Srivastav, A. L., & Soni, D. K. (2022). An overview for biomedical waste management during pandemic like COVID-19. *International Journal of Environmental Science and Technology*. https://doi.org/10.1007/s13762-022-04287-5

Karak, T., Bhagat, R. M., & Bhattacharyya, P. (2012). Municipal solid waste generation, composition, and management: The world scenario. *Critical Reviews in Environmental Science and Technology, 42*(15), 1509–1630. https://doi.org/10.1080/10643389.2011.569871

Katleho, R. T., Omondi, A. B., & Mekbib, S. B. (2022). Sewage water from a hospital and its impact to microbial profiles: A case study. *International Journal of Environment and Waste Management, 30*(4), 440–452. https://doi.org/10.1504/IJEWM.2022.128211

Kaur, R., Yadav, B., & Tyagi, R. D. (2020). Microbiology of hospital wastewater. *Current Developments in Biotechnology and Bioengineering*, 103–148. https://doi.org/10.1016/B978-0-12-819722-6.00004-3

Kemper, N. (2008). Veterinary antibiotics in the aquatic and terrestrial environment. *Ecological Indicators, 8*(1), 1–13. https://doi.org/10.1016/j.ecolind.2007.06.002

Ketema, S., Melaku, A., Demelash, H., G/Mariam, M., Mekonen, S., Addis, T., & Ambelu, A. (2023). Safety practices and associated factors among healthcare waste handlers in four public hospitals, Southwestern Ethiopia. *Safety, 9*(2), 41. https://doi.org/10.3390/safety9020041

Khan, B. A., Cheng, L., Khan, A. A., & Ahmed, H. (2019)Healthcare waste management in Asian developing countries: A mini review. *Waste Management and Research, 37*(9), 863–875. https://doi.org/10.1177/0734242X19857470

Khan, M. T., Shah, I. A., Ihsanullah, I., Naushad, Mu, Ali, S., Shah, S. H. A., & Mohammad, A. W. (2021). Hospital wastewater as a source of environmental contamination: An overview of management practices, environmental risks, and treatment processes. *Journal of Water Process Engineering*, *41*, 101990. https://doi.org/10.1016/j.jwpe.2021.101990

Kilaru, P., Hill, D., Anderson, K., Collins, M. B., Green, H., Kmush, B. L., & Larsen, D. A. (2023). Wastewater surveillance for infectious disease: A systematic review. *American Journal of Epidemiology*, *192*(2), 305–322. https://doi.org/10.1093/aje/kwac175

Kohn, T., Decrey, L., & Vinnerås, B. (2017). Chemical disinfectants. In J. B. Rose & B. Jiménez-Cisneros (Eds.), *Global Water Pathogens Project*, http://www.waterpathogens.org (C. Haas (eds) Part 4 Management of Risk from Excreta and Wastewater). http://www.waterpathogens.org/book/chemical-disinfectants Michigan State University, E. Lansing, MI, UNESCO.

Kühling, J.-G., & Pieper, U. (2012). Management of healthcare waste: Developments in Southeast Asia in the twenty-first century. *Waste Management & Research: The Journal for a Sustainable Circular Economy*, *30*(9_suppl), 100–104. https://doi.org/10.1177/0734242X12452907

Kuppusamy, P. P., Bhatia, A., Verma, A., Shah, N. R., Pratyush, P., Shanmugarajan, V., Kim, S. C., Poongavanam, G., & Duraisamy, S. (2022). Accumulation of biomedical waste during the COVID-19 pandemic: Concerns and strategies for effective treatment. *Environmental Science and Pollution Research International*, *29*(37), 55528–55540. https://doi.org/10.1007/s11356-022-21086-0

Kutralam-Muniasamy, G., Pérez-Guevara, F., & Shruti, V. C. (2022). A critical synthesis of current peer-reviewed literature on the environmental and human health impacts of COVID-19 PPE litter: New findings and next steps. *Journal of Hazardous Materials*, *422*, 126945. https://doi.org/10.1016/j.jhazmat.2021.126945

Lauer, J. L., Battles, D. R., & Vesley, D. (1982). Decontaminating infectious laboratory waste by autoclaving. *Applied and Environmental Microbiology*, *44*(3), 690–694. https://doi.org/10.1128/aem.44.3.690-694.1982

Lebarbenchon, C., Brown, S. P., Poulin, R., Gauthier-Clerc, M., & Thomas, F. (2008). Evolution of pathogens in a man-made world. *Molecular Ecology*, *17*(1), 475–484. https://doi.org/10.1111/j.1365-294X.2007.03375.x

Lunag, M. N., Duran, J. Z., & Buyucan, E. D. (2019). Waste analysis and characterisation study of a hill station: A case study of Baguio City, Philippines. *Waste Management and Research*, *37*(11), 1102–1116. https://doi.org/10.1177/0734242X19866249

Madigan, M., Bender, K., Buckley, D., Sattley, W. M., & Stahl, D. (2021). *Brock Biology of Microorganisms* (16th ed.). Pearson Education Limited

Magalang, A. A. (2014). Municipal solid waste management in the Philippines. In A. Pariatamby & M. Tanaka (Eds.), *Municipal Solid Waste Management in Asia and the Pacific Islands: Challenges and Strategic Solutions* (pp. 281–297). Springer. https://doi.org/10.1007/978-981-4451-73-4_14

Maina, S. M., Nyerere, A. K., & Ngugi, C. W. (2018). Isolation of bacterial diversity present in medical waste and health care settings in hospitals in Kenya. *African Journal of Microbiology Research*, *12*(26), 606–615. https://doi.org/10.5897/AJMR2018.8876

Mamahlodi, M. T. (2019). Potential benefits and harms of the use of UV radiation in transmission of tuberculosis in South African health facilities. *Journal of Public Health in Africa*, *10*(742). https://doi.org/10.4081/jphia.2019.742

Manzoor, J., & Sharma, M. (2019). Impact of biomedical waste on environment and human health. *Environmental Claims Journal*, *31*(4), 311–334. https://doi.org/10.1080/10406026.2019.1619265

Mara, D., Lane, J., Scott, B., & Trouba, D. (2010). Sanitation and health. *PLOS Medicine*, *7*(11), e1000363. https://doi.org/10.1371/journal.pmed.1000363

Maria, M., Hussain, T., Pervez, M. T., Babar, M. E., Bibi, R., & Maqbool, A. (2019). Application of metagenomics in the analysis of microbial diversity in soil of hospitals. *Pakistan Journal of Biochemistry and Molecular Biology*, *52*(2), Article 2.

Mecklem, R. L. (n.d.). *Defining and Managing Biohazardous Waste in Research-Based Universities in the United States: A Survey of Environmental Health and Safety Professionals*. Oregon State University.

Mecklem, R. L., & Neumann, C. M. (2003). Defining and managing biohazardous waste in US research-oriented universities; A survey of environmental health and safety professionals. *Journal of Environmental Health*, *66*(1), 17–22.

Mesfin, A., Worku, W., & Gizaw, Z. (2014). Assessment of health care waste segregation practice and associated factors of health care workers in Gondar University Hospital, North West Ethiopia, 2013. *Universal Journal of Public Health*, *2*(7), 201–207. https://doi.org/10.13189/ujph.2014.020703

Mohan, H., Rajput, S., Jadhav, E., Singh Sankhla, M., Sonone, S., Jadhav, S., & Kumar, R. (2021). Ecotoxicity, occurrence, and removal of pharmaceuticals and illicit drugs from aquatic systems. *Biointerface Research in Applied Chemistry*, *11*(5), 12530–12546. https://doi.org/10.33263/BRIAC115.1253012546

Mor, S., Ravindra, K., Dahiya, R. P., & Chandra, A. (2006). Leachate characterization and assessment of groundwater pollution near municipal solid waste landfill site. *Environmental Monitoring and Assessment*, *118*(1), 435–456. https://doi.org/10.1007/s10661-006-1505-7

Morley, N. J. (2009). Environmental risk and toxicology of human and veterinary waste pharmaceutical exposure to wild aquatic host–parasite relationships. *Environmental Toxicology and Pharmacology*, *27*(2), 161–175. https://doi.org/10.1016/j.etap.2008.11.004

Möse, J. R., & Reinthaler, F. (1985). Microbiological studies of the contamination of hospital waste and household refuse. *Zentralblatt Fur Bakteriologie, Mikrobiologie Und Hygiene. 1. Abt. Originale B, Hygiene*, *181*(1–2), 98–110

Munson, E., Bowles, E. J., Dern, R., Beck, E., Podzorski, R. P., Bateman, A. C., Block, T. K., Kropp, J. L., Radke, T., Siebers, K., Simmons, B., Smith, M. A., Spray-Larson, F., & Warshauer, D. M. (2018). Laboratory focus on improving the culture of biosafety: Statewide risk assessment of clinical laboratories that process specimens for microbiologic analysis. *Journal of Clinical Microbiology*, *56*(1), e01569-17. https://doi.org/10.1128/JCM.01569-17

Néstor, M. C., & Mariana, C. (2019). Impact of pharmaceutical waste on biodiversity. In L. M. Gómez-Oliván (Ed.), *Ecopharmacovigilance: Multidisciplinary Approaches to Environmental Safety of Medicines* (pp. 235–253). Springer International Publishing. https://doi.org/10.1007/698_2017_151

Nguyen, T. D. T., Nakakubo, T., & Kawai, K. (2022). Analysis of COVID-19 waste management in Vietnam and recommendations to adapt to the 'new normal' period. *Journal of Material Cycles and Waste Management*. https://doi.org/10.1007/s10163-022-01563-x

Olalo, K. F., Nakatani, J., & Fujita, T. (2022). Optimal Process Network for Integrated Solid Waste Management in Davao City, Philippines. *Sustainability*, *14*(4), Article 4. https://doi.org/10.3390/su14042419

Olanrewaju, O. O., & Fasinmirin, R. J. (2019). Design of medical wastes incinerator for health care facilities in Akure. *Journal of Engineering Research and Reports*, 1–13. https://doi.org/10.9734/jerr/2019/v5i216919

Osafo, R., Balali, G. I., Amissah-Reynolds, P. K., Gyapong, F., Addy, R., Nyarko, A. A., & Wiafe, P. (2022). Microbial and parasitic contamination of vegetables in developing countries and their food safety guidelines. *Journal of Food Quality*, *2022*, 4141914. https://doi.org/10.1155/2022/4141914

Othman, N. H., Alias, N. H., Fuzil, N. S., Marpani, F., Shahruddin, M. Z., Chew, C. M., David Ng, K. M., Lau, W. J., & Ismail, A. F. (2021). A review on the use of membrane technology systems in developing countries. *Membranes*, *12*(1), 30. https://doi.org/10.3390/membranes12010030

Panta, G., Richardson, A. K., & Shaw, I. C. (2019). Effectiveness of autoclaving in sterilizing reusable medical devices in healthcare facilities. *The Journal of Infection in Developing Countries*, *13*(10), 858–864. https://doi.org/10.3855/jidc.11433

Parvin, F., & Tareq, S. M. (2021). Impact of landfill leachate contamination on surface and groundwater of Bangladesh: A systematic review and possible public health risks assessment. *Applied Water Science*, *11*(6), 100. https://doi.org/10.1007/s13201-021-01431-3

Pathak, D. R., Nepal, S., Thapa, T., Dhakal, N., Tiwari, P., & Sinha, T. K. (2021). Capacity assessment and implementation analysis of common treatment facility for the management of infectious healthcare waste in rapidly urbanising city of Nepal. *Waste Management & Research: The Journal for a Sustainable Circular Economy*, *39*(1_suppl), 64–75. https://doi.org/10.1177/0734242X211013910

Paul, J. G., Arce-Jaque, J., Ravena, N., & Villamor, S. P. (2012). Integration of the informal sector into municipal solid waste management in the Philippines – What does it need? *Waste Management*, *32*(11), 2018–2028. https://doi.org/10.1016/j.wasman.2012.05.026

Pawar, S. D., Khare, A. B., Keng, S. S., Kode, S. S., Tare, D. S., Singh, D. K., More, R. L., & Mullick, J. (2021). Selection and application of biological safety cabinets in diagnostic and research laboratories with special emphasis on COVID-19. *Review of Scientific Instruments*, *92*(8), 081401. https://doi.org/10.1063/5.0047716

Pedley, S., & Howard, G. (1997). The public health implications of microbiological contamination of groundwater. *Quarterly Journal of Engineering Geology*, *30*(2), 179–188. https://doi.org/10.1144/GSL.QJEGH.1997.030.P2.10

Perry, M. R., Lepper, H. C., McNally, L., Wee, B. A., Munk, P., Warr, A., Moore, B., Kalima, P., Philip, C., de Roda Husman, A. M., Aarestrup, F. M., Woolhouse, M. E. J., & van Bunnik, B. A. D. (2021). Secrets of the hospital underbelly: Patterns of abundance of antimicrobial resistance genes in hospital wastewater vary by specific antimicrobial and bacterial family. *Frontiers in Microbiology*, *12*. https://www.frontiersin.org/articles/10.3389/fmicb.2021.703560

Pienpatanakij, N., Armin, N., & Niyomdecha, N. (2016). Comparing the results of microbiological sterilization by autoclaving in different waste packaging formats. *Journal of Pure and Applied Microbiology*, *10*(2), 1033+. Gale Academic OneFile. https://link.gale.com/apps/doc/A481650189/AONE?u=anon~2f763e3f&sid=bookmark-AONE&xid=1bdc2132

Qamar, M. K., Shaikh, B. T., & Afzal, A. (2020). What do the dental students know about infection control? A cross-sectional study in a teaching hospital, Rawalpindi, Pakistan. *BioMed Research International*, *2020*, 1–5. https://doi.org/10.1155/2020/3413087

Rahman, M. M., Bodrud-Doza, M., Griffiths, M. D., & Mamun, M. A. (2020a). Biomedical waste amid COVID-19: Perspectives from Bangladesh. *The Lancet. Global Health*, *8*(10), e1262. https://doi.org/10.1016/S2214-109X(20)30349-1

Raimi, M. O., Iyingiala, A.-A., Sawyerr, O. H., Saliu, A. O., Ebuete, A. W., Emberru, R. E., Sanchez, N. D., & Osungbemiro, W. B. (2022a). Leaving no one behind: Impact of soil pollution on biodiversity in the global south: A global call for action. In S. Chibueze Izah (Ed.), *Biodiversity in Africa: Potentials, Threats and Conservation* (pp. 205–237). Springer Nature. https://doi.org/10.1007/978-981-19-3326-4_8

Raimi, M. O., Iyingiala, A.-A., Sawyerr, O. H., Saliu, A. O., Ebuete, A. W., Emberru, R. E., Sanchez, N. D., & Osungbemiro, W. B. (2022b). Leaving no one behind: Impact of soil pollution on biodiversity in the global south: A global call for action. In S. Chibueze Izah (Ed.), *Biodiversity in Africa: Potentials, Threats and Conservation* (pp. 205–237). Springer Nature. https://doi.org/10.1007/978-981-19-3326-4_8

Rau, E. H., Alaimo, R. J., Ashbrook, P. C., Austin, S. M., Borenstein, N., Evans, M. R., French, H. M., Gilpin, R. W., Hughes, J., Hummel, S. J., Jacobsohn, A. P., Lee, C. Y., Merkle, S., Radzinski, T., Sloane, R., Wagner, K. D., & Weaner, L. E. (2000). Minimization and management of wastes from biomedical research. *Environmental Health Perspectives, 108*(suppl 6), 953–977. https://doi.org/10.1289/ehp.00108s6953

Reyes, P. B., & Furto, M. V. (2013). Greening of the Solid Waste Management in Batangas City. *Journal of Energy Technologies and Policy*, 3(11), 187–194.

Rhind, S. M. (2009). Anthropogenic pollutants: A threat to ecosystem sustainability? *Philosophical Transactions of the Royal Society B: Biological Sciences, 364*(1534), 3391–3401. https://doi.org/10.1098/rstb.2009.0122

Riaz, L., Yang, Q., Sikandar, A., Safeer, R., Anjum, M., Mahmood, T., Rehman, M. S. U., Rashid, A., & Yuan, W. (2020). Antibiotics use in hospitals and their presence in the associated waste. In M. Z. Hashmi (Ed.), *Antibiotics and Antimicrobial Resistance Genes: Environmental Occurrence and Treatment Technologies* (pp. 27–49). Springer International Publishing. https://doi.org/10.1007/978-3-030-40422-2_2

Robertson, D., Gnanaraj, J., Wauben, L., Huijs, J., Samuel, V. M., Dankelman, J., & Horeman-Franse, T. (2021). Assessment of laparoscopic instrument reprocessing in rural India: A mixed methods study. *Antimicrobial Resistance and Infection Control, 10*(1), 109. https://doi.org/10.1186/s13756-021-00976-x

Rocha, I. C. N., Ramos, K. G., & Crispino, K. T. (2022). Food and waterborne disease outbreaks after a super Typhoon Hit the Southern Philippines during the COVID-19 pandemic: A triple public health emergency. *Prehospital and Disaster Medicine, 37*(3), 421–422. https://doi.org/10.1017/S1049023X2200053X

Rønning, A., Hanssen, O. J., & Nyland, C. A. (2000). *Product-Oriented Environmental Management Strategies*. pp. 1–33

Rosenberg Goldstein, R. E., Micallef, S. A., Gibbs, S. G., Davis, J. A., He, X., George, A., Kleinfelter, L. M., Schreiber, N. A., Mukherjee, S., Sapkota, A., Joseph, S. W., & Sapkota, A. R. (2012). Methicillin-resistant Staphylococcus aureus (MRSA) detected at four U.S. wastewater treatment plants. *Environmental Health Perspectives, 120*(11), 1551–1558. https://doi.org/10.1289/ehp.1205436

Sahiledengle, B. (2019). Self-reported healthcare waste segregation practice and its correlate among healthcare workers in hospitals of Southeast Ethiopia. *BMC Health Services Research, 19*(1), 591. https://doi.org/10.1186/s12913-019-4439-9

Samreen, Ahmad, I., Malak, H. A., & Abulreesh, H. H. (2021). Environmental antimicrobial resistance and its drivers: A potential threat to public health. *Journal of Global Antimicrobial Resistance, 27*, 101–111. https://doi.org/10.1016/j.jgar.2021.08.001

Sapkota, B., Gupta, G. K., & Mainali, D. (2014). Impact of intervention on healthcare waste management practices in a tertiary care governmental hospital of Nepal. *BMC Public Health, 14*(1), 1005. https://doi.org/10.1186/1471-2458-14-1005

Sawant, A. D., Warrier, S., & Pawar, G. (2021). Socio-economic implications of waste management. In A. Kateja & R. Jain (Eds.), *Urban Growth and Environmental Issues in India* (pp. 239–252). Springer Singapore. https://doi.org/10.1007/978-981-16-4273-9_15

Sayadi, M. H., Pathak, R. K. T., & R. K. (2010). Pollution of pharmaceuticals in environment. *I Control Pollution*, *26*(1), 89–94

Sembiring, E., & Nitivattananon, V. (2010). Sustainable solid waste management toward an inclusive society: Integration of the informal sector. *Resources, Conservation and Recycling*, *54*(11), 802–809. https://doi.org/10.1016/j.resconrec.2009.12.010

Sethi, Y., Kaka, N., Patel, N., Roy, D., Chopra, H., & Emran, T. B. (2023). Environmental correlates of infectious diseases in South-East Asia: A perspective on the missed link. *New Microbes and New Infections*, *53*, 101118. https://doi.org/10.1016/j.nmni.2023.101118

Sharma, K., & Kaushik, G. (2021). Urbanization and pharmaceutical waste: An upcoming environmental challenge. In A. Kateja & R. Jain (Eds.), *Urban Growth and Environmental Issues in India* (pp. 287–300). Springer. https://doi.org/10.1007/978-981-16-4273-9_18

Shayo, G. M., Elimbinzi, E., Shao, G. N., & Fabian, C. (2023). Severity of waterborne diseases in developing countries and the effectiveness of ceramic filters for improving water quality. *Bulletin of the National Research Centre*, *47*(1), 113. https://doi.org/10.1186/s42269-023-01088-9

Shekdar, A. V. (2009). Sustainable solid waste management: An integrated approach for Asian countries. *Waste Management*, *29*(4), 1438–1448. https://doi.org/10.1016/j.wasman.2008.08.025

Shinee, E., Gombojav, E., Nishimura, A., Hamajima, N., & Ito, K. (2008). Healthcare waste management in the capital city of Mongolia. *Waste Management*, *28*(2), 435–441. https://doi.org/10.1016/j.wasman.2006.12.022

Sia Su, G. (2005). Water-borne illness from contaminated drinking water sources in close proximity to a dumpsite in Payatas, The Philippines. *Journal of Rural and Tropical Public Health*, *4*, 43–48.

Singh, N., Ogunseitan, O. A., & Tang, Y. (2022). Medical waste: Current challenges and future opportunities for sustainable management. *Critical Reviews in Environmental Science and Technology*, *52*(11), 2000–2022. https://doi.org/10.1080/10643389.2021.1885325

Some, S., Mondal, R., Mitra, D., Jain, D., Verma, D., & Das, S. (2021). Microbial pollution of water with special reference to coliform bacteria and their nexus with environment. *Energy Nexus*, *1*, 100008. https://doi.org/10.1016/j.nexus.2021.100008

Srikanth, P. K., Srimanoi, W., Kashyap, P., & Visvanathan, C. (2022). Overview of infectious healthcare waste management in Thailand in Pre- and during COVID-19 context. In S. K. Ghosh & P. Agamuthu (Eds.), *Health Care Waste Management and COVID 19 Pandemic: Policy, Implementation Status and Vaccine Management* (pp. 121–150). Springer Nature. https://doi.org/10.1007/978-981-16-9336-6_6

Sun, Y., Meng, Y., Ou, Z., Li, Y., Zhang, M., Chen, Y., Zhang, Z., Chen, X., Mu, P., Norbäck, D., Zhao, Z., Zhang, X., & Fu, X. (2022). Indoor microbiome, air pollutants and asthma, rhinitis and eczema in preschool children – A repeated cross-sectional study. *Environment International*, *161*, 107137. https://doi.org/10.1016/j.envint.2022.107137

Swenson, V. A., Stacy, A. D., Gaylor, M. O., Ushijima, B., Philmus, B., Cozy, L. M., Videau, N. M., & Videau, P. (2018). Assessment and verification of commercially available pressure cookers for laboratory sterilization. *PLOS One*, *13*(12), e0208769. https://doi.org/10.1371/journal.pone.0208769

Tadesse, M. L., & Kumie, A. (2014). Healthcare waste generation and management practice in government health centers of Addis Ababa, Ethiopia. *BMC Public Health*, *14*(1), 1221. https://doi.org/10.1186/1471-2458-14-1221

Tickner, D., Opperman, J. J., Abell, R., Acreman, M., Arthington, A. H., Bunn, S. E., Cooke, S. J., Dalton, J., Darwall, W., Edwards, G., Harrison, I., Hughes, K., Jones, T., Leclère, D., Lynch, A. J., Leonard, P., McClain, M. E., Muruven, D., Olden, J. D., … Young, L. (2020). Bending the curve of global freshwater biodiversity loss: An emergency recovery plan. *BioScience*, *70*(4), 330–342. https://doi.org/10.1093/biosci/biaa002

Tong, Y. D., Huynh, T. D. X., & Khong, T. D. (2021). Understanding the role of informal sector for sustainable development of municipal solid waste management system: A case study in Vietnam. *Waste Management*, *124*, 118–127. https://doi.org/10.1016/j.wasman.2021.01.033

Tortora, G., Funke, B., & Case, C. (2019). *Microbiology an Introduction*. Pearson Education Limited.

US-Bio-Clean. (2014, June 27). *10 Industries That Need Proper Bio-waste Disposal*. US Bio-Clean. https://usbioclean.com/10-industries-need-proper-bio-waste-disposal/

Van Minh, H., & Hung, N. V. (2011). Economic aspects of sanitation in developing countries. *Environmental Health Insights*, *5*, EHI.S8199. https://doi.org/10.4137/EHI.S8199

Velis, C. A., & Cook, E. (2021). Mismanagement of plastic waste through open burning with emphasis on the global south: A systematic review of risks to occupational and public health. *Environmental Science and Technology*, *55*(11), 7186–7207. https://doi.org/10.1021/acs.est.0c08536

Visvanathan, C. (2006). *Medical waste management issues in Asia*. Asia 3R Conference, Tokyo, Japan.

Vivar, P. C., Salvador, P., & Abocejo, F. (2015). Village-level solid waste management in Lahug, Cebu City, Philippines. *Countryside Development Research Journal*, *3*(01), Article 01.

Wen, X., Luo, Q., Hu, H., Wang, N., Chen, Y., Jin, J., Hao, Y., Xu, G., Li, F., & Fang, W. (2014). Comparison research on waste classification between China and the EU, Japan, and the USA. *Journal of Material Cycles and Waste Management*, *16*(2), 321–334. https://doi.org/10.1007/s10163-013-0190-1

Wen, Y., Schoups, G., & van de Giesen, N. (2017). Organic pollution of rivers: Combined threats of urbanization, livestock farming and global climate change. *Scientific Reports*, *7*(1), Article 1. https://doi.org/10.1038/srep43289

White, A., & Hughes, J. M. (2019). Critical importance of a one health approach to antimicrobial resistance. *EcoHealth*, *16*(3), 404–409. https://doi.org/10.1007/s10393-019-01415-5

WHO. (2020a). Laboratory biosafety manual (4th ed.). World Health Organization.

WHO. (2020b). *Decontamination and waste management (Laboratory biosafety manual, fourth edition and associated monographs)*. World Health Organization.

Willey, J. M., Sandman, K., Wood, D. H., & Prescott, L. M. (2023). *Prescott's Microbiology* (12th ed., International student edition). McGraw Hill.

Wilson, D. C., Velis, C., & Cheeseman, C. (2006). Role of informal sector recycling in waste management in developing countries. *Habitat International*, *30*(4), 797–808. https://doi.org/10.1016/j.habitatint.2005.09.005

Windfeld, E. S., & Brooks, M. S.-L. (2015). Medical waste management – A review. *Journal of Environmental Management*, *163*, 98–108. https://doi.org/10.1016/j.jenvman.2015.08.013

Wynne, A. L., Nieves, P. M., Vulava, V. M., Qirko, H. N., & Callahan, T. J. (2018). A community-based approach to solid waste management for riverine and coastal resource sustainability in the Philippines. *Ocean & Coastal Management*, *151*, 36–44. https://doi.org/10.1016/j.ocecoaman.2017.10.028

Yekkalar, M., Panahi, S., & Nikravan, M. (2015). Evaluation of current laboratory waste management: A step towards green campus at Amirkabir University of Technology. In W. Leal Filho, N. Muthu, G. Edwin, & M. Sima (Eds.), *Implementing Campus Greening Initiatives: Approaches, Methods and Perspectives* (pp. 215–227). Springer International Publishing. https://doi.org/10.1007/978-3-319-11961-8_17

Yu, J. (2020). *How Wuhan copes with Mountains of Medical Waste.* https://news.cgtn.com/news/2020-03-17/How-Wuhan-copes-with-its-mountains-of-medical-waste--OUxhr4jW1i/index.html

Zaki, A. N., & Campbell, J. R. (1997). Infectious waste management and laboratory design criteria. *American Industrial Hygiene Association Journal*, *58*(11), 800–808. https://doi.org/10.1080/15428119791012306

Zhang, S., Huang, J., Zhao, Z., Cao, Y., & Li, B. (2020). Hospital wastewater as a reservoir for antibiotic resistance genes: A meta-analysis. *Frontiers in Public Health*, *8*. https://www.frontiersin.org/articles/10.3389/fpubh.2020.574968

# 7 Biosafety and Biosecurity in the Academe

*Marian P. De Leon*
Museum of Natural History
University of the Philippines
Los Baños College
Los Baños Laguna, Philippines

*Michael O. Baclig*
Trinity University of Asia
St. Luke's Medical Center
College of Medicine-William H. Quasha Memorial
Quezon City, Philippines

## 7.1 INTRODUCTION

The most recent data from the Johns Hopkins Coronavirus Resource Center showed that there are nearly 700 million confirmed cases of COVID-19 and nearly seven million deaths worldwide as of March 10, 2023. Johns Hopkins has discontinued its operations after three years of around-the-clock tracking of COVID-19. Locally, there are over 4 million confirmed COVID-19 cases, with more than 4 million recoveries and over 65,000 deaths. Early on, COVID-19 continued to spread with alarming speed, bringing instruction and research activities to a near, if not total, standstill, as many countries, including the Philippines, impose restrictions and mitigation measures to halt the further spread of the virus.

Higher education institutions (HEIs) in the Philippines transition into a research university from a teaching university. In a research university or college, high investments in research provide better opportunities for students to be engaged in research that serves as their training ground for learning and discovery. The Commission on Higher Education (CHED) of the Republic of the Philippines recognizes the importance of universities not only as a generator of knowledge, an educator of young minds, and transmitters of culture but also as a major agent of economic growth, a Research and Development laboratory and a mechanism through which the nation builds its human capital to enable it to participate in the global economy actively (CHED Memorandum Order No. 52 Series of 2016).

In an article by Witze (2020), it was mentioned that even after the COVID-19 pandemic, the effects brought about by the global pandemic could permanently change how researchers work, what scientists study, and how much funding they receive. Despite these challenges, the scientific community's response to the global pandemic


DOI: 10.1201/9781003426219-7

presents an opportunity for scientists to innovate and align with the realities happening in various research and academic institutions, locally and abroad (Korbel et al., 2020). For example, the experiences described by Gillum and colleagues at Arizona State University exemplify the challenge to balance inquiry and scientific process against the urgent need for understanding pandemic dynamics and developing new best practices (Gillum 2022).

It is noteworthy to mention that for scientists, the pandemic is an opportune time, rather than a pause, to go ahead and carry on what needs to be done. The intermittent lockdowns and limited access to research laboratories posed an unprecedented challenge to humanity and science and displaced research not related to COVID-19 (Riccaboni & Verginer, 2022). The rapid shift in the research focus in the first three months of the onset of COVID-19, according to Riccaboni and Verginer (2022), result in a fivefold increase in the number of scientific papers on COVID-19 compared to the number of articles on H1N1 swine influenza. In the Philippines, more than a hundred research on COVID-19 are being funded by the Department of Science and Technology (DOST) under its health research and development arm, the Philippine Council for Health Research and Development (PCHRD) (https://www.pchrd.dost.gov.ph/about-pchrd/).

Considering the global pandemic, the number of articles published about COVID-19 in peer-reviewed journals, for instance, are rapidly increasing regarding the epidemiology, pathogenesis, clinical diagnosis, and treatment strategies. However, much remains to be elucidated about this disease, given its multifaceted presentation, viral characteristics, and clinical course. Experts are carefully analyzing the long-term effects of the disease. Thus, this serves as an avenue for further research not just in the clinics but also from a biosafety point of view. Newer risk mitigation strategies may need to be implemented based on local risk assessments but also consider other emerging and re-emerging infectious diseases.

The COVID-19 pandemic resulted in hurdles students and researchers face in the academe. For instance, it has caused a significant disruption in the enrollment of study participants, data collection, laboratory experimentation, maintenance and preservation of microbial cultures in microbial culture collections, biospecimens in biobanks, and tissues and historical DNAs in museums and laboratory animals. Research centers and laboratories in the academe and other stakeholders operate with minimal personnel and limited activities. On the other hand, academic institutions transition to flexible learning, including face-to-face, online, offline, or any combination of these modalities. Flexible learning has become part of the new norm. While innovative solutions have been established for flexible learning, recent problems have arisen, including but not limited to technical glitches and virtual fatigue.

In the local setting, a Joint Memorandum Circular No. 2021-004 by CHED and the Department of Health (DOH) have provided guidelines on the implementation of limited face-to-face classes for all programs of Higher Education Institutions (HEIs) in areas under Alert Levels System for COVID-19 Response since the local cases still have occasional surges (Joint Memorandum Circular No. 2021-004 and CHED Memorandum Order No. 1 Series of 2022). In particular, the minimum public health standards set forth by the CHED Memorandum are shown in Table 7.1, and some are illustrated in Figure 7.1.

**TABLE 7.1**
**Minimum Health Standards**

| Practices | Description |
|---|---|
| Classroom capacity | Maximum of 100% under Alert 1* |
| Physical distance in HEIs | There are no restrictions, but HEIs may provide distancing protocols as they deem appropriate |
| Engineering controls | Ensure adequate air quality (6 air changes per hour, $CO_2$ less than 1,000 ppm*) and ventilation (open windows, exhaust fans) |
| Hand hygiene and sanitation facilities | Adequate water supply, hand washing stations, soap, alcohol, hands-free trash receptacles |
| Visual cues, reminders, and guides | Mask wearing, hand hygiene, DOH* hotline, and other safety reminders |
| Wearing of face masks | Well-fitted masks, proper disposal of PPE* |
| Retrofitting and disinfection of facilities, classrooms, and laboratories | Use of plastic or acrylic barriers or dividers |
| Cyclical shifting model/regular class schedule | Shifting schedule is optional, given maximum indoor capacity of 100% |
| Safety seal certification program | Maintain safety seal certification |
| Student dormitories | Coordinate with local government unit before opening |
| Physical education classes | Reduce capacity based on capability to comply with the minimum health standards |
| Medical insurance coverage for students | Registered with PhilHealth* or equivalent health insurance |
| Mechanisms for the provision of care | Referral of patients or other emergency health conditions |

**Note:* Alert Level 1 = there is full indoor and outdoor venue capacity for fully vaccinated individuals (Joint Memorandum Circular No. 2021-004); HEIs = Higher Education Institutions; ppm = parts per million; DOH = Department of Health; PhilHealth = Philippine Health Insurance Corporation

The minimum public health standards agree with the practices in other countries and cooperation with health experts. As of this writing, the National Capital Region (NCR) and the Cordillera Administrative Region (CAR) are classified under moderate risk for COVID-19. The DOH Undersecretary reported that NCR and CAR's average daily attack rates have exceeded the threshold of five cases per 100,000 population (Philippine News Agency). The rest of the regions showed a downward trend of cases in the country. On May 4, 2023, *the International Health Regulation (IHR) Emergency Committee of the WHO highlighted the decreasing trend in COVID-19 deaths, the decline in COVID-19-related hospitalizations and intensive care unit admission, and the high levels of population immunity to SARS-CoV2.* The following day, *the WHO Director-General concurred with the advice offered by the committee regarding the ongoing COVID-19 pandemic. It determined that COVID-19 is now an established and ongoing health issue that no longer constitutes a public health emergency of international concern. It advised that it is time to transition to long-term management of the COVID-19 pandemic.* On May 8, 2023,

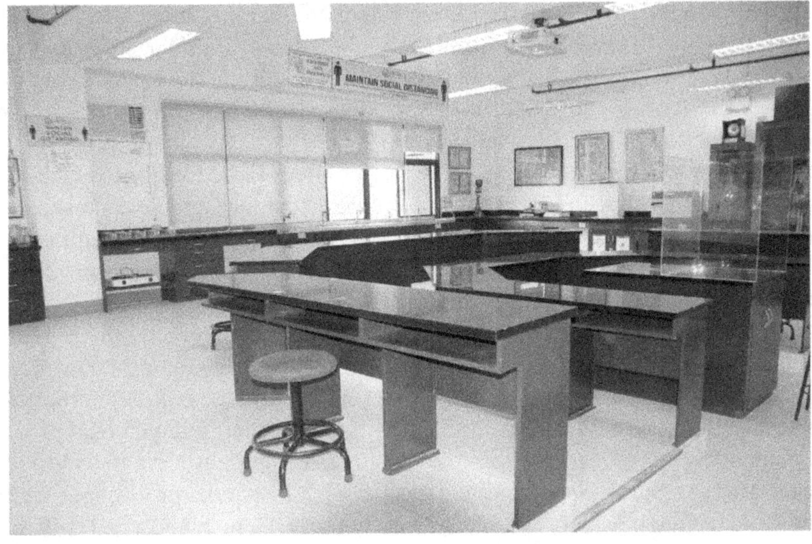

(a)

(b)

**FIGURE 7.1** Minimum standards implemented in HEIs in the Philippines. Use of acrylic barriers in school facilities (Pictures courtesy of Trinity University of Asia).

*the Inter-Agency Task Force for the Management of Emerging Infectious Diseases, in its Resolution No. 08 (2023), requested lifting the State of Public Health Emergency throughout the Philippines.* More recently, on July 21, 2023, the Malacanang Palace issued Proclamation No. 297 that lifted the public health emergency throughout the Philippines due to COVID-19 (Malacanang).

It is worth mentioning that prior to the reopening of HEIs for limited face-to-face classes for all programs under the Alert Levels System for COVID-19 response, the HEIs conducted a self-assessment on the readiness of HEIs to reopen campuses. The Chair of the Crisis Management Committee certifies the assessment checklist. The areas of assessment are shown in Table 7.2.

It is well-recognized that the backbone of the biosafety practice is risk assessment, a risk- and evidence-based approach. Thus, the HEIs' self-assessment allows the stakeholders to identify, evaluate, mitigate, and control the risk. A combination of engineering and administrative controls, new best practices, and personal protective equipment mitigate the risk of exposure of vaccinated and unvaccinated individuals.

Aside from research opportunities, the COVID-19 pandemic brought forth the essentials of having biosafety and biosecurity in the academic landscape, which were almost neglected in favor of other priorities in the HEIs, and the lack of orientation and training delayed its implementation and appreciation.

On August 14, 2023, the World Health Organization (WHO) reported the rising global COVID-19 cases. The WHO mentioned for Member States to maintain and not dismantle the established COVID-19 infrastructures. In addition, early reporting, surveillance, variant tracking, early clinical care provision, booster shots, especially to high-risk groups, and adequate and proper ventilation cannot be overemphasized.

The succeeding sections will discuss the requirements vis-à-vis the practices and mitigation measures implemented in most HEIs in the Philippines.

## 7.2 FACILITY DESIGN AND ENGINEERING CONTROLS IN THE ACADEME

The COVID-19 pandemic has posed unprecedented challenges to reopening colleges and universities. Some proven strategies for controlling transmission include physical

**TABLE 7.2**
**Areas of Assessment**

| Checklist | Description |
|---|---|
| Management and oversight | The crisis management committee oversees the implementation and monitoring of compliance with the conduct of limited face-to-face classes. |
| Institutional policies and protocols | Screening, containment, and lockdown protocols are in place. |
| Controls | Engineering controls such as proper ventilation and hand washing facilities are installed strategically. |

distancing, body temperature measurement, health declaration, and COVID-19 screening. However, facility design and engineering controls are crucial in preventing the spread of infections like COVID-19 and laboratory-acquired infection (LAI) (Table 7.3).

According to Destura et al. (2022), between 1982 and 2016, 27 LAIs were reported in the Asia-Pacific, and 52% happened in the research laboratories. Fortunately or not, no LAIs have been reported despite numerous agricultural, microbiological, medical, and biotechnological research universities and centers in the country. LAIs can be prevented or minimized by appropriate facility design and engineering controls, along with practices and controls discussed in succeeding sections of this Chapter.

Requirements and details on facility design and engineering controls are presented in Chapter 15. At the same time, those applicable to the academic setting will be mentioned in this Chapter.

In research universities, sealed centrifuge rotors are used for loading and unloading safety cups within the biosafety cabinet to prevent aerosol formation. The biosafety cabinets are maintained and certified by authorized agencies. HEPA filter leaks, downflow velocity, face velocity, and air flow smoke pattern are some of the tests performed annually by third-party agencies to ensure user, product, and environment protection (Baclig, 2021).

Laboratory facilities are designated biosafety levels (BSLs) based on the laboratory type, practices and equipment. There are four BSLs, namely BSL-1, BSL-2, BSL-3, and BSL-4. Table 7.4 shows some of the equipment needed in a BSL-1,

**TABLE 7.3**
**Facility Design and Engineering Controls**

| Controls | Description |
|---|---|
| Ventilation | There is controlled and adequate ventilation with the prescribed air change per hour. Recommended air change is at least six changes per hour. |
| Air quality | Windows shall be open while there are occupants in the workplace. |
| Exhaust fans | This should be operational, especially while there are occupants in the area. |
| Hand hygiene and sanitation facilities | There is adequate water supply and disinfectant. The washing station or sink is hands-free. |
| Trash receptacles | Foot-pedal-operated trash bins. |
| Barriers | Use of plastic or acrylic dividers inside the classrooms. |
| Restricted work areas | There are visible signages of health and safety reminders and markings on the floor to direct the movement of personnel. |
| Entry gate | There is a screening station for body temperature monitoring at the entry point. There is also a station for hand hygiene. |

Adopted from Joint Memorandum Circular No. 2021- 01

**TABLE 7.4**
**Common Equipment and Instruments Used in Teaching and Research Laboratories**

| Equipment | BSL-1 Common Teaching Laboratory | BSL-2 Research Laboratory |
|---|---|---|
| Biosafety cabinet | Not all | Class II, A1 or A2 |
| Pipettors | Glass, plastic, automated | Mostly automated; filtered pipette tips |
| Centrifuge | Yes | Operated inside the biosafety cabinet |
| Vortex mixer | Yes | Operated inside the biosafety cabinet |
| Autoclave | No | Yes |

which is common for teaching, and BSL-2 in research laboratories. For research laboratories, it is desirable to have a certified biosafety cabinet and an onsite autoclave for decontamination

## 7.3 GOOD LABORATORY WORK PRACTICES AND PROCEDURES

Good Laboratory Practices (GLP) refer to a system of management controls that ensures researchers' tests and operations provide reliable, consistent, and reproducible results under safe conditions (Bornstein-Forst, 2017). GLP shall play an important role in ensuring laboratory test consistency, reliability, reproducibility, and quality (Alatgi & Chougule, 2015). Further, Alatgi and Chougule (2015) emphasized that GLP provides a framework for laboratories to plan, perform, monitor, record, and report their activities. It also reflects the quality of laboratory services during assessments and accreditation processes.

Good Laboratory Work Practice (GLWP), on the one hand, is a practice, technique, or procedure that, when followed, ensures the protection of the laboratory workers and the environment and reduces the risk of exposure to hazardous agents (Global Biorisk Management Curriculum Library, 2016) while a Good Microbiological Practice and Procedure (GMPP) is specific for microbiology laboratory (WHO Laboratory Safety Manual, 2020). For discussion, GLWP and GMPP will differentiate GLP and make it more specific for biosafety and biosecurity in the academe (GLWP) and microbiology laboratories (GMPP). The developing architecture of HEIs as research catalysts requires an established GLWP to successfully implement the laboratory biosafety and biosecurity (biorisk) program to ensure the safety of the personnel and security of valuable biological materials or (VBMs).

The onset of emerging and re-emerging diseases of man and animal in the past years triggered research initiatives in the academe to shift gears into biomedical, biological, and veterinary sciences. Discoveries of new and alternative sources of therapeutics generated biologics (e.g., microorganisms, cells, tissues, genes, proteins) have resulted in diagnostic and state-of-the-art equipment in the laboratories in the universities (Bolon et al., 2018). The onset of the genomics era, where molecular biology techniques and combined omics approaches advance the understanding and increased titer of biological agents.

Bolon and colleagues (2018) also presented the key features of the GLP for nonclinical laboratories that may be undertaken for the GLWP for the biorisk program in an academic setting. These include but are not limited to the following: (1) organizational structure, (2) personnel responsibilities, (3) personnel training practices, (4) facilities, (5) equipment, (6) Standard Operating Procedures (SOP), (7) study documentation (record keeping), and (8) record and sample retention.

In the research academic setting, the organizational structure provides the hierarchy of commands and responsibilities in implementing the biorisk program and monitoring compliance with the system standards. Individuals who will assume leadership, also known as Responsible Person (RP) (e.g., laboratory manager, division head, center director, curator, project or study leaders, etc.), are expected to have the authority to make important personnel decisions (e.g., hiring or reassignment) and expertise on the review of research protocols and procedures, experimental designs, and laboratory facilities together with equipment use and care. Responsible personnel (RPs) must be identified and proper orientation and training must be provided to avoid safety hazards, equipment damage, infrastructure loss, and confusion regarding operations and oversight (Bornstein-Forst, 2017). Management and leadership are also responsible for providing and arranging the appropriate training of all personnel based on their functional roles and responsibilities in support of the biorisk program (BMBL, 2020).

The laboratory personnel like researchers, students, and technicians assume the responsibility of strict implementation of the principles of the GLWP, report any infringement or nonobservance of the GLWP, and review and revise the SOP if necessary. Figure 7.2 shows some of the GLWP, such as the use of floor signs and visual cues to direct traffic and high-efficiency particulate air (HEPA) filtration air purifiers.

Regular training on GLWP is necessary to update the laboratory personnel, especially on newly installed equipment, and adapt robust methodology and techniques and research and laboratory-related training for proficiency. Established facility-specific best practices and procedures are essential to support the implementation and sustainability of a successful biorisk program (BMBL, 2020). As discussed in the preceding section, facilities have ample space and adequate separation of the elements or activities of the study that will prevent contamination of the VBMs (Chougule, 2015). Equipment and apparatus must be regularly or periodically inspected, cleaned, maintained, calibrated, and validated according to the internal SOPs. Records of these must be kept and regularly updated based on the schedule of activities.

Among the key features of a GLWP, adopting a common SOP provides an academic research laboratory to assess whether or not the researchers, students, laboratory technicians, teaching assistants, and faculty members consistently perform operations appropriately. SOPs *should be developed that minimize the potential exposure of responding personnel to potentially hazardous biological materials* (BMBL, 2020). When GLWP is implemented, the workflow from task performance to assessment of performance supports efficiency in time and budget allocation, as well as academic excellence. As industrial QC is applied post-process, the equivalent

(a)

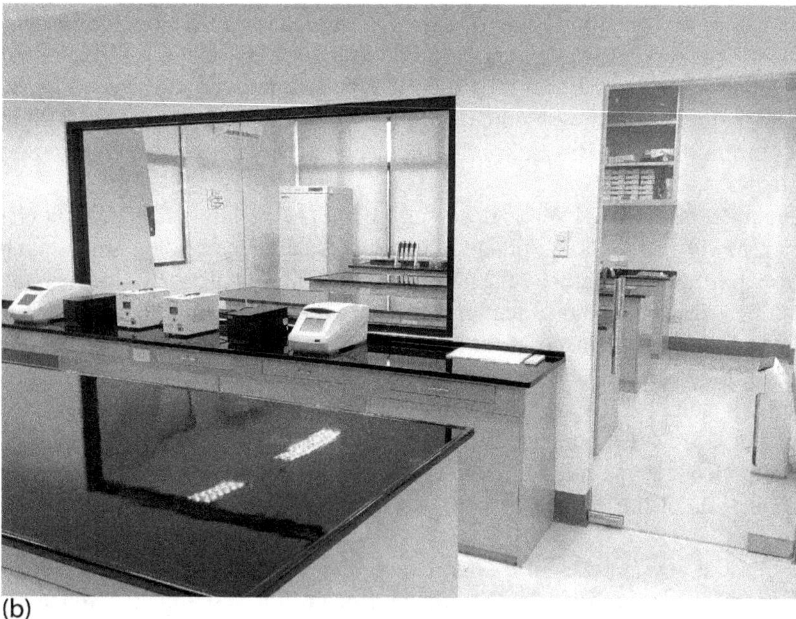

(b)

**FIGURE 7.2** Floor signs, visual cues to direct traffic, and HEPA air purifiers.

educational assessment would entail the summative review of student, staff, and faculty performance noted above. Benchmark indicators may be encompassed by course reviews, faculty promotion and tenure documents, staff performance appraisals, grades, and exams. GLWP and SOPs are strictly implemented and observed by the students, interns, researchers, and laboratory technicians. Aides in teaching and research laboratories in HEIs ensure safe and secure research activities. Some good microbiological practices that may reduce, if not eliminate exposure to biological hazards related to COVID-19 testing, are shown in Table 7.5, and definite examples of GLWPs for teaching and research laboratories are in Table 7.6.

It is noteworthy to mention that some universities have been granted Safety Seal Certification by The Commission on Higher Education (CHED). The Safety Seal is a voluntary certification scheme that affirms that an establishment is compliant with the minimum public health standards set by the government.

## 7.4 ADMINISTRATIVE CONTROLS

Administrative controls include policies, standards, guidelines, and procedures. Policies are the highest form of administrative control, resulting in a safe work environment. According to the Environment, Health and Safety of Cornell University, the colleges and departments are responsible for developing policies to protect laboratory workers against exposure to hazardous materials (e.g., biological and chemical). Meanwhile, it is the laboratory supervisor's responsibility to ensure that laboratory personnel are informed of campus-wide policies and procedures related to laboratory biosafety. Specific examples of administrative controls are shown in Table 7.7.

**TABLE 7.5**
**Good Microbiological Practices**

| Practices | Description |
|---|---|
| Signages | There are readable signages of health and safety reminders placed in strategic areas. |
| Standard operating procedures | SOPs on handling, storing, and disposal of wastes are in place. SOPs on laboratory procedures and use of equipment are available. |
| Facility and staff training | Proper and adequate training of laboratory workers before actual work. |
| Face mask | Masks should be worn as mitigation strategies to suppress transmission and save lives. |
| Fit testing of NIOSH-approved respirator | Fit testing of respirators such as N95 or higher prior to actual use. |
| Disinfection of surface | Freshly prepared disinfects such as sodium hypochlorite are used to clean surfaces. For spills, appropriate contact time at the correct dilutions is strictly observed. |
| Proper segregation and disposal of wastes | Biohazardous materials are autoclaved prior to disposal. |

### TABLE 7.6
### Examples of Good Laboratory Work Practices in Teaching and Research Laboratories

| Teaching Laboratories | Research Laboratories (in addition to those GLWPs in the Teaching Laboratories) |
|---|---|
| Wash hands after entering and before exiting the laboratory. | Keep note-taking and discussion practices separate from work with hazardous or infectious material. |
| Do not bring food, gum, drinks (including water), or water bottles into the laboratory. | Use micro/bacticinerators or disposable loops rather than Bunsen burners. |
| Do not mouth pipette. | Use proper transport vessels (test tube racks) for moving cultures in the laboratory, and store vessels containing cultures in leak-proof containers when working with them is complete. |
| Label all containers clearly. | Use leak-proof containers for storage and transport of infectious materials. |
| Disinfect the bench before and after the laboratory session with a disinfectant known to kill the organisms handled. Use disinfectants according to manufacturer instructions. | Proper identification and segregation of contaminated wastes should be implemented prior to decontamination and disposal. |
| | Samples must be shipped following the triple packaging system. |
| Notify the instructor of all spills or injuries. Document all injuries according to school, university, or college policy. | Packaging and transport of samples must comply with local and/or international standards. |
| Do not handle broken glass with fingers; use a dustpan and broom. | A qualified service engineer inspects and replaces the HEPA filter of a biosafety cabinet and performs a leak test, among others. |
| Teach, practice, and enforce proper wearing, removal, and use of gloves. No wearing of lab coats or PPE outside of the lab. | |
| Arrange for proper (safe) decontamination and disposal of contaminated material (e.g. in a properly maintained and validated autoclave) or arrange for licensed waste removal in accordance with local, and national guidelines. | |
| Use only institution-provided marking pens and writing instruments. These remain in the lab at all times and must be disinfected regularly. | |

(*Continued*)

**TABLE 7.6 (CONTINUED)**
**Examples of Good Laboratory Work Practices in Teaching and Research Laboratories**

| Teaching Laboratories | Research Laboratories (in addition to those GLWPs in the Teaching Laboratories) |
|---|---|
| In the laboratory, do not handle personal items (cosmetics, cell phones, laptops, calculators, pens, pencils, etc.). Cell phone usage is sometimes permitted during the lab; however, cell phones MUST be kept in a resealable plastic bag. After use, the bag should be disposed of in the proper receptacle. | |
| Do not wear valuable electronics (smartwatches, Fitbits, etc.) or dangling jewelry in the laboratory. | |
| Tie back long hair. Do not touch the face, apply cosmetics, adjust contact lenses, or bite nails. | |

Adopted from the Guidelines for Biosafety in Teaching Laboratories by the American Society for Microbiology, 2019.

**TABLE 7.7**
**Administrative Control Measures**

| Administrative Controls | Description |
|---|---|
| Health checks | Measurement of body temperature and health declaration of students and teaching and non-teaching staff. |
| Safety officer | Appointment of safety officers to ensure compliance with occupational health and safety guidelines in the workplace. |
| Training | Biosafety and biosecurity training or workshops such as spill drills are conducted regularly. |
| Institutional Biosafety Committee | The committee is responsible for reviewing, approving, and monitoring university-wide research projects involving biological materials that may pose differing safety, health, or environmental risks to plants, animals, or humans. |

The principal investigator, instructor, and laboratory supervisor are responsible for ensuring that everyone working in the laboratory, like the students, researchers, and laboratorians, are aware and strictly implement and adhere to policies and procedures imposed by the university, college or department, and the laboratory. These include guidelines and requirements in the laboratory safety and security manuals and standard operating procedures (SOP). In every laboratory, a well-written and easy-to-understand SOP is a MUST. These are sets of written guidelines, documents, or supplemental information that contain experimental processes; procedural use of equipment following manufacturer's instructions; use and care of engineering controls (e.g., biosafety cabinet and fume hood); emergency procedures and responses (e.g., biological spill management); use, fit, and seal tests and disposal of PPEs; and decontamination and disposal of wastes. These are just a few examples that require a written and updated SOP.

Access to laboratories and storage facilities of VBM and chemical supplies must be restricted from unauthorized individuals, visitors, and children. Access controls, security locks, and cameras may be installed to deter loss and/or theft of VBMs and hazardous chemicals.

In order to put into place appropriate administrative controls, all laboratories handling VBMs must conduct risk assessments to determine safety and security risks in the laboratory, including routine work and unexpected situations (Sandia Laboratories, 2023). Biorisk assessment is a critical component in all activities conducted in a biological laboratory. The process involved in biorisk assessment is discussed in detail in Chapter 5. For emphasis, laboratories in academe require routine risk assessment to ensure proper administrative controls are in place. Biorisk assessment will help the top management heads determine the following: (1) effective allocation of resources to mitigate risks; (2) identification of training needs and supervision; (3) planning for renovation; (4) evaluation of procedural changes; (5) compliance with governmental regulations; (6) justification for space and equipment needs; (7) evaluation of emergency plans; (8) planning for preventative maintenance; (9) evaluation of exchanges and workflow with other laboratories/units (Sandia Laboratories, 2023); (10) draft, implement, and review policies; (11) compliance in upgraded standards; and (12) implementation and adaptability of protocols and procedures, among others.

Biorisk assessment of daily laboratory activities or operations provides a guide for selecting appropriate biological safety measures (including microbiological practices and safety equipment), security measures, and other facility safeguards to mitigate the determined risks to an acceptable or manageable level.

## 7.5 ESTABLISHMENT OF AN INSTITUTIONAL BIOSAFETY COMMITTEE

Establishing an Institutional Biosafety Committee (IBC) is still in its infancy in the Philippines. To date, only a few colleges or universities review, approve, and monitor research projects involving biological agents or their toxins. However, it is noteworthy to mention that following the successful conduct of the Philippine Advanced Biorisk Officers' Training (PhABOT) by the National Training Center for Biosafety and Biosecurity (NTCBB), University of the Philippines-National Institutes of Health (UP-NIH) in collaboration with the US Biosecurity Engagement Program (BEP) and the US Civilian Research and Development Foundation (CRDF) Global, there is now an increased awareness for the creation of the IBC in the country. Indeed, the PhABOT graduates from various academic or research institutions, both private and public, initiated the establishment of the IBC in close coordination with the upper management. To date, 102 Certified Biosafety Officers (CBOs) are working in clinical academic, military, and research laboratories, all connected to the NTCBB through the PhABOT program.

The important considerations in conducting the IBC review include but are not limited to risk identification, transmission routes, engineering controls such as safety equipment and facilities, administrative controls, good laboratory practices and procedures, as well as proper and adequate use of personal protective equipment. The IBC will determine approval of the research proposal, and the Principal Investigator will be notified of the decision. An example of a workflow in the review process is shown in Figure 7.3.

In general, the IBC is composed of at least five members. The University President or Chancellor appoints the Chair and members. The IBC consists of faculty members, research personnel and one or two non-affiliated institution members who will represent the community. The members of the IBC should be individuals who are knowledgeable in institutional policies and professional standards. Some of the qualifications of IBC members are as follows: (1) basic training on biosafety and biosecurity, (2) good moral character, and (3) willing to undergo further training related to the practice of biosafety and biosecurity.

## 7.6 CHALLENGES AND OPPORTUNITIES: MOVING FORWARD

The emergence and re-emergence of infectious diseases, including but not limited to COVID-19, Zika and MERS-CoV, have significantly changed how we live and do things (Table 7.8). In addition, the rapid developments in modern biotechnology,

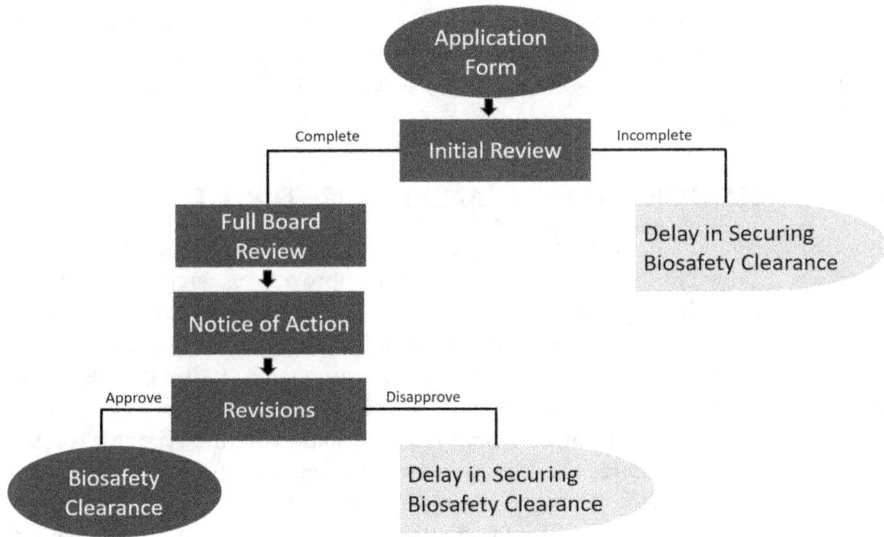


**FIGURE 7.3** Institutional Biosafety Committee review process.

**TABLE 7.8**
**Emerging and Re-emerging Infections in the Philippines**

| Year | Infection |
|---|---|
| 2021 | Nipah |
| 2020 | SAR-CoV2 |
| 2016 | Zika |
| 2016 | MERS-CoV, Ebola Reston |
| 2016 | MDR TB |
| 2014 | Measles, MERS-CoV |
| 2013 | Pertussis, Measles |
| 2012 | Leptospirosis |
| 2011 | Chikungunya |
| 2010 | Dengue |
| 2009 | Pandemic A H1N1 |
| 2008 | Ebola Reston, Leptospirosis, *Salmonella typhi* |
| 2007 | Resistant *Shigella flexneri* |

microbiology, and synthetic biology presented challenges and opportunities to advance science and improve quality of life. Thus, establishing the biosafety and biosecurity landscape around the globe should be a top priority to avert any unnecessary occurrences while ensuring optimal health for people, animals, and our environment.

From a biosecurity perspective, strengthening the rules and regulations of various stakeholders is of utmost importance. Table 7.9 shows the ideal roles of the various stakeholders as described by Destura et al. (2021).

The Department of Health, Office of Health Laboratories has recently published the Laboratory Biosafety and Biosecurity Standards (2023). Specifically, eight biosecurity standards were discussed to ensure valuable biological materials' safe use and security (Table 7.10).

## 7.7 CONCLUDING REMARKS AND FUTURE DIRECTIONS

The CHED Memorandum Order 52 (2016) listed several concerns that must be addressed by the Philippine HEIs, namely, (1) improved research capabilities of the faculty, research staff, and students; (2) instilling research culture and vocation among the faculty and graduate students; (3) upgrading of the physical resources and research infrastructure; (4) building up, retraining, and retaining a sustainable stream of new generation of researchers; (5) increasing research productivity and raising research quality and impact; and (6) having and institutionalizing a research code of ethics that maintains the integrity, openness, and transparency of a research process and safeguards intellectual property. In addition to the concerns highlighted, HEIs must adhere to the ethical research guidelines, including compliance with the Philippine Biosafety Guidelines. The Philippine Biosafety Guidelines focus on

**TABLE 7.9**
**Roles of Stakeholders in the Philippines**

| Stakeholders | Ideal Roles |
|---|---|
| Department of the Interior and Local Government | Investigate alleged bioterrorism |
| Department of Health | Maintain a list of biological materials of concern |
| Anti-Terrorism Council | Create a national action plan |
| Armed Forces of the Philippines | Respond in case of emergency |
| Department of National Defense | |
| National Disaster Risk Reduction and Management Control | |
| Office of Civil Defense | |
| Philippine National Police | |
| Bureau of Customs | Regulate entry and exit of biological materials of concern |
| Bureau of Immigration | |
| Department of Transportation | Regulate transport of biological materials of concern by land, water, and air |

**TABLE 7.10**
**Biosecurity Standards**

| Standard | Security Controls | Description |
|---|---|---|
| Standard 1 | Biosecurity Awareness | Conduct biosecurity training for laboratory personnel and other stakeholders. |
| Standard 2 | Personnel Reliability | Human factors can significantly impact the success of biorisk management, and insider or outsider threats cannot be ruled out. |
| Standard 3 | Information Security | Any sensitive information (logbooks, SOPs, computer files) that could lead to misuse of dangerous materials should be carefully secured. |
| Standard 4 | Material Control and Accountability | Awareness of what exists in the workplace, where it is, and who is responsible. |
| Standard 5 | Emergency/Incident Response | Reporting near misses, incidents, and accidents. |
| Standard 6 | Management | Top management support is crucial to ensure safe and secure work practices. |
| Standard 7 | Physical Security | Mechanisms for early detection, delay, rapid response, and limited access control should be in place. |
| Standard 8 | Transport Security | Assurance that the same rigorous process that protects valuable biological materials in the laboratory follows the stringent process when the materials are transported outside the workplace. |

genetically modified organisms and do not include microbial resources from environmental samples characterized for producing bioactive compounds with agricultural, food, medical, and industrial applications and as carriers for antimicrobial genes and toxins. There is a need for a regulatory agency to monitor, regulate, and accredit teaching and research laboratories in the country. This will ensure that biosafety and biosecurity practices in teaching and research laboratories in Philippine HEIs are strictly adhered to for the safety and security of everyone working in the laboratory and the community.

## ACKNOWLEDGMENTS

The authors would like to thank the National Training Center for Biosafety and Biosecurity (NTCBB) of the University of the Philippines-Manila for the comprehensive training given to the authors through the Philippine Advanced Biorisk Officers' Training (PhABOT) and the Trinity University of Asia through Vice President for Academic Affairs Dr. Howell T. Ho for providing us the pictures used in this chapter.

## REFERENCES

Alatgi, A. C., Chougule, S. (2015). Good Laboratory Practice. *Journal of Evolution of Medical and Dental Sciences* 4(103), 16901–16906. https://doi.org/10.14260/jemds/2015/2543.

Baclig, M. O. (2021). Biosafety in the Time of Severe Acute Respiratory Syndrome Coronavirus 2 Pandemic: The Philippine Experience. *Applied Biosafety: Journal of the American Biological Safety Association* 26(Suppl 1), S10–S15. https://doi.org/10.1089/apb.20.0069.

Biosafety in Microbiological and Biomedical Laboratories (BMBL) 5th edition. https://www.cdc.gov/labs/pdf/CDC-BiosafetymicrobiologicalBiomedicalLaboratories-2009-P.pdf.

Bolon, B., W. Baze, C. J. Shilling, K. L. Keatley, D. J. Patrick, K. A. Schafer. (2018). Good Laboratory Practice in the Academic Setting: Fundamental Principles for Nonclinical Safety Assessment and GLP-Compliant Pathology Support When Developing Innovative Biomedical Products. *ILAR Journal* 59(1), 18–28.

Bornstein-Forst, S. M. (2017). Establishing Good Laboratory Practice in Small Colleges and Universities. *Journal of Microbiology and Biology Education* 18(1), 18.1.10.

CHED Memorandum Order No. 1 Series of 2022. Supplemental Guidelines to CHED-Joint Memorandum Circular (JMC) No. 2021-004 on the Additional Guidelines for the Operations of Limited Face-to-Face Classes of Higher Education Institutions (HEIs) in Areas Under Alert Level 1. https://ched.gov.ph/wp-content/uploads/CMO-NO.-1-S.-2022.pdf.

CHED Memorandum Order No. 52 Series of 2016. Pathways to Equity, Relevance and Advancement in Research, Innovation and Extension in Philippine Higher Education. https://ched.gov.ph/wp-content/uploads/2017/10/CMO-52-s.-2016.pdf.

COVID-19 Dashboard by the Center for Systems Science and Engineering (CSSE) at Johns Hopkins University. (2022). https://coronavirus.jhu.edu/map.html.

Destura, R. V., Lam, H. Y., Navarro, R. C., Lopez, J. C. F., Sales, R. K. P., Gomez, M. I. F. A., Dela Tonga, A., Ulanday, G. E. (2021). Assessment of the Biosafety and Biosecurity Landscape in the Philippines and the Development of the National Biorisk Management Framework. *Applied Biosafety: Journal of the American Biological Safety Association* 26(4), 232–244. https://doi.org/10.1089/apb.20.0070.

Gillum, D. R., Rice, A. D., Mendoza, I. A. (2022). The COVID-19 Pandemic Response: Biosafety Perspectives from a Large Research and Teaching Institution. *Applied Biosafety*, 64–78. https://doi.org/10.1089/apb.2022.0001.

Global Biorisk Management Curriculum Library. (2016). https://gcbs.sandia.gov/tools/gbrm.

Guidelines on Biosafety in Teaching Laboratories. (2019). American Society for Microbiology.

Joint Memorandum Circular No. 2021-004. Guidelines on the Implementation of Limited Face-To Face Classes for All Programs of Higher Education Institutions (HEIs) in Areas Under Alert Levels System for COVID-19 Response. https://ched.gov.ph/wp-content/uploads/CHED-DOH-JMC-No.-2021-004.pdf.

Korbel, J. O., Stegle, O. (2020 May 11). Effects of the COVID-19 Pandemic on Life Scientists. *Genome Biology* 21(1), 113. https://doi.org/10.1186/s13059-020-02031-1.

Malacanang Palace. Proclamation No. 297. Lifting of the State of Public Health Emergency Throughout the Philippines Due to COVID 19(July 21).

National Institutes of Health (US); Biological Sciences Curriculum Study. NIH Curriculum Supplement Series [Internet]. Bethesda (MD): National Institutes of Health (US). 2007. Understanding Emerging and Re-emerging Infectious Diseases. https://www.ncbi.nlm.nih.gov/books/NBK20370/.

Philippine News Agency. 2022. NCR, Cordillera Under Moderate Risk for COVID-19: DOH. https://www.pna.gov.ph/articles/1180735.

Riccaboni, M., Verginer, L. (2022). The Impact of the COVID-19 Pandemic on Scientific Research in the Life Sciences. *PLOS One* 17(2), e0263001. https://doi.org/10.1371/journal.pone.0263001.

Sandia National Laboratories. https://gcbs.sandia.gov/what-we-do/risk-management/.

Statement on the Fifteenth Meeting of the IHR Emergency Committee on the COVID-19 Pandemic. World Health Organization.

The Philippine Council for Health Research and Development. https://www.pchrd.dost.gov.ph/about-pchrd/.

WHO Laboratory Safety Manual 3rd edition. https://iris.who.int/bitstream/handle/10665/42981/9241546506_eng.pdf?sequence=1.

Witze, A. (2020). Universities Will Never Be the Same After the Coronavirus Crisis. *Nature* 582(7811), 162–164. https://doi.org/10.1038/d41586-020-01518-y.

*Laboratory Biosafety and Biosecurity Standards* (2023). Department of Health, Office of Health Laboratories.

# 8 Hazards to Harmony
## *Biosafety and Biosecurity Practice in Healthcare Settings*

*Gianne Eduard L. Ulanday*
National Training Center for Biosafety and Biosecurity
National Institutes of Health University of the Philippines
Manila, Philippines
Department of Medical Microbiology
College of Public Health
University of the Philippines
Manila, Philippines

*Toni Rose Lamata-Porras*
Office for Health Laboratories
Department of Health
Manila, Philippines
Department of Pathology and Laboratory
Quirino Memorial Medical Center
Quezon City, Philippines
College of Allied Health Professions
University of the East Ramon Magsaysay
Memorial Medical Center, Inc.
Quezon City, Philippines

## 8.1 SYNERGISM OF BIOSAFETY AND INFECTION CONTROL PRACTICE

### 8.1.1 The Practice of Infection Control and Biosafety in Healthcare Settings

Infection control and biosafety are critical components of healthcare and laboratory practices aimed at preventing the spread of infections and ensuring the safety of personnel and the environment. Biosafety refers to the application of laboratory practices, safety equipment, and procedures to protect individuals, the public, and the environment when working with potentially infectious microorganisms (Medina et al., 2017). Biosafety measures focus on the safe handling and containment of

 DOI: 10.1201/9781003426219-8

infectious microorganisms and hazardous biological materials to prevent unintentional exposure to pathogens and toxins. These measures are particularly crucial in laboratory settings where the handling of biological materials and infectious agents is routine. Furthermore, biosafety mitigation measures in clinical microbiology laboratories have been found to have a good prevention and control effect on preventing infections among medical staff, thereby reducing the risk of infection and promoting good working habits (Muhammad et al., 2017; Munson et al., 2018).

On the other hand, infection control is a discipline central to healthcare. It focuses on preventing and managing the spread of infections within healthcare settings, including the use of standardized clinical procedures, containment devices, and protective barriers (Coelho & García-Díez, 2015). In healthcare settings, infection control practices involve activities aimed at preventing the spread of infections by assessing infection-related risk factors and implementing preventive interventions. This includes measures such as hand hygiene, proper waste disposal, and the use of personal protective equipment to minimize the risk of transmission of infectious agents (Wang et al., 2020; Zhou et al., 2022).

The objectives of the two practice areas are similar in many ways and overlap. The notion of containment is a shared characteristic between biosafety and infection control. Within the hospital environment, the customary practice involves the segregation of the patient, isolating a potential reservoir of a contagion that might be transmitted to other patients or healthcare personnel. Within the laboratory setting, the implementation of engineering controls and the utilization of personal protective equipment, including respiratory protection, play a crucial role in safeguarding laboratory personnel against unwarranted exposure to infectious organisms. One notable contrast exists between the healthcare context and the laboratory environment, wherein the patient assumes the role of the biological agent in the former, while in the latter, it typically refers to a sample or specimen containing infectious biological elements.

Notwithstanding the considerable convergence of areas of expertise, infection prevention and control fail to incorporate biosecurity adequately. Initially, infection prevention and control encompass domains such as surveillance, epidemiology, and patient care that are not directly related to biosafety or biorisk management. These specific domains are seldom linked to a security-oriented culture or mindset. However, with the significant impact of pandemics and cutting-edge scientific advancements, biosecurity will likely be included in infection prevention (Burnette & SpringerLink, 2021).

### 8.1.2 Foundations of Biosafety in Biomedical Education

Biosafety in medical education is a critical aspect of training future healthcare professionals to manage and mitigate biohazards in healthcare settings effectively. The importance of biosafety training has been underscored by various studies emphasizing the need to expand biosafety training to individuals without appropriate backgrounds in biohazard mitigation, especially in the context of disease outbreaks such as COVID-19. Similarly, studies highlighted the limited information regarding the knowledge, awareness, and practices of medical students regarding infection prevention and control (IPC) and the need to evaluate the learning approaches to improve their knowledge in this area (Ibrahim & Elshafie, 2016; Mayer et al., 2022). This underscores the necessity of comprehensive biosafety training for future medical personnel to ensure their competency in managing biological risks in the workplace.

Furthermore, the importance of simulation training courses for healthcare workers to improve their risk perception and preparedness in caring for patients potentially infected with biological hazards such as Ebola and COVID-19 has been mentioned previously (Carvalho et al., 2019). This highlights the value of practical, hands-on training in enhancing biosafety competencies among healthcare professionals. Additionally, emphasis should be given to evaluating knowledge and practices about infection control guidelines among medical students and exploring their educational needs as perceived by both students and faculty. It further emphasized the importance of tailored biosafety education in medical curricula (Ayub et al., 2013).

The significance of promoting knowledge, attitude, and practice regarding infection control measures among medical undergraduates through practical infection control training indicates the positive impact of targeted training interventions on biosafety competencies (Saati & Alkalash, 2022). Having mentioned that biosafety and infection control share a common goal requires collaboration and cross-training between these professions, especially during disease outbreaks (Emery et al., 2016; Emery et al., 2022). Both professions share common competencies such as disease transmission and prevention, risk assessment, and sterilization. However, they also acknowledge that biosafety professionals focus on protecting laboratory workers from exposure to infectious agents, while infection preventionists primarily focus on patient safety in healthcare settings. In some areas where resources might be scarce, the implementation of biosafety in hospitals falls within the scope of hospital infection prevention and control committees. They are also responsible for biosafety policies and mitigation within their laboratories.

## 8.2 ANALYSIS OF BIOSAFETY IN MEDICAL AND PUBLIC HEALTH PRACTICE IN LOW RESOURCE SETTINGS

### 8.2.1 The Impact of Biological Hazards and Risks on Low- to Middle-Income Healthcare Settings

The trends in the biological safety of emerging infections have highlighted the importance of monitoring and preventing the re-emergence of traditional infectious diseases, as well as the outbreak and spread of emerging infectious diseases. The current biological hazards affecting public health and medical practice encompass a range of challenges that pose risks to both healthcare workers and the general population. These hazards include exposure to infectious diseases, antibiotic resistance, and occupational hazards in healthcare settings. Low- and middle-income countries (LMICs) are at a heightened risk for biological catastrophes due to factors such as the emergence of new zoonotic diseases and inadequate healthcare facilities (Luby & Arthur, 2019).

Often overlooked, other infectious diseases such as community-acquired bacterial bloodstream infections have been a concern in the Philippines, with a study in Metro Manila showing a significant percentage of blood culture-confirmed infections among hospitalized patients (Saito et al., 2022). Moreover, zoonotic malaria prevalence has been reported in non-human primates across Southeast Asia, including in Indonesia, the Philippines, Vietnam, Brunei, and Myanmar, with hotspots in areas that have experienced extensive deforestation and landscape modification (Johnson et al., 2022). Additionally, the emergence of drug-resistant infections has become a growing concern, necessitating transdisciplinary collaborations involving social, biological, medical, and public health practitioners to address the escalating health burdens imposed by antibiotic resistance (Smith et al., 2015).

In the context of hospital infection control, the management of biosafety risks is crucial to prevent healthcare worker infections and potential disruptions to medical services. Healthcare workers are particularly vulnerable to biological hazards, as they are frequently exposed to chemical, biological, physical, and psychosocial occupational hazards (Huang et al., 2021; Li et al., 2022; Ndejjo et al., 2015). This exposure occurs during various activities such as collecting or processing biological materials, disinfecting, cleaning, and transporting contaminated equipment, or working in contaminated areas (Alshalani & Salama, 2019). Strategies to intercept the chain of infection and protect vulnerable populations have been emphasized in emergency departments, demonstrating the importance of biosafety in healthcare settings during epidemics. Additionally, biosafety training has been identified as a valuable tool in improving infection prevention and control practices, highlighting the role of education in enhancing biosafety measures (Davwar et al., 2023; Hu et al., 2020).

Medical laboratory staff are at high risk of exposure to hazardous biological substances and are susceptible to laboratory-associated infections (Chiwande et al., 2021; Tait et al., 2018). Hence, the significance of biosafety is even more evident in laboratory settings, where the prevention of laboratory-acquired infections is a priority. Research on bioaerosol distribution and deposition in biosafety level-3 laboratories underscores the urgency of understanding and mitigating biosafety risks in high-risk environments (Liu et al., 2020; Zhu et al., 2020). Moreover, biosafety measures are crucial in the context of epidemic outbreaks and pandemics, where alternative approaches, such as minimally invasive tissue sampling, have been proposed as an alternative to autopsies to mitigate biosafety risks associated with investigating emerging pathogens (Bassat et al., 2021).

Moreover, the risks extend beyond healthcare facilities, as evidenced by the occupational hazards faced by immigrant restaurant workers, highlighting the broader societal impact of biological hazards (Tsai & Salazar, 2007). These hazards include infectious or biological agents transmitted between individuals via air, droplets, or contact. In addition to infectious disease-related biosafety hazards, foodborne hazards also pose a significant health burden in low- and middle-income countries, with milk-borne hazards being particularly prevalent in LMICs in Africa and Asia (Prakashbabu et al., 2020). These hazards can have detrimental effects on public health and necessitate the implementation of effective biosafety measures in food production and handling.

Furthermore, the management of regulated medical waste poses unique challenges, as it is significantly more expensive to treat and dispose of compared to non-regulated, non-hazardous waste which may attract risk-tolerant stakeholders to deviate from standard procedures (Gorman et al., 2013). Additionally, the occupational health and safety practices at government mortuaries have been identified as a critical area for assessment, given the diverse hazards present, including sharp objects, electrical, physical, biological, radiation, and chemical hazards (Molewa et al., 2021).

These challenges in low-income countries are compounded by resource limitations, making it crucial to develop strategic interventions and capacity-building initiatives to address these hazards effectively. The biosafety risk management systems (BRMs) in LMICs are often weak due to fragmented implementation strategies. This weakness in biosafety management systems can lead to increased risks of occupational exposures to hazards among healthcare workers in LMICs (Orelle et al., 2021; Rai et al., 2021). Additionally, the lack of institutional structures such as Institutional Biosafety Committees (IBCs) in LMICs further exacerbates these biosafety challenges (Guissou et al., 2022). The establishment of biosafety plans and the implementation of quality management systems are essential components in managing biosafety hazards in clinical laboratories and research facilities in low-income countries (Alam et al., 2022; Back et al., 2022).

### 8.2.2 Implications of COVID-19: Insights Gained in Preparation for the Next Pandemic

The COVID-19 pandemic has indeed brought to light pressing biosafety and biosecurity challenges, especially in LMICs. The handling of infectious agents such as SARS-CoV-2 in laboratories in these countries requires stringent biosafety measures to prevent accidental release and exposure. However, the pandemic has strained already limited laboratory capacities, which are crucial for both diagnosing the disease and conducting essential research toward therapeutics while maintaining adequate safety standards. Furthermore, the design and implementation of laboratory biosafety and biosecurity programs are inconsistent globally, especially in low- and middle-income countries, with major gaps influenced by differences in national and local infrastructures, available funding, and priorities (Orelle et al., 2021).

The lack of comprehensive surveillance and laboratory testing in many low-resource settings has made it difficult to monitor the spread of COVID-19. The use of portable RT-PCR systems has been proposed as a rapid and scalable diagnostic tool for COVID-19 testing, which can help decentralize testing and offer rapid results for patients in remote and low-resource settings (Claude et al., 2020; Zowawi et al., 2021). Moreover, the development of in-house and innovative protective devices made from projector transparency sheets and cardboard boxes fitted with hypochlorite spraying devices has been suggested as an alternative to biosafety cabinets for protecting laboratory personnel in dedicated COVID-19 hospitals with inadequate resources (Dubey et al., 2023). Therefore, this has underscored the need for a relative risk-based framework for safer, more secure, and sustainable medical capacity building, especially in LMICs, as was previously thought of (Dickmann et al., 2015).

The significance of implementing infection control and biosafety measures inside healthcare environments was also highlighted during the pandemic. The fragile healthcare systems in these settings make it challenging to provide specialized high-resource clinical settings with highly trained healthcare professionals and access to mechanical ventilators, which are essential for the hospitalization of COVID-19 patients (Elhadi et al., 2020).

Notably, the pandemic led to the implementation of new biosafety protocols and improvements in laboratory facilities to ensure the safety of healthcare workers and prevent disease spread. It has prompted an increase in the adoption of biosafety mitigation measures by the public, such as the use of surgical masks, N95 masks, face shields, and protective coats, to protect not only healthcare workers and patients but also the general population which was almost unheard of prior to this catastrophic event.

### 8.2.3 Focused Discussion: The 2022 Monkeypox Outbreak

The re-emergence of monkeypox virus and its rising incidence in non-endemic countries pose a new challenge to global health. Monkeypox has emerged as a significant biosafety concern due to its recent spread to non-endemic regions, including Europe, North America, and beyond, which led to the declaration of the World Health Organization as a Public Health Emergency of International Concern in 2022 (Chang et al., 2022; León-Figueroa et al., 2022). The 2022 outbreak of monkeypox has led to a paradigm shift in the association between humans and poxviruses, with new modes of transmission and concerns about viral evolution, including the potential for it to become a sexually transmitted disease (Srinivasan & Rao, 2022). Likewise, it has raised concerns about its potential for global spread and the need for preparedness in infectious disease management at a time when the world is still recovering from the effects of COVID-19. In anticipation of the imminent threat posed by the infectious disease, Malaysia implemented a five-point strategy comprising the following: enhancement of public awareness, consolidation of laboratory diagnostic facilities, case management and treatment, cluster management, and early detection of monkeypox (Chang et al., 2022). However, this was not the first instance wherein the same strategies have been identified and used in a similar event. The 2017 outbreak of human monkeypox in Nigeria revealed gaps in outbreak response that led other hospitals to strengthen epidemic preparedness and response activities in the hospital setting. Early and sustained response to the epidemic by all stakeholders was positively affected by exemplary clinical and political leadership, capacity building of healthcare personnel, allocation of enough mitigation resources, and incentives. Moreover, the study also brought to light the necessity for hospitals to establish a multidisciplinary team, in addition to an infection control team, in order to effectively address future outbreaks. An optimal team composition for effective case management in this context would encompass a diverse range of clinical professionals, such as physicians specializing in various fields, pharmacists, and nurses. Additionally, the team should include a laboratory team consisting of pathologists and laboratory scientists, a safe burial team comprising pathologists and morticians, as well as a counseling and social support team consisting of psychiatrists, social workers, and health counselors. Furthermore, support services such as cleaners and health attendants should be included to provide the necessary assistance (Ogoina et al., 2019).

### 8.2.4 Focused Discussion: The Virology Institute of the Philippines and the Philippine Centers for Disease Control

During the peak of the COVID-19 outbreak in the Philippines, a proposal was submitted with the aim of establishing the Philippine Center for Disease Control and Prevention (CDC) and the Virology and Vaccine Institute of the Philippines (VIP). The CDC will serve as the lead agency for developing communicable disease control and prevention initiatives. The organization will be tasked with investigating potential public health emergencies, ensuring compliance with regulations to prevent the transmission of communicable diseases, procuring and distributing medical supplies such as vaccines and antibiotics, and working in collaboration with international organizations and nations to enhance systems and practices for disease prevention and control. The proposal included the streamlining of procurement and coordination among experts and laboratory networks to ensure a more rapid response during public health emergencies. It addresses the challenges encountered by the Philippine health sector during the initial stages of its response to the COVID-19 outbreak.

Leading the Virology and Vaccine Institute of the Philippines (VIP) program, the Department of Science and Technology (DOST) hopes to establish VIP as the nation's preeminent virology research and development center. This will aid in the nation's readiness for potential pandemics, bioterrorism events, and the avoidance of endemic disease resurgence. Additionally, the institution will facilitate the establishment of a platform for scientists, both domestic and international, to engage in joint research endeavors pertaining to viruses that hold significance in the fields of agriculture, industry, clinical medicine, and the environment. This initiative aims to foster strategic collaborations with renowned scientists, virology centers, vaccine laboratories, and other esteemed institutes on a global scale. The institute will serve as the designated national reference laboratory for all viral etiology-related zoonotic and botanical illnesses. Moreover, the plan contained provisions for imposing fines and legal repercussions in relation to breaches of biosafety and biosecurity within the country. Albeit being in the early plenary stages, the decision made by policymakers to pursue these initiatives demonstrates their dedication to advancing the safety and security of the nation (*Virology Science and Technology Institute of the Philippines Act of 2021.*, 2021).

## 8.3 BIOSECURITY IN HEALTHCARE FACILITIES

Biosecurity refers to the protection, control of, and accountability for high-consequence biological agents and toxins, and critically relevant biological materials and information within facilities to prevent unauthorized possession, loss, theft, misuse, diversion, and intentional release.

Biosecurity measures are traditionally applied to laboratories handling and manipulating highly infectious agents. While this is true, biosecurity may also be usefully applied in clinical settings and other healthcare facilities, such as isolation facilities and critical care units for the management of patients with highly infectious diseases. Infectious materials from the patients, in particular diagnostic specimens, may be stolen during storage or during transport within or outside of the clinical area.

In the Philippines, biosecurity in clinical healthcare facilities is generally not routinely applied and considered. Biosafety, on the other hand, is given much more emphasis in these areas since the safety of both patients and healthcare workers is given priority.

### 8.3.1 Potential Threats in Healthcare Facilities

It is assumed that if a person or a group of people with intent to harm will always consciously keep an eye on its target. One having intentions to use biological agents for various reasons such as sabotage or terrorism may have the knowledge and technological capability to transport, store, replicate, and transform these into a biological weapon. Thus, it is not impossible that these people may obtain these Valuable Biological Materials (VBMs) from the point of source, the patient; hence, biosecurity should find an application in clinical healthcare facilities, especially those that are managing patients with highly infectious diseases.

### 8.3.2 Implementing Effective Biosecurity Measures

Similar to biosafety where understanding the characteristics of the pathogen being handled greatly impacts mitigation measures to be implemented, in biosecurity, the same principle applies. Recognizing their characteristics and their potential use as biological weapons will also impact how stringent the biosecurity measures are to be deployed.

Pathogens are generally classified into risk groups based on their level of virulence and the severity of the diseases they cause.

- Risk Group 1 (RG1) includes microorganisms that are unlikely to cause disease in healthy individuals. Examples include non-pathogenic strains of *E. coli* and *Bacillus subtilis*. These organisms can be handled safely with standard practices.
- Risk Group 2 (RG2) pathogens have the potential to cause human disease, but effective treatments or preventive measures are available. Examples include *Salmonella*, hepatitis B virus, and influenza A virus. Proper containment measures, such as personal protective equipment and biological safety cabinets, must be used when working with RG2 pathogens.
- Risk Group 3 (RG3) pathogens can cause severe diseases that may spread within a community or population but can be effectively treated or prevented by vaccines or other medical interventions. Tuberculosis and HIV fall into this category.

- Risk Group 4 (RG4) includes highly virulent pathogens that pose a high risk of causing severe diseases without any effective treatments or preventive measures available. Examples include Ebola virus and Marburg virus.

It is imperative to identify whether healthcare facilities possess biological agents, in their case, are patients harboring these biological agents that may be attractive to those who may wish to use them maliciously. The confidentiality and integrity of sensitive information held in the clinics that could be used with malicious intent should also be considered in implementing biosecurity measures.

Planning for biosecurity management programs in healthcare facilities should consider all dimensions of biosecurity pillars: physical security, personnel management, material control and accountability, transport security, and information security. However, to achieve an effective program, biosecurity risks should be thoroughly identified through recognition of healthcare facility threats and vulnerability. Identifying vulnerabilities is akin to answering the question, 'How could harm occur?' 'Are our clinical facilities open for anyone?' while identifying threats is comparable to answering the question, 'Who or what could cause harm?' In a broad sense, a threat is anything that could exploit a vulnerability and hinder the confidentiality, integrity, and availability of these VBMs.

A huge support service for clinical research of highly infectious agents and novel microorganisms is the diagnostic/research laboratory. The importance of discovering microorganisms that play an etiological role in infectious disease outbreaks and clinical syndromes transcends academic interest. If direct causality between the microorganisms and the disease syndrome is established by clinical associations and animal models, then further study is indicated for defining specific diagnostic, therapeutic, and preventative measures. This is especially true in novel microorganisms. To carry out these, high containment laboratories are recommended to handle these highly infectious microorganisms. Setting up high containment laboratories (BSL-3 and BSL-4) facilities in the Philippines presents numerous obstacles where one of the major hurdles is the lack of infrastructure and resources. These facilities require specialized equipment, such as negative pressure rooms, HEPA filters, and dedicated waste management systems, which are often not readily available in the country. Another challenge is the shortage of trained personnel. Operating BSL-3 and BSL-4 laboratories requires highly skilled scientists, technicians, and support staff who are knowledgeable about biocontainment procedures. However, there is a limited number of individuals with this expertise in the Philippines. Furthermore, regulatory compliance poses a significant obstacle. Establishing biosafety standards and guidelines that adhere to international protocols is crucial for ensuring safety within these facilities. However, developing comprehensive regulations that encompass all aspects of biosafety can be a complex process that requires coordination among various government agencies.

At the core, hospitals and/or any clinical healthcare facilities to include support services play a vital role in the biosecurity system in the country as they also handle VBMs that are potentially attractive to those who have undesirable intentions.

## 8.4 INTERNATIONAL STANDARDS FOR MEDICAL BIOSAFETY AND BIOSECURITY

International standards for medical biosafety and biosecurity are essential to ensure the safe handling and management of biological materials, including infectious agents and toxins, in healthcare settings, research laboratories, and other facilities. These standards are developed and maintained by international organizations, governmental bodies, and expert groups to establish guidelines and best practices for minimizing the risks associated with the handling of biological materials.

The World Health Organization (WHO) plays a key role in providing guidelines and recommendations for medical biosafety and biosecurity. This organization collaborates with others to create standards and guidelines for various aspects of biosafety and biosecurity. It is important to note that many countries have their own national regulations and standards related to biosafety and biosecurity. These may be based on international guidelines but can include additional requirements or variations to suit their local specific needs. The Philippines is the first country in Southeast Asia to adopt a national biosafety guideline; however, this focuses on genetic engineering and other activities that require the importation, introduction, field release, and breeding of nonindigenous organisms. The country signed into the Cartagena Protocol on Biosafety to the Convention on Biological Diversity in 2000. Despite the progress made in the mentioned areas, there are still issues included within the field of biosafety and biosecurity that needs to be given attention in the Philippines such as in research, clinical, and diagnostic areas.

### 8.4.1 Biosafety and Biosecurity Challenges in Asia Pacific Region and in the Philippines

Biosafety and biosecurity challenges in the Asia Pacific region have become a growing concern in recent years. This became evident with the COVID-19 pandemic when we realized our healthcare gaps. Moreover, with the rapid advancements in biotechnology and the increasing number of facilities working with dangerous pathogens and catering to patients with highly infectious diseases, there is an urgent need to address these challenges to protect public health and ensure global security.

One of the major biosafety challenges in the region is the lack of standardized regulations and guidelines for handling dangerous pathogens for diagnostic, research, and clinical settings. The Philippines Executive Order 514, the National Biosafety Framework prescribes that biosafety policies may apply to multiple sectors related to biosafety. However, the scope of the framework is limited to products of modern biotechnology, exotic species, and invasive alien species, leaning heavily toward the agriculture and environment sector. Hence, an expanded concept may be necessary to balance the perspectives of science, security, prevention, and preparedness beyond laboratory work.

In addition to the lack of comprehensive national biosafety and biosecurity policies and guidelines, many countries in the Asia Pacific Region have facilities handling infectious agents that were built more than 10 or 20 years ago with limited provision for biosafety and biosecurity measures in their design and practices. This could also pose a significant risk, particularly when dealing with infectious agents that require control and specialized equipment and infrastructure for proper handling (Chua, 2008).

Biosafety and biosecurity are crucial aspects of laboratory operations, especially when dealing with hazardous biological materials. In the Philippines, small laboratories play a significant role in diagnostic activities supporting clinics with their patient management. However, ensuring proper regulation of biosafety and biosecurity measures in these facilities remains a challenge. To address this issue, the Philippine government has implemented various regulations to promote biosafety and biosecurity practices. The Department of Health (DOH) through the Administrative Order 2021-0037 has incorporated guidelines for laboratory biosafety, which cover risk assessment, biosafety and biosecurity officers, containment measures, and training requirements. These guidelines aim to protect laboratory personnel, the community, and the environment from potential hazards associated with biological agents. Despite these efforts, there are still gaps in implementing regulations effectively. Limited resources pose a challenge for small laboratories to comply fully with biosafety standards. Furthermore, lack of awareness among laboratory personnel about existing regulations hinders their ability to implement necessary safety protocols. While progress has been made through government initiatives, further efforts are needed to bridge gaps between regulation implementation and compliance within these facilities.

## 8.5 CONCLUSION AND FINAL THOUGHTS

Handling biological materials in healthcare settings is of utmost importance to ensure the safety and well-being of both healthcare professionals and patients. Biological materials, such as blood, tissues, and bodily fluids, can carry infectious agents that pose a significant risk if not handled properly. Firstly, it is crucial to establish strict protocols for handling biological materials in an institution. A thorough institution-based risk assessment plays a key role in planning for healthcare measures. The procedures being performed in the clinics and hospitals should be considered as well as the risk grouping of microorganisms of cases being handled and managed.

Healthcare professionals should also be trained on different mitigation measures such as proper techniques for handling, collection, transportation, and disposal of these materials. It is also prudent that healthcare professionals are aware of the nature and characteristics of possible infectious agents present in patients. This will determine appropriate personal protective equipment (PPE) to be used to prevent contact with potentially infectious substances.

Healthcare facilities must also have appropriate design to accommodate biosafety and biosecurity measures such as designated areas for the storage and handling of biological materials. These areas should be equipped with proper ventilation systems to minimize the risk of airborne transmission. Additionally, rooms or units handling cases with these materials should be labeled clearly with biohazard symbols to alert individuals about potential risks.

Furthermore, proper disposal and decontamination of contaminated materials is critical to prevent the spread of infections and protect public health. Regular cleaning and disinfection of all equipment used during procedures involving biological materials should be thoroughly cleaned and sterilized before and after use.

In conclusion, handling biological materials in healthcare settings requires strict adherence to protocols and guidelines. By implementing proper training programs for healthcare professionals, ensuring that healthcare infrastructure has appropriate provisions for biosafety and biosecurity measures, establishing designated areas for storage and handling, and ensuring regular cleaning and disinfection practices are followed diligently, we can effectively minimize the risks associated with these potentially hazardous substances. Ultimately, prioritizing safety measures when dealing with biological materials will contribute significantly to maintaining a safe environment within healthcare facilities.

## REFERENCES

Alam, M. S., Abdalla, S. E. B., & Jabeen, F. (2022). Assessment of Biosafety Practices in Clinical Laboratories in Khartoum State, Sudan. *Journal of Biosciences and Medicines*. https://doi.org/10.4236/jbm.2022.1011008

Alshalani, A. J., & Salama, K. F. (2019). Assessment of Occupational Safety Practices Among Medical Laboratory Staff in Governmental Hospitals in Riyadh, Saudi Arabia. *Journal of Safety Studies*, *5*(1), 1. https://doi.org/10.5296/jss.v5i1.14992

Ayub, A., Goyal, A., Kotwal, A., Kulkarni, A., Kotwal, A., & Mahen, A. (2013). Infection Control Practices in Health Care: Teaching and Learning Requirements of Medical Undergraduates. *Medical Journal Armed Forces India*, *69*(2), 107–112. https://doi.org/10.1016/j.mjafi.2012.07.021

Back, J. B., Martinez, L., Nettenstrom, L., Sheerar, D., & Thornton, S. (2022). Establishing a Biosafety Plan for a Flow Cytometry Shared Resource Laboratory. *Cytometry Part A*. https://doi.org/10.1002/cyto.a.24524

Bassat, Q., Varo, R., Hurtado, J. C., Marimon, L., Ferrando, M., Ismail, M. R., Carrilho, C., Fernandes, F., Castro, P., Maixenchs, M., Rodrigo-Calvo, M. T., Guerrero, J. A. S., Martinez, A., Lacerda, M. V. G., Mandomando, I., Menéndez, C., Martínez-Priego, L., Ordi, J., & Rakislova, N. (2021). Minimally Invasive Tissue Sampling as an Alternative to Complete Diagnostic Autopsies in the Context of Epidemic Outbreaks and Pandemics: The Example of Coronavirus Disease 2019 (COVID-19). *Clinical Infectious Diseases*. https://doi.org/10.1093/cid/ciab760

Burnette, R. N., & SpringerLink. (2021). *Applied Biosecurity: Global Health, Biodefense, and Developing Technologies* (1st ed.). Springer International Publishing: Imprint: Springer.

Carvalho, E., Castro, P., León, E., Del Río, A., Crespo, F., Trigo, L., Fernández, S., Trilla, A., Varela, P., & Nicolás, J. M. (2019). Multi-professional Simulation and Risk Perception of Health Care Workers Caring for Ebola-Infected Patients. *Nursing in Critical Care*, *24*(5), 256–262. https://doi.org/10.1111/nicc.12396

Chang, C. T., Lim, X.-J., Nordin, A. A., Thum, C. C., Sararaks, S., Periasamy, K., & Rajan, P. (2022). Health System Preparedness in Infectious Diseases: Perspective of Malaysia, a Middle-Income Country, in the Face of Monkeypox Outbreaks. *Tropical Medicine and Health*. https://doi.org/10.1186/s41182-022-00479-4

Chiwande, E. S., Shinde, V., & Gudhe, V. (2021). Assessment of Awareness about Occupational Safety among Laboratory Staff in Healthcare Settings: A Study Protocol. *Journal of Pharmaceutical Research International*, 262–267. https://doi.org/10.9734/jpri/2021/v33i64a35365

Chua, T. (2008). *Biosafety & Biosecurity Challenges in the Asia Pacific Region*. Geneva: Asia Pacific Biosafety Association.

Claude, K. M., Serge, M. S., Alexis, K. K., & Hawkes, M. (2020). Prevention of COVID-19 in Internally Displaced Persons Camps in War-Torn North Kivu, Democratic Republic of the Congo: A Mixed-Methods Study. *Global Health Science and Practice*. https://doi.org/10.9745/ghsp-d-20-00272

Coelho, A. C., & García-Díez, J. (2015). Biological Risks and Laboratory-Acquired Infections: A Reality That Cannot Be Ignored in Health Biotechnology. *Frontiers in Bioengineering and Biotechnology*. https://doi.org/10.3389/fbioe.2015.00056

Davwar, P. M., Luka, P. D., Dami, F. D., Pam, D., Weldon, C., Brocard, A. S., Paessler, S., Weaver, S. C., & Shehu, N. Y. (2023). One Health Epidemic Preparedness: Biosafety Quality Improvement Training in Nigeria. *International Journal of One Health*. https://doi.org/10.14202/ijoh.2023.10-14

Dickmann, P., Sheeley, H., & Lightfoot, N. F. (2015). Biosafety and Biosecurity: A Relative Risk-Based Framework for Safer, More Secure, and Sustainable Laboratory Capacity Building. *Frontiers in Public Health*. https://doi.org/10.3389/fpubh.2015.00241

Dubey, A., Bansal, A., Sonkar, S. C., Goswami, B., Makwane, N., Manchanda, V., & Koner, B. C. (2023). In-House Assembled Protective Devices in Laboratory Safety in Clinical Biochemistry Laboratory of a COVID-19 Dedicated Hospital. *Acta Scientific Microbiology*. https://doi.org/10.31080/asmi.2023.06.1220

Elhadi, M., Msherghi, A., Alkeelani, M., Alsuyihili, A., Khaled, A., Buzreg, A., Boughididah, T., Abukhashem, M., Alhashimi, A., Khel, S., Gaffaz, R., Saleim, N. B., Bahroun, S. G., Elharb, A., Eisay, M., Alnafati, N., Almiqlash, B., Biala, M., & Alghanai, E. (2020). Concerns for Low-Resource Countries, With Under-Prepared Intensive Care Units, Facing the COVID-19 Pandemic. *Infection, Disease and Health*. https://doi.org/10.1016/j.idh.2020.05.008

Emery, R. J., Patlovich, S. J., King, K., Lowe, J. J., & Rios, J. (2016). Comparing the Established Competency Categories of the Biosafety and Infection Prevention Professions: A Possible Roadmap for Addressing Professional Development Training Needs for a New Era. *Applied Biosafety*, *21*(2), 79–83. https://doi.org/10.1177/1535676016651250

Emery, R. J., Patlovich, S. J., King, K. G., Lowe, J. M., & Rios, J. (2022). Assessing the Established Competency Categories of the Biosafety, Infection Prevention, and Public Health Professions: A Guide for Addressing Needed Professional Development Training for the Current and Next Pandemic. *Applied Biosafety*, *27*(2), 53–57. https://doi.org/10.1089/apb.2022.0002

*Establishing the National Biosafety Framework, Prescribing Guidelines for Its Implementation, Strengthening the National Committee on Biosafety of the Philippines, and for Other Purposes*. (2006, March 17). The Official Gazette. Retrieved November 25, 2023, from https://www.officialgazette.gov.ph/2006/03/17/executive-order-no-514-s-2006/

Gorman, T., Dropkin, J., Kamen, J., Nimbalkar, S., Zuckerman, N., Lowe, T., Szeinuk, J., Milek, D., Piligian, G., & Freund, A. (2013). Controlling Health Hazards to Hospital Workers. *New Solutions*, *23*(Suppl), 1–167. https://doi.org/10.2190/NS.23.Suppl

Guissou, C., Quinlan, M. M., Sanou, R., Ouedraogo, R. K., Namountougou, M., & Diabaté, A. (2022). Preparing an Insectary in Burkina Faso to Support Research in Genetic Technologies for Malaria Control. *Vector Borne and Zoonotic Diseases*. https://doi.org/10.1089/vbz.2021.0041

Hu, X., Liu, S., Wang, B., Xiong, H., & Wang, P. (2020). Management Practices of Emergency Departments in General Hospitals Based on Blockage of Chain of Infection During a COVID-19 Epidemic. *Internal and Emergency Medicine.* https://doi.org/10.1007/s11739-020-02499-6

Huang, F., Chen, W.-T., Sun, W., Zhang, l., & Lu, H. (2021). The Impact of Policy and COVID-19 Prevention Strategies in China: A Qualitative Study. https://doi.org/10.21203/rs.3.rs-710681/v1

Ibrahim, A. A., & Elshafie, S. S. (2016). Knowledge, Awareness, and Attitude regarding Infection Prevention and Control among Medical Students: A Call for Educational Intervention. *Advances in Medical Education and Practice, 7,* 505–510. https://doi.org/10.2147/amep.S109830

Johnson, E., Sharma, R. S. K., Cuenca, P. R., Byrne, I., Salgado-Lynn, M., Shahar, Z. S., Lin, L. C., Zulkifli, N., Saidi, N. D. M., Drakeley, C., Matthiopoulos, J., Nelli, L., & Fornace, K. (2022). *Landscape Drives Zoonotic Malaria Prevalence in Non-human Primates.* https://doi.org/10.1101/2022.06.17.496284

León-Figueroa, D. A., Bonilla-Aldana, D. K., Pachar, M., Romani, L., Saldaña-Cumpa, H. M., Anchay-Zuloeta, C., Diaz-Torres, M., Franco-Paredes, C., Suárez, J. A., Ramírez, J. D., Paniz-Mondolfi, A., & Rodriguez-Morales, A. J. (2022). The Never-Ending Global Emergence of Viral Zoonoses After COVID-19? The Rising Concern of Monkeypox in Europe, North America and Beyond. *Travel Medicine and Infectious Disease.* https://doi.org/10.1016/j.tmaid.2022.102362

Li, X., He, M., Lin, X., & Lin, Y. (2022). *Biosafety Management Risk Analysis for Clinical Departments of Military Central Hospitals in the Fujian Province of China.* Sage Open. https://doi.org/10.1177/21582440221085270

Liu, Z., Zhuang, W., Hu, X., Zhi-heng, Z., Rong, R., Ding, W., Li, J., & Li, N. (2020). Effect of Equipment Layout on Bioaerosol Temporal-Spatial Distribution and Deposition in One BSL-3 Laboratory. *Building and Environment.* https://doi.org/10.1016/j.buildenv.2020.107149

Luby, S., & Arthur, R. (2019). Risk and Response to Biological Catastrophe in Lower Income Countries. *Current Topics in Microbiology and Immunology, 424,* 85–105. https://doi.org/10.1007/82_2019_162

Mayer, O., Pfundt, T., Fortenberry, G. Z., Harcourt, B. H., & Bower, W. A. (2022). Lessons Learned From a COVID-19 Biohazard Spill During Swabbing at a Quarantine Facility. *Disaster Medicine and Public Health Preparedness, 16*(3), 1279–1281. https://doi.org/10.1017/dmp.2020.436

Medina, P., Padua, A., Casagan, M., Olpindo, R., Tandoc Iii, A., & Lupisan, S. (2017). Development and Pilot Implementation of a Ladderized Biosafety Training Program in a Specialty Infectious Disease Hospital and Research Institute. *PJP, 2*(1), 5. https://philippinejournalofpathology.org/index.php/PJP/article/view/48

Molewa, M. L., Mbonane, T. P., Shirinde, J., & Masekameni, D. M. (2021). Assessment of Occupational Health and Safety Practices at Government Mortuaries in Gauteng Province: A Cross-Sectional Study. *Pan African Medical Journal, 38,* 76. https://doi.org/10.11604/pamj.2021.38.76.21699

Muhammad, F. J., Siddiqui, N. U., Ali, N., & Kazmi, S. U. (2017). Analysis of Biosafety Performance in Selected Hospital Medical Laboratories in Karachi, Pakistan. *Applied Biosafety.* https://doi.org/10.1177/1535676017742378

Munson, E., Bowles, E. J. A., Dern, R., Beck, E. T., Podzorski, R. P., Bateman, A. C., Block, T., Kropp, J. L., Radke, T., Siebers, K., Simmons, B. J., Smith, M. A., Spray-Larson, F., & Warshauer, D. M. (2018). Laboratory Focus on Improving the Culture of Biosafety:

Statewide Risk Assessment of Clinical Laboratories That Process Specimens for Microbiologic Analysis. *Journal of Clinical Microbiology*. https://doi.org/10.1128/jcm.01569-17

Ndejjo, R., Musinguzi, G., Yu, X., Buregyeya, E., Musoke, D., Wang, J. S., Halage, A. A., Whalen, C., Bazeyo, W., Williams, P., & Ssempebwa, J. (2015). Occupational Health Hazards among Healthcare Workers in Kampala, Uganda. *Journal of Environmental and Public Health, 2015*, 913741. https://doi.org/10.1155/2015/913741

*New Rules and Regulations Governing the Regulation of Clinical Laboratories in The Philippines* Philippine Department of Health, (DOH AO 2021-0037). (2021, June 11).

Ogoina, D., Izibewule, J. H., Ogunleye, A., Ederiane, E., Anebonam, U., Aworabhi, N., Oyeyemi, A., Etebu, E. N., & Ihekweazu, C. (2019). The 2017 Human Monkeypox Outbreak in Nigeria—Report of Outbreak Experience and Response in the Niger Delta University Teaching Hospital, Bayelsa State, Nigeria. *PLOS One*. https://doi.org/10.1371/journal.pone.0214229

Orelle, A., Nikiema, A., Zakaryan, A., Albetkova, A., Rayfield, M., Peruski, L. F., Pierson, A., & Kachuwaire, O. (2021). National Biosafety Management System: A Combined Framework Approach Based on 15 Key Elements. *Frontiers in Public Health*. https://doi.org/10.3389/fpubh.2021.609107

Philippine Center for Disease Prevention and Control (CDC) Act. (2023). Retrieved November 24, 2023, from https://legacy.senate.gov.ph/lis/bill_res.aspx?congress=19&q=SBN-1869

Prakashbabu, B. C., Cardwell, J. M., Craighead, L., Ndour, A. P. N., Yempabou, D., Ba, E., Alambedji, R. B., Akakpo, A. J., & Guitian, J. (2020). "We Never Boil Our Milk, It Will Cause Sore Udders and Mastitis in Our Cows"- Consumption Practices, Knowledge and Milk Safety Awareness in Senegal. *BMC Public Health*. https://doi.org/10.1186/s12889-020-08877-1

Rai, R., El-Zaemey, S., Dorji, N., Doj, B., & Fritschi, L. (2021). Exposure to Occupational Hazards Among Health Care Workers in Low- and Middle-Income Countries: A Scoping Review. *International Journal of Environmental Research and Public Health*. https://doi.org/10.3390/ijerph18052603

Saati, A. A., & Alkalash, S. H. (2022). Promotion of Knowledge, Attitude, and Practice Among Medical Undergraduates Regarding Infection Control Measures During COVID-19 Pandemic. *Frontiers in Public Health, 10*, 932465. https://doi.org/10.3389/fpubh.2022.932465

Saito, N., Solante, R., & Guzman, F. Telan, d., E. O., Umipig, D. V., Calayo, J., Frayco, C. H., Lazaro, J., Ribo, M. R., Dimapilis, A. Q., Dimapilis, V. O., Villanueva, A. M. G., Mauhay, J. L., Suzuki, M., Yasunami, M., Koizumi, N., Kitashoji, E., Sakashita, K., Yasuda, I., … Parry, C. M. (2022). A Prospective Observational Study of Community-Acquired Bacterial Bloodstream Infections in Metro Manila, The Philippines. *PLOS Neglected Tropical Diseases*. https://doi.org/10.1371/journal.pntd.0010414

*Science Safety Security – Risk Groups*. S3: Science, Safety, Security. (2015). https://www.phe.gov/s3/BioriskManagement/biosafety/Pages/Risk-Groups.aspx

Smith, R. A., M'Ikanatha, N. M., & Read, A. F. (2015). Antibiotic Resistance: A Primer and Call to Action. *Health Communication, 30*(3), 309–314. https://doi.org/10.1080/10410236.2014.943634

Sridhar, S., To, K. K., Chan, J. F., Lau, S. K., Woo, P. C., & Yuen, K. Y. (2015). A Systematic Approach to Novel Virus Discovery in Emerging Infectious Disease Outbreaks. *The Journal of Molecular Diagnostics : JMD, 17*(3), 230–241. https://doi.org/10.1016/j.jmoldx.2014.12.002

Srinivasan, K., & Rao, M. (2022). Poxvirus-Driven Human Diseases and Emerging Therapeutics. *Therapeutic Advances in Infectious Disease*. https://doi.org/10.1177/20499361221136751

Tait, F. N., Mburu, C., & Gikunju, J. (2018). Occupational Safety and Health Status of Medical Laboratories in Kajiado County, Kenya. *Pan African Medical Journal, 29*, 65. https://doi.org/10.11604/pamj.2018.29.65.12578

Tsai, J. H., & Salazar, M. K. (2007). Occupational Hazards and Risks Faced by Chinese Immigrant Restaurant Workers. *Family and Community Health, 30*(2 Suppl), S71–79. https://doi.org/10.1097/01.Fch.0000264882.73440.20

Virology Science and Technology Institute of the Philippines Act of 2021. (2021). Retrieved November 24, 2023, from https://legacy.senate.gov.ph/lis/bill_res.aspx?congress=18&q=SBN-2155

Wang, Y.-H., Bychkov, A., Chakrabarti, I., Jain, D., Liu, Z., He, S., Hanum, S. F., Bakrin, I. H., Templo, F. S., Nguyen, T., Kumarasinghe, P., Jung, C. K., Kakudo, K., & Chen, C. L. (2020). Impact of the COVID-19 Pandemic on Cytology Practice: An International Survey in the Asia-Pacific Region. *Cancer Cytopathology*. https://doi.org/10.1002/cncy.22354

Zhou, W., Zou, L., Zhu, F., & Yang, J. (2022). Biosafety Protection and Workflow of Clinical Microbiology Laboratory Under COVID-19: A Review. *Medicine*. https://doi.org/10.1097/md.0000000000031740

Zhu, X. H., Zhao, Z., Wang, c., Chen, H., Wang, M., Wei, K., Li, Z., & Liu, Z. (2020). *Laboratory-Acquired Brucella Infection and S2 Vaccine Infection Events in China*. https://doi.org/10.22541/au.159819290.02533746

Zowawi, H. M., Alenazi, T. H., Alomaim, W., Wazzan, A., Alsufayan, A., Hasanain, R. A., Aldibasi, O., Althawadi, S., Altamimi, S. A., Mutabagani, M. S., Alamri, M., Almaghrabi, R. S., Al-Abdely, H. M., Memish, Z. A., & Alqahtani, S. A. (2021). Portable RT-PCR System: A Rapid and Scalable Diagnostic Tool for COVID-19 Testing. *Journal of Clinical Microbiology*. https://doi.org/10.1128/jcm.03004-20

# 9 The Threats of Transboundary Plant Pathogens

## *A Perspective on Philippines' Plant Biosecurity*

*Mark Angelo Balendres*
Department of Biology, College of Science
De La Salle University, Manila, Philippines

*Christian Joseph Cumagun*
Parma Research and Extension Center
University of Idaho
Idaho, USA

*Paul Taylor*
Faculty of Science
University of Melbourne
Victoria, Australia

## 9.1 INTRODUCTION

The Philippine economy has expanded with a record Gross Domestic Product (GDP) increase of 7.8% (Department of Agriculture 2023). Agriculture shares 8.6% of the country's GDP, trailing behind the service (61.4%) and industry (29.7%) sectors in 2022. There was a slight decrease in the annual GDP of the agriculture sector from 9.6% (2021) to 8.9% (2022), which could be mainly attributed to the 'vulnerability of the sector to natural disasters' and the increase in costs of inputs (Department of Agriculture 2023). More than half (51.9%) of the Gross Value Added (GVA) is contributed by the crop subsector, which makes this subsector the most significant contributor to the country's GVA in agriculture as compared with the livestock (10.3%), poultry (10.5%), and fisheries (12.1%) subsectors (Department of Agriculture 2023). The country's top agricultural exports include plants (fruits, nuts, etc.) and plant products (preparations of vegetables, fruits, nuts, and other plant parts). Given the significant contribution of the crops subsector to the country's agricultural sector, which is a key growth driver to the Philippines' GDP and economy, the government

 DOI: 10.1201/9781003426219-9

must protect its crops against factors limiting crop growth and contributing to production losses.

Agricultural damage and losses in the country in 2022 were accounted for by four major factors: typhoons (90%, valued at Php 21.7 billion), geological disasters (4.4%, valued at Php 1.06 billion), other climate-related disasters (4.3%, valued at Php 1.04 billion), and pest infestation and disease incidence (1.3%, valued at Php 313.8 million) (Department of Agriculture 2023). In terms of commodity, most of the losses were accounted for from the crop's subsector, which includes rice, corn, cassava, and high-value crops (Department of Agriculture 2023). Of these four contributors to agricultural damage and losses, only pest infestation and disease incidence are caused by biological and infectious agents. In addition, typhoons and other climate-related disasters may also contribute to increased disease incidence.

The country imports more agricultural products than exporting products to other countries. The export value is less than half of its import values, accounting for a USD 11.8 billion deficit, resulting in a negative agricultural trade balance (Department of Agriculture 2023). The value and volume of products traded in 2022 indicate frequent movement of products into the country. While importation contributes to filling in the gap of production deficit or toward creating new products, the movement of agricultural products and materials may introduce and spread biological and infectious pathogens that could be destructive when these pathogens reach local agricultural production. While many of the fresh fruits and vegetables imported into the country are for re-sale and high-end market distribution only (NPQSD 2023a), there is still a possibility that they carry pathogens that could be hiding in these fresh, healthy-looking produce. For instance, a recent study detected the fungus *Colletotrichum fructicola* in imported persimmon fruits re-sold by a local vendor (Evallo et al. 2022). Although the production of persimmon in the country is relatively insignificant, the isolated fungus may cross-infect other plant species, as is the case of many plant-pathogenic fungi (Freeman & Shabi 1996; Phoulivong et al. 2012; Eaton et al. 2021). Aside from fresh produce, plants and planting materials are also regulated for importation based on existing international and national phytosanitary legislation. These plant and planting materials may be used for commercial feeds, feedstock, pasture development, personal collection, planting, processing, propagation, and research (NPQSD 2023b). Thus, there is an increasing threat to the country's agriculture sector by the possible entry of virulent and highly infectious plant pathogens through the movement of fresh produce, plants, and planting materials that may harbor these pathogens. Indeed, it was known that the pathogen *Sclerospora sacchari* (currently known as *Peronosclerospora sacchari* T. Miyake) was brought into the country by the introduction of some sugarcane varieties from Japan in the early 1920s (Lee and Medalla 1921). In 1957, leaf mottling or greening disease was reported in Batangas and believed to have spread to other provinces and many parts of the country through infected planting materials. The disease was not known until the introduction of citrus planting materials from China and India (Altamirano et al. 1976). Consequently, a plant biosecurity program is needed to protect the country's agriculture (and even forestry) sector. Moreover, given the archipelagic geography of the Philippines, plant biosecurity becomes even more challenging as it needs to address and prevent the intra-island movement of prohibited plants and planting materials, which may harbor unwanted plant pathogens.

The UK's Department for Environment, Food, and Rural Affairs (2023) defines 'biosecurity' as a 'set of precautions that aim to prevent the introduction and spread of harmful organisms. These include non-native pests, insects, and disease-causing organisms called pathogens, such as some viruses, bacteria, and fungi'. In the context of plant biosecurity, these precautions are aimed at those pathogens causing plant diseases. However, the term 'biosecurity' is not usually used in the Philippine plant quarantine laws. The term 'plant quarantine' is usually used to describe measures to prevent the introduction and spread of pests. In this chapter, we will use the term 'plant biosecurity' to refer to these measures to prevent introduction and spread of pests (plant quarantine). This chapter discusses the threats of plant pathogens to agricultural production and their importance in food supply shortage and food insecurity. This chapter also provides a broad overview of key plant pathogens that have quarantine importance to the Philippines, based on pest risk analysis by the Bureau of Plant Industry-National Plant Quarantine Services Division, both foreign (Table 9.1) and domestic (Figure 9.1) plant pathogens. Most of these pathogens, herein referred to as transboundary plant pathogens or plant diseases, have been included in the databases of international organizations (EPPO or European and Mediterranean Plant Protection Organization and the IPPC or International Plant Protection Convention) and are subject of the works of scientists in their scholarly publications; we highlight essential information that is relevant to the possible pathway of entry into the country, establishment, spread, and potential economic impact. Hence, we focused on the pathogen's current reported geographic location close to the Philippines and its possible host range. This chapter also describes phytosanitary measures implemented by the country's National Plant Protection Organization (NPPO) to prevent the entry and spread of foreign and domestic plant pathogens and diseases. Portales et al. (2023) outlined several major activities that the national plant quarantine organization implemented as part of the mandate of the plant quarantine law, international conventions and agreements. The legal bases of the country's plant biosecurity programs and quarantine measures are also presented (Figure 9.2).

## 9.2 THREATS AND IMPACT OF TRANSBOUNDARY PLANT PATHOGENS

The Food and Agriculture Organization of the United Nations (2023) defined transboundary plant pests and diseases (TPPDs) as 'migratory pests that pose a significant threat to food security, trade, and livelihoods of people in the affected countries, and generate huge losses of crops and pastures. Preventive measures, early action, and long-term solutions are essential for protecting crops and pastures from TPPDs'. They are regulated by organizations to prevent their entry into a country or area, free from that pathogen, contain their establishment and spread upon entry into the country/area, and mitigate their direct and indirect effect on crop yield, food supply, and directly impacted community. According to Prasanna et al. (2022), the increase in

## TABLE 9.1
## Plant Pathogens in the Philippines List of Regulated Pests

| Pathogen | Species | Disease Name |
|---|---|---|
| Bacteria | *Burkholderia caryophilli* | Bacterial wilt of carnation, Bacterial stem crack of carnation |
| Bacteria | *Candidatus* Liberibacter solanacearum | African greening, Greening of Citrus, Huanglongbing, Yellow branch, Greening |
| Bacteria | *Clavibacter michiganensis* subsp. *sepedonicus* | Potato ring rot, Bacterial ring rot of potato, Vascular wilt of potato |
| Bacteria | *Pectobacterium raphontici* | Rhubarb crown rot, Wheat pink seed, Cereal pink grain |
| Bacteria | *Pseudomonas cichorii* | Bacterial blight of endive, Varnish spot of lettuce, Bacterial rot of lettuce, Leaf spot of ornamentals, Brown stem of celery, Bud blight of chrysanthemum, Leaf spot of chrysanthemum, Bacterial spot of ornamentals, Lettuce varnish spot, Bacterial blight of celery, Endive bacterial blight, Leaf rot of pepper, Leaf spot of *Geranium* spp., Drippy gill of mushrooms, Leaf spot of vegetables |
| Bacteria | *Pseudomonas marginalis* pv. *marginalis* | Lettuce marginal leaf blight, Lettuce kansas disease, Bacterial leaf rot of lettuce, Butt rot of lettuce, *Pseudomonas* soft rot, Pink rib of lettuce, Bacterial soft rot |
| Bacteria | *Pseudomonas syringae* pv. *syringae* | Bacterial canker or blast (stone and pome fruits), Bacterial brown spot (beans), Bacterial sheath rot, Bacterial eye spot, Blast of citrus, Blister spot of apple, Pear blossom blight, Bacterial leaf spot, Bacterial black spot, Apoplexy of apricots, Peach-tree short-life, Bacterial necrosis on kiwifruit |
| Bacteria | *Pseudomonas viridiflava* | Bacterial leaf blight of tomato (USA), Hydrangea bud blight (USA), Bacterial leaf necrosis of basil (USA), Bacterial rot of lettuce (Korea Republic), Bacterial rot of Chinese cabbage (Korea Republic), Bacterial soft rot of tomato (USA, Crete), Bacterial blossom blight of kiwi (New Zealand) |
| Bacteria | *Rhodococcus fascians* | Fasciation: leafy gall, Witches' broom syndrome, Cauliflower disease (ornamentals and strawberry), Fasciation (strawberry), Leafy gall (tobacco) |
| Bacteria | *Xanthomonas fragariae* | Angular leaf spot, Leaf blight of strawberry, Vascular collapse of strawberry |

(*Continued*)

**TABLE 9.1 (CONTINUED)**
**Plant Pathogens in the Philippines List of Regulated Pests**

| Pathogen | Species | Disease Name |
|---|---|---|
| Phytoplasma | Aster yellows phytoplasma group | Tomato big bud, Western aster yellows, Periwinkle little leaf, Eastern aster yellows, Purple coneflower yellows, Multiplier disease, Maryland aster yellows, Dwarf western aster yellows, Marguerite yellows, Periwinkle yellows, Periwinkle witches' broom and virescence, Italian lettuce yellows, Hydrangea phyllody and virescence, American aster yellows, Onion yellows and virescence, *Gladiolus* virescence, Anemone virescence, Chrysanthemum yellows |
| Phytoplasma | *Candidatus Phytoplasma asteris* | Aconitum proliferation, Aconitum virescence, Alberta aster yellows, Alfalfa stunt, Alstroemeria decline, American aster yellows, Anemone virescence, Apple sessile leaf, Apricot chlorotic leaf roll, Azalea little leaf, Basil little leaf, Bermuda grass white leaf, Black currant reversion, Black pepper yellows, Blueberry stunt, Broccoli phyllody, Bunias phyllody, Cactus virescence, Cactus witches' broom, Calendula virescence, etc. |
| Phytoplasma | Clover phyllody phytoplasma | Phyllody of clover, green petal of strawberry |
| Phytoplasma | *Candidatus Phytoplasma aurantifolia* | Lime witches' broom phytoplasma, Witches' broom disease of lime |
| Fungi | *Alternaria japonica* | Alternaria black spot (and wirestem), Alternaria black spot of radish, Leaf spot of radish, Black pod blotch of radish |
| Fungi | *Alternaria radicina* | Black rot of carrots |
| Fungi | *Balansia oryzae-sativae* | Udbatta disease, Black choke, Incense rot, False ergot, Black ring, Sterility disease |
| Fungi | *Botryotinia allii* | Neck rot of onion |
| Fungi | *Diaporthe phaseolorum* var. *sojae* | Soyabean pod and stem blight, pod and stem blight of soybean |
| Fungi | *Didymella ligulicola* | Ray blight of chrysanthemum |
| Fungi | *Golovinomyces cichoracearum* | White mold, powdery mildew |
| Fungi | *Phytophthora fragariae* | Strawberry red stele root rot, Raspberry root rot, Red core disease of strawberry, Red stele disease of strawberry |
| Fungi | *Puccinia horiana* | White rust of chrysanthemum |

(*Continued*)

**TABLE 9.1 (CONTINUED)**
**Plant Pathogens in the Philippines List of Regulated Pests**

| Pathogen | Species | Disease Name |
|---|---|---|
| Fungi | *Sclerophthora macrospora* | Downy mildew, Crazy top of maize, Yellow wilt of rice, Witches' broom on maize |
| Fungi | *Sporisorium cruentum* | Loose kernel smit, Loose smut |
| Fungi | *Stromatinia cepivora* | White rot of onion and garlic, Onion white root rot, Bulb rot of onion, Allium white rot, Onion white rot |
| Fungi | *Synchytrium endobioticum* | Wart disease of potato, Potato wart disease, Black wart of potato, Potato black scab |
| Virus | *Alfalfa mosaic virus* | Alfalfa yellow spot, Tomato necrotic tip curl |
| Virus | *Arabis mosaic virus* | Hop bare-bine, Forsythia yellow net, Strawberry mosaic virus, Hop split leaf blotch virus |
| Virus | *Beet western yellows virus* | Turnip (mild) yellows, TuMYV, BWYV |
| Virus | *Clover yellow vein virus* | Clover yellow vein potyvirus, Pea necrosis virus, Bean yellow mosaic virus-S (severe strain), Bean yellow mosaic virus-N (necrotic strain), Statice virus Y |
| Virus | *Leek yellow stripe virus* | Leek yellow stripe potyvirus |
| Virus | *Lily mottle virus* | Lily mottle potyvirus, Tulip breaking virus, Lily mosaic virus |
| Virus | *Lily virus X* | Lily X potexvirus |
| Virus | *Onion yellow dwarf virus* | Garlic mosaic virus, Onion yellow dwarf potyvirus, Garlic yellow streak virus, Garlic yellow stripe |
| Virus | *Potato yellow dwarf virus* | Potato yellow dwarf, Nucleorhabdovirus, Yellow dwarf of potato |
| Virus | *Prunus necrotic ringspot virus* | Almond bud failure, Almond calico, Almond line pattern, Almond necrotic ringspot, Apricot line pattern, Apricot necrotic ringspot, Cherry ringspot, Cherry rugose mosaic, Cherry tatter leaf, Cherry lace leaf, Cherry necrotic ringspot, Peach mule's ear, Peach necrotic leafspot, Peach necrotic ringspot, Peach willow leaf, Plum decline, Plum oak leaf, Peach ringspot, Sour cherry fruit necrosis, Sour cherry line mosaic, Cherry stecklenberger disease, Sour cherry neurotic ringspot, Cherry line pattern, Cherry (sour) necrotic ringspot virus, Neurotic ringspot virus |
| Virus | *Strawberry latent C virus* | Strawberry latent C disease |
| Virus | *Strawberry latent ringspot virus* | Latent ringspot of strawberry |

*(Continued)*

**TABLE 9.1 (CONTINUED)**
**Plant Pathogens in the Philippines List of Regulated Pests**

| Pathogen | Species | Disease Name |
|---|---|---|
| Virus | *Strawberry mild yellow edge virus* | Strawberry mild yellow edge disease, Strawberry yellow edge, Strawberry mild yellow edge potexvirus, Strawberry mild yellow edge associated virus, Strawberry xanthosis |
| Virus | *Strawberry vein banding virus* | Vein banding of strawberry, Strawberry leaf curl |
| Virus | *Tobacco rattle virus* | Spraing of potato, Corky ringspot of potato, Internal rust of potato |
| Virus | *Tobacco ringspot virus* | Tobacco ringspot nepovirus, Anemone necrosis virus, Tobacco ringspotvirus No.1, Nicotiana virus 12, Blueberry necrotic ringspot virus, Soybean bud blight virus, Tobacco Brazilian streak virus |
| Virus | *Tobacco streak virus* | Tobacco streak ilarvirus, Asparagus stunt virus, New logan virus, Asparagus stunt virus, Datura quercina virus, Nicotiana virus 8, Nicotiana virus vulnerans, Strawberry necrotic shock virus, Black raspberry latent virus, Black raspberry latent ilarvirus, Bean red node strain, Black raspberry latent strain |
| Virus | *Tomato black ring virus* | Ringspot of beet, Yellow vein of celery, Ringspot of lettuce, Bouquet of potato, Pseudo-aucuba of potato, Black ring of tomato, Ringspot of bean |
| Virus | *Tomato spotted wilt virus* | Tomato spotted wilt, Bronze wilt leaf, Tomato bronze leaf virus |
| Nematodes | *Ditylenchus destructor* | Potato tuber nematode<br>Potato rot nematode, Potato eelworm |
| Nematodes | *Ditylenchus dipsaci* | Stem and bulb nematode, Onion bloat, Brown ring disease of hyacinth, Bulb eelworm, Ring disease of bulbs |
| Nematodes | *Heterodera schachtii* | Beet cyst eelworm, Sugarbeet nematode |
| Nematodes | *Heterodera carotae* | Carrot cyst nematode, Carrot root eelworm |
| Nematodes | *Meloidogyne minor* | |
| Nematodes | *Meloidogyne chitwoodi* | Columbia root-knot nematode |
| Nematodes | *Nacobbus aberrans* | False root-knot nematode, Potato rosary nematode |
| Nematodes | *Pratylenchus penetrans* | Meadow nematode<br>Root-lesion nematode, Cobb's root-lesion nematode, Northern root lesion nematode |
| Nematodes | *Pratylenchus thornei* | |
| Nematodes | *Pratylenchus vulnus* | Walnut root lesion nematode, Root lesion nematode, Meadow nematode |

*Source:* National Plant Quarantine Services Division (DA Department Circular No 18 Series of 2018). Access via https://www.ippc.int/en/countries/philippines/reportingobligation/2020/04/philippines-list-of-regulated-plant-pests/, access on August 25, 2023).

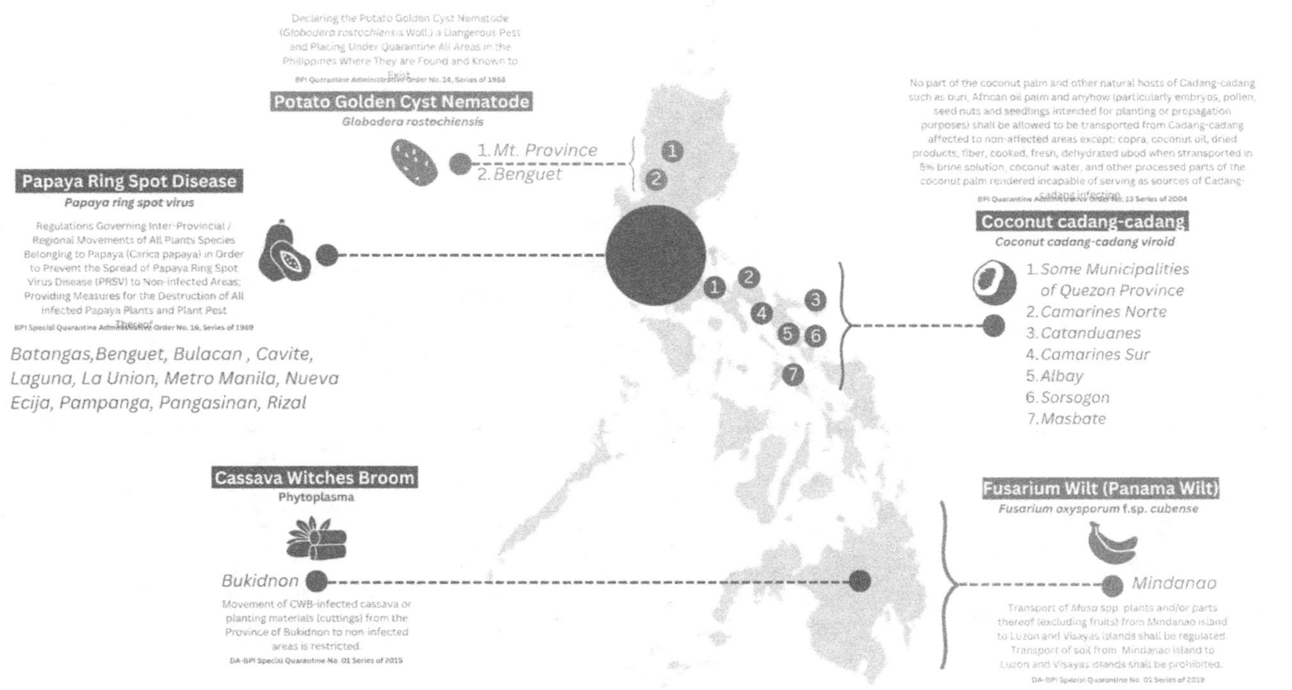


**FIGURE 9.1** Plant pathogens that have been detected in the Philippines that are regulated in the country. Data based on existing Department of Agriculture and Bureau of Plant Industry administrative orders (AO) and special quarantine orders (SQO). Data from published papers that were not used by the regulators as basis of the AOs and SQOs are not included.

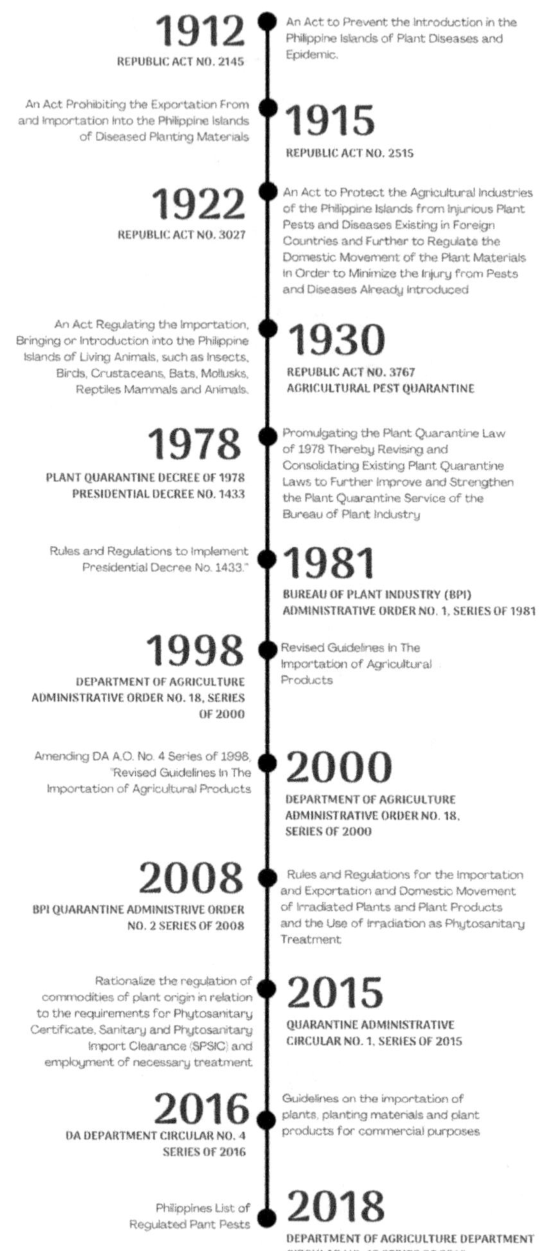


**FIGURE 9.2** Some legal bases of plant biosecurity programs, measures, and approaches in the Philippines.

agricultural trade, travel, and weak phytosanitary systems in some countries accelerates the global spread of transboundary plant pests and pathogens.

The direct, indirect, and potential impacts of regulated plant pathogens are significant and devastating. The estimated cost of plant diseases to the global economy is around USD 220 billion (IPPC Secretariat 2021), and these diseases threaten the goal of achieving global food security (Strange & Scott 2005). We can look back to what happened in Ireland during the Irish potato famine in the 1840s caused by the oomycete pathogen *Phytophthora infestans* (this pathogen remains a regulated or quarantined pest in some countries). As the yield and crop losses continued due to the severe infection of potato plants, people had no food or money to pay for their rent. Some people moved to workhouses, where they got sick, and millions eventually died due to hunger and disease (Ristaino 2021). Millions of people from Ireland also moved to other countries, which reduced Ireland's workforce. The potato blight caused by *P. infestans* is among the many other plant diseases, e.g., coffee leaf rust caused by *Hemileia vastatrix* (Bowden 1971), citrus huanglongbing (HLB) caused by '*Candidatus* Liberibacter asiaticus' (Blaustein et al. 2018), that showcased the negative impact of plant diseases to the society and a country.

Nevertheless, we do not need to go far to see the negative impact of plant diseases and their associated pathogens. One of the Philippines' most important regulated plant pathogens is *Fusarium oxysporum* f. sp. *cubense* tropical race 4 (FocTR4), responsible for the dreaded Panama wilt disease. Panama wilt caused by FocTR4 is an increasingly serious banana disease affecting the popularly exported Cavendish strain (Ploetz 2021). In the country, the FocTR4 is only present in several provinces on the island of Mindanao (BPI Special Quarantine Administrative Order No. 01 Series of 2012). In 2011, the disease destroyed at least a thousand hectares of bananas. They threatened the survival of at least 300,000 families whose members worked in the banana industry (Lacorte & Quiros 2011). In the 2015 survey, 15,507 hectares of banana plantations in Davao were infested with FocTR4 (Lumawag 2019). It was also reported that at least 150 banana growers shifted from banana to corn planting due to Panama wilt (Palicte 2023; Hinutan 2023). The potential export loss due to FocTR4 was estimated at US$ 544 million without effective management measures (Pattison et al. 2020).

Serious plant disease outbreaks are food security issues which are national issues (Ristaino 2021). Plant disease outbreaks and the occurrence of infection in previously non-infested areas require government involvement to contain the outbreaks, minimizing the pathogens' spread and mitigating diseases' impact on the community and relevant industries. This involvement is evidenced by the release of special quarantine orders, memoranda, and administrative orders from the relevant authorities, national organizations, and regulators (Figure 9.2). The actions from the national and local organizations also require funding and coordination that require human resources and labor force, as well as providing support to people and their families who lost their jobs due to the closure of factories or establishments. Hence, the impact of plant pathogens and associated plant diseases is beyond crop yield loss and reduced profit. In the Philippines, more than half of plant disease reports since the 1920s are caused by plant pathogenic fungi (Balendres 2022). Incidence and detection of fungi infecting various plant species have also been documented (e.g., Aumentado & Balendres 2023a, Aumentado & Balendres 2023b, Evallo et al. 2023, Aumentado & Balendres 2022a, Dimayacyac & Balendres 2022; Aumentado & Balendres 2022b).

## 9.3 QUARANTINED AND REGULATED PLANT PATHOGENS

As of 2018, there are 56 quarantined and regulated plant pathogens (Table 9.1) that are included in the list of regulated pests in the Philippines based on pest risk analysis reported by the National Plant Quarantine Services Division (NPQSD) (Department of Agriculture Department Circular no. 18 series of 2018 and Annex 1 DA DC no.18 s.2018) to the International Plant Protection Convention (IPPC) (https://www.ippc.int/en/countries/philippines/reportingobligation/2020/04/philippines-list-of-regulated-plant-pests/). Nevertheless, some pathogens may or may not be included in the list due to limited information The NPQSD is a division under the Bureau of Plant Industry (BPI), which regulates the transport of plant and planting materials, and monitors and controls the movement of plant pests and diseases. The BPI-NPQSD has become the country's frontline in the fight against foreign plant pathogens and protects Philippine agriculture 'by preventing the introduction of new pests and its further spread' (NPQSD 2023a) according to the Presidential Decree No. 1433 or the Plant Quarantine Decree of 1978. Of the 56 regulated plant pathogens in the current version of the list, 10 are bacteria, 4 are phytoplasma, 13 are fungi and fungi-like, 19 are viruses, and 10 are nematode species (Table 9.1). Visual inspection, bioassays, serology (ELISA-based), and molecular (PCR-based) approaches are used to detect and identify these plant pathogens.

### 9.3.1 Regulated Bacterial Pathogens

*Burkholderia caryophilli* (listed as *Paraburkholderia caryophylli*) in the EPPO Global Database (2023) is responsible for bacterial wilt and bacterial stem crack of carnation (*Dianthus caryophyllus*). This pathogen has been reported in some countries in Asia, Europe, and North and South America. In Asia, the bacterium's presence has been recorded in Taiwan (Liu et al. 1990), Japan (Onazaki et al. 2002), India (EPPO Global Database 2023), and Jilin, China (EPPO Global Database 2023).

Another regulated pathogen in the Philippine list is *Candidatus* Liberibacter solanacearum. Although it was noted as the causative agent of 'African greening, Greening of Citrus, Huanglongbing, Yellow branch, Greening' in the Annex 1 DA DC no.18 s.2018 listing, this pathogen is widely known to cause zebra chip disease of potatoes. This bacterium is vectored by *Bactricera cockerilli* (also known as the potato or tomato psyllid) (Bextine et al. (2013), *B. trigonica* (Ben et al. 2018), and *Trioza apicalis* (also known as the carrot psyllid) (Haapalainen et al. 2018). The pathogen is unculturable and was first identified from *B. cockerelli* as '*Candidatus* Liberibacter psyllaurous' (Hansen et al. 2008) and in several crops as '*Candidatus* Liberibacter solanacearum' (Liefting et al. 2009). This pathogen has been reported as 'present' with 'few occurrences' to 'restricted distribution' in Europe, the US, and

Mexico. This pathogen is not present in the Asian region but has been detected in New Zealand (Pitman et al. 2011).

*Clavibacter michiganensis* subsp. *sepedonicus* is the causative agent of potato ring rot, bacterial ring rot of potato, ring rot of potato, and vascular wilt of potato. Within the neighboring region, it has been reported in Taiwan (Hsu et al. 1986), Japan (EPPO Global Database 2023), Korea (EPPO Global Database 2023), Nepal (EPPO Global Database 2023), Pakistan (Zehra et al. 2022), and some parts of China (Zhang et al. 1990). Potato is the primary host of this bacterium. However, in an experimental assay, it is known to infect *Beta vulgaris*, tomato, and various plant species, most belonging to the genus *Solanum.*

*Pectobacterium raphontici* is listed as *Erwinia raphontici* in the EPPO Global Database (2023). This pathogen is associated with crown rot of rhubarb, pink grain of cereals, and the pink seed of wheat. This pathogen has been reported in Malaysia (EPPO Global Database 2023), Japan (Tsuji et al. 2019), Korea (Huang et al. 2003), some parts of China (Chu et al. 2010), as well as in some European countries and the US (EPPO Global Database 2023). Several plant species host this bacterium, including onion, *Citrus* spp., carrot, tomato, wheat, and pea. In field of peas (*Pisum sativum*), the symptoms are pale pinkish brown-to-bright pink discoloration that is permanent and cannot be removed by thoroughly washing the seed coat (Schroeder et al. 2002).

Four *Pseudomonas* species are included in the list of regulated pests in the country. These are *P. cichorii and P. marginalis* pv. *marginalis*, *P. syringae* pv. *syringae*, and *P. viridiflava.* The bacterial blight and leaf spot pathogen, *P. cichorii*, has been recorded in Australia, New Zealand, Taiwan, Japan, and some parts of China (EPPO Global Database 2023). This bacterium infects basil, pepper (*Capsicum* sp.), lettuce, tobacco, and tomato. The lettuce pathogen, *P. marginalis* pv. *marginalis*, has been reported in Australia, New Zealand, Japan, and Korea (EPPO Global Database 2023). Aside from lettuce, it has also been reported to infect other vegetables. The bacterial canker pathogen, *P. syringae* pv. *syringae* is associated with the bacterial canker of lilac, pear, stone fruits, trees, sorghum and alfalfa (EPPO Global Database 2023). However, it also causes leaf spots and blast. Lastly, *P. viridiflava* is a bacterium associated with bacterial blight of kiwi, root rot of lucerne, greasy canker of poinsettia (EPPO Global Database 2023), and a variety of diseases in different plant organs (Lipps & Samac 2022). It also infects cauliflower, pumpkin, lupin, passionfruit, poppy, grape, pea, and Chinese gooseberry (Wilkie et al. 1973). All four regulated *Pseudomonas* species are important in agriculture due to their impact on plant growth and crop yield.

*Rhodococcus fascians* is associated with the fasciation of strawberries, leafy galls of ornamentals and tobacco, and cauliflower disease of ornamentals and strawberries. Within the region, this bacterium has been reported in India (Andhra Pradesh), Australia (New South Wales), and New Zealand (EPPO Global Database 2023). This pathogen is a limiting factor in profitable ornamental crop production (Putnam & Miller 2007).

The strawberry pathogen, *Xanthomonas fragariae*, is associated with angular leaf spot, leaf blight and vascular collapse. The bacterium has been reported in

Taiwan (EPPO Global Database 2023), Korea (Kim et al. 2016), and some provinces of China (Zhang et al. 2022). This pathogen was first reported in the US (Kennedy and King 1962) and is prevalent in North America. While it is prevalent in some parts of the world where strawberries are grown, it remains a quarantine pest in several countries, including Morocco, Mexico, China, Israel, and New Zealand (EPPO Global Database 2023).

### 9.3.2 Regulated Phytoplasma Pathogens

Four Phytoplasma are on the list of regulated pests in the Philippines. These are Aster yellows phytoplasma group, Candidatus Phytoplasma asteris, Clover phyllody phytoplasma, and Candidatus Phytoplasma aurantifolia. Aster yellows phytoplasma group has been associated with more than 300 kinds of plants. It is spread via an unidentified insect vector (IPPC 2023). One particular, important member is 'Candidatus Phytoplasma asteris', which is associated with aster yellows and other related diseases (Lee et al. 2004). This pathogen has been reported in Myanmar (EPPO Global Database 2023) and is present in Japan and China with a restricted distribution (EPPO Global Database 2023). Another phytoplasma in the regulated list is Clover phyllody phytoplasma, which is associated with the phyllody of clover (Firrao et al. 1996) and green petal of strawberry (HonetŜlegrová et al. 1996). Finally, Candidatus Phytoplasma aurantifolia is associated with Oman witches' broom disease, witches' broom of lime (Bové et al. 2000), and potato purple top (Caicedo et al. 2015). Candidatus Phytoplasma aurantifolia is hosted by several plant species, including some of those in the *Citrus* family (EPPO Global Database 2023), and is vectored by the leafhopper *Hishimonus phycitis* (Salehi et al. 2007).

### 9.3.3 Regulated Fungi and Fungi-Like Pathogens

*Alternaria japonica* is associated with black pod blotch, black spot, and leaf spot of radish (EPPO Global Database 2023), black spot in turnip (Bassimba et al. 2013), and can also infect seeds of wild and cultivated rocket (Gilardi et al. 2015). On the other hand, *A. radicina* has been associated with black rot in carrots (Meier et al. 1922), which is a seed-borne pathogen previously detected on imported carrot seeds (Scott and Wenham 1973). The pathogen has been detected in South Australia on carrot seeds, seedlings, and roots (Coles & Wicks 2003) and in New Zealand infecting carrot leaves (Soteros 1979).

*Balansia oryzae-sativae* (syn. *Ephelis oryzae*) is the false ergot pathogen and is also associated with a black choke of rice, incense rot, and udbatta disease of rice (EPPO Global Database 2023). This pathogen has been reported in Indonesia (Amir 1986), some parts of China (Booth 1979; Zhang & Huang 1990), and some regions of India (Murty and Alimeer 1978). The weed species, *Echinochloa crusgalli*, has been also reported as an alternative host of this fungal pathogen (Das et al. 2000).

*Botryotinia allii* is listed as *Ciborinia allii* in the EPPO Global Database (2023). This pathogen is associated with mycelial neck rot of onion and sclerotinia disease of Welsh onion. Another fungal pathogen in the list of regulated pests is *Diaporthe phaseolorum* var. *sojae*, which has been associated with pod blight, seed decay (Kmetz et al. 1978), and stem blight of soybean. Severe infection can affect soybean yield (Pedersen & Grau 2010). This fungus is widely distributed but is not yet recorded in the Philippines. It is present or has been recorded in the neighboring countries of Taiwan (Chen et al. 2009), Malaysia (Mahmodi et al. 2013), Thailand, China, and Korea (EPPO Global Database 2023). Soybean (*Glycine max*) is the major host but can also infect *Phaseolus lunatus*, *P. vulgaris*, and *Vigna unguiculata*.

*Didymella ligulicola*, currently known as *Stagonosporopsis chrysanthemi* (EPPO Global Database 2023, Mycobank), is associated with flower and ray blight of chrysanthemum (Baker et al. 1949; Romero & Romero 1990). This pathogen has been reported in Korea on *Chrysanthemum boreale* (Kim et al. 2001), Papua New Guinea, New Zealand, and Japan and is prevalent in Europe (EPPO Global Database 2023).

*Golovinomyces cichoracearum* (syn. *Erysiphe cichoracearum*) is associated with powdery mildews of various ornamental and vegetable plants, including gerbera, lettuce (Lebeda & Mieslerová 2011), potato, tobacco, and cucurbits (EPPO Global Database 2023). Although this pathogen is on the list of regulated pests in the country, *G. cichoracearum* is widely distributed and has been recorded in many countries, including Australia, Thailand, Korea, India, Mexico, the US, China, Argentina, France, Russia, Israel, Japan, and many more (Farr & Rossman 2023). In the Philippines, *G. cichoracearum* has been detected in *Zinnia* sp., causing defoliation (Balendres et al. 2021). The pathogen was identified based on morphocultural characterization and its association with *Ampelomyces quisqualis*. In 2022, Aumentado and Balendres detected powdery mildews on *Vigna* spp., grown within a 50-meter radius, where the powdery mildews on *Zinnia* sp. were detected. Based on the molecular characteristics, the powdery mildew pathogen associated with *Vigna* spp. was identified as *Podosphaera xanthii* (Aumentado & Balendres 2022). Most recently, *P. xanthii* was again associated with eggplant powdery mildews (Aumentado et al. 2023a) in the same area (within 50 50-meter radius) where the *Zinnia* sp. were previously grown. There is a possibility that the powdery mildew detected in *Zinnia* sp. could be *P. xanthii*, given the distance of the *Zinnia* sp. plants to *Vigna* spp. and eggplant. Nevertheless, the morphological characteristics of powdery mildew isolated from *Zinnia* sp. (Balendres et al. 2021) resemble those of *G. cichoracearum*. Further sampling and characterization of powdery mildew from *Zinnia* sp. would be needed to validate the previous study (Balendres et al. 2021). Both *G. cichoracearum* and *P. xanthii* are important powdery mildew pathogens of vegetables (Lebeda et al. 2011). Further, *Erysiphe cichoracearum*, the previous name of *G. cichoracearum*, has already been reported in the Philippines on several plant species (Tangonan 1999). For instance, Quimio and Nadayao (1980) reported powdery mildew of tomato caused by *E. cichoracearum*.

*Stromatinia cepivora* is associated with white rot of onion and garlic and bulb rot of onion. This pathogen has been reported in China, Japan, Australia, New Zealand, the USA, Europe, South America, and Africa (EPPO Global Database 2023) ). Aside from onion and garlic, this fungal pathogen has also been reported to cause white rot in Korean wild chives (Kim et al. 2023). The pathogen produces sclerotia that allows the fungus to survive in the soil without its host (Marcuzzo & Schmoeller 2017).

There are two basidiomycetes in the list of regulated pests in the Philippines. *Puccinia horiana* is associated with white rust of chrysanthemum (*Chrysanthemum × morifolium*) (Firman and Martin 1968 ; Whipps 1993). This rust pathogen has been reported in neighboring countries like Brunei Darussalam (Peregrine and Ahmad 1982), Malaysia (EPPO Global Database 2023), Taiwan (Leu et al. 1982), Thailand (Lohsomboon et al. 1986), Japan (Abiko et al. 1977), Korea (Cho & Shin 2004), some regions of China (Zhuang 2001), New Zealand, and Australia (Catley 1987). This pathogen remains a quarantine pest in Canada, Mexico, the US, Israel, Morocco, Tunisia, Norway, Belarus, and Moldova (EPPO Global Database 2023). The other basidiomycetes is *Sporisorium cruentum*, associated with loose kernel smut of sorghum or loose smut of sorghum. This pathogen has been reported in several African countries, some parts of Europe, and some regions of China, Taiwan, and Japan (EPPO Global Database 2023).

Two oomycetes are included in the list of regulated pests. These are the rot pathogen, *Phytophthora fragariae*, and the downy mildew pathogen, *Sclerophthora macrospora*. *P. fragariae* is associated with the red core of strawberry (Hickman 1941) and the red stele of strawberry. In the Asian region, this pathogen has been recorded only in Japan (Morita 1965). Aside from strawberries, this oomycete can also cause root rot in raspberries (Laun & Zinkernagel 1997). On the other hand, *Sclerophthora macrospora* is associated with the crazy top of maize and sorghum, downy mildew of cereals (e.g., maize, wheat, sorghum, and pearl millet) (EPPO Global Database 2023), and sugarcane (Magarey 2006). This downy mildew pathogen has been reported in Taiwan (Chang & Lai 1970), Japan, Korea (Jee et al. 2002), Australia (Ryley et al. 2022), New Zealand, and some parts of China (EPPO Global Database 2023).

The only Chytridiomycetous pathogen in the regulated pest list is *Synchytrium endobioticum*, the potato wart pathogen (Vossenberg et al. 2022). This pathogen can survive in the soil for several years through winter sporangia (Przetakiewicz 2015). The pathogen is widespread in Europe and has been reported in specific areas in India (Sharma & Chakrabarti 2020), and in the Guizhou province of China (Bi & Hu 2005). Potato is the major host of this pathogen but is also known to be hosted, experimentally, by *Datura metel*, tomato, *Duboisia* sp., *Nicandra* sp., and *Solanum dulcamara* (EPPO Global Database 2023).

### 9.3.4 Regulated Plant Viruses

Plant viruses had the most number (19) of plant pathogens in the list of regulated pests in the Philippines (Table 9.1). These plant viruses are *Alfalfa mosaic virus*, *Arabis mosaic virus*, *Beet western yellows virus*, *Clover yellow vein virus*, *Leek yellow stripe virus*, *Lily mottle virus*, *Lily virus X*, *Onion yellow dwarf virus*, *Potato yellow dwarf virus*, *Prunus necrotic ringspot virus*, *Strawberry latent C virus*, *Strawberry latent ringspot virus*, *Strawberry mild yellow edge virus*, *Strawberry vein banding virus*, *Tobacco rattle virus*, *Tobacco ringspot virus*, *Tobacco streak virus*, *Tomato black ring virus*, and *Tomato spotted wilt virus*. Several viruses infect vegetables, particularly those in the solanaceous group (e.g., tomato, tobacco, potato).

### 9.3.5 Regulated Plant-Parasitic Nematodes

Ten plant parasitic nematodes are listed as regulated pests in the Philippines (DA Department Circular No. 18 Series of 2018). These are *Ditylenchus destructor*, *Ditylenchus dipsaci*, *Heterodera schachtii*, *Heterodera carotae*, *Meloidogyne minor*, *Meloidogyne chitwoodi*, *Nacobbus aberrans*, *Pratylenchus penetrans*, *Pratylenchus thornei*, and *Pratylenchus vulnus*. The current list remains significant as these nematodes are yet to be reported in the Philippines. For instance, *H. schachtii* remains undetected in the Southeast Asian region but has been detected in neighboring countries like China (Huan et al. 2022) and Australia (Khair 1981). Another species, *D. destructor*, has been detected in many provinces in China (EPPO Global Database 2023), but its presence in the Southeast Asian region is not known. *P. penetrans*, *P. thornei*, and *P. vulnus* have never been scientifically reported in the Philippines.

Nevertheless, other plant parasitic nematode species may also be considered in addition to the above-listed species. These nematode species appear in the list of regulated pests in other countries. However, they are not present in the Philippines and are yet to appear in the country's regulated pests list. For instance, *Aphelenchoides fragariae* causes disease in *Allium* sp., potato, and its major host, strawberry, which has been reported in some provinces in China and Java, Indonesia (Chaerani 2022, EPPO Global Database 2023). Another regulated nematode is *Xiphinema rivesi*, hosted by several plant species, including *Allium* sp., *Citrus sinensis*, and mango (EPPO Global Database 2023). In the Asian region, it has only been reported in Pakistan (Khan & Tereen 2012; EPPO Global Database 2023). Nonetheless, thorough surveys across the Philippine archipelago are warranted to determine the country's current status of plant parasitic nematode species. The information generated from these surveys would be useful in informed- and science-based decisions in updating the country's current list of regulated pests (nematodes).

## 9.4 PLANT PATHOGENS WITH QUARANTINE IMPORTANCE (DOMESTIC)

The country has five regulated plant pathogens (Figure 9.1) based on administrative orders (AOs) released by relevant regulating agencies (key AOs are presented in Figure 9.2). These are the Papaya ringspot virus, Potato Golden Cyst nematode (*Globodera rostachiensis*), *Coconut cadang cadang* (viroid), *Cassava Witches Broom* (phytoplasma), and *Fusarium oxysporum* f. sp. *cubense* (FocTR4). In 1923, an AO (No. 39) was also released to regulate the eradication of coconut bud-rot. These pathogens and associated diseases have been reported in some regions/localities. Given that the country is an archipelago composed of more than 7,000 islands, the movement of the infected materials has been regulated to avoid the spread of the pathogens.

Papaya ringspot virus (PRSV) is a serious disease of papaya causing conspicuous ringspots on fruits and mottle or mosaic symptoms on leaves (Sta Cruz et al. 2009). This virus has been detected in some provinces in Luzon (Sta Cruz et al. 2009). Under the BPI Special Quarantine Administrative Order No. 16, Series of 1989, '*Regulations Governing Inter-Provincial / Regional Movements of All Plants Species Belonging to Papaya (Carica papaya) in Order to Prevent the Spread of Papaya Ring Spot Virus Disease (PRSV) to Non-infected Areas, Providing Measures for the Destruction of All Infected Papaya Plants and Plant Pest Thereof*', movement of fruits, seeds, and seedlings of papaya is strictly prohibited from identified infested areas (Cavite, Batangas, Laguna, Rizal, Metro Manila, Bulacan, Pampanga, Benguet, La Union, Nueva Ecija, and Pangasinan) to non-infested areas in Visayas and Mindanao islands. However, in 2001, PRSV was reported for the first time on the island of Mindanao. Symptoms of PRSV were reported in papaya and validated by ELISA tests in three municipalities in South Cotabato (Polomolok, Tupi, and Tampakan) and in Sta. Cruz Davao Del Sur (Herradura et al. 2001). No further scientific reports have been made after the 2001 report (Herradura et al. 2001). Survey and monitoring in the island of Mindanao for PRSV is warranted to determine if previous PRSV occurrence has been eradicated or PRSV is still present on the island.

In 1988, the Potato Golden Cyst (GCN) nematode (*Globodera rostachiensis*) was declared by the Bureau of Plant Industry as a dangerous pest and placed the Province of Benguet and Mountain Province as under quarantine (BPI Quarantine Administrative Order No. 14, Series of 1988). Under this order, the movement, transfer, or transport of seed potatoes or parts, therefore capable of propagation from these two provinces to other areas in the country, was prohibited. Sections 3 and 4 also state that 'areas confirmed infested with [GCN], the owner thereof shall be required to observe alternate planting of his field with non-solanaceous vegetables for at least two years … and all potato seed pieces shall pass the seed certification scheme of the Bureau of Plant Industry' (BPI Quarantine Administrative Order No. 14, Series of 1988). Since its first declaration, no records or scientific reports have been released regarding PGC; therefore, there is a need to revisit the status of this pathogen through surveys in potato-growing regions in the country.

'Cadang cadang' (Coconut cadang cadang viroid) was first reported in Southern Luzon in the Philippines (Randles et al. 1980). An estimated 30 million coconut palms have been destroyed due to the disease (Zelanzy et al. 1982). The latest legal basis for the regulatory status of cadang-cadang is the BPI Quarantine Administrative Order No. 13, Series of 2004. Under this AO, some areas of Quezon province, all cities and municipalities of Albay, Camarines Sur, Camarines Norte, Catanduanes, Masbate, Sorsogon, some areas of Eastern Samar, all municipalities of Northern Samar, some towns of Samar, and two towns of Biliran were declared as cadang-cadang infected areas.

In 2015, through the Special Quarantine Order No. 1, series of 2015, DA-BPI declared cassava witches' broom (caused by a phytoplasma) (CWB) was declared as a dangerous disease of cassava with the Province of Bukidnon placed under quarantine. A 100% yield loss can occur if the symptom occurs during crop establishment and up to 60% yield loss when symptoms occur five months post-crop establishment. CWB-infected planting materials or stem cuttings spread the disease from field to field. In section 2 of the special quarantine order, movement and transfer of cassava plants/planting materials from Bukidnon to other areas were prohibited. CWB is a devastating cassava disease in Southeast Asia in Thailand, Cambodia, Vietnam, Laos (Pardo et al. 2023), and the Philippines (Dolores et al. 2023). Just recently, the provinces where CWB is present have been expanded. CWB is now present in ten provinces in Luzon, six provinces in Visayas, and three provinces in Mindanao (including Bukidnon) (Dolores et al. 2023). The cause of CWB in the Philippines was identified as a 'Candidatus Phytoplasma luffa'-related strain (16SrVIII) based on 16S rRNA gene analysis (Dolores et al. 2023). With the recent detailed reports by Dolores et al. (2023), the Special Quarantine Order No. 1, series of 2015 could be revised by keeping its status as a 'dangerous disease of cassava' but lifting the part where Bukidnon is placed as 'under quarantine' and include other areas where CWB has been detected. All measures to avoid disease spread must remain in place.

*Fusarium oxysporum* f. sp. *cubense* (FocTR4) is a more virulent strain of *Fusarium oxysporum* that occurred and severely affected farms in Mindanao (Molina et al. 2008; Molina et al. 2011). Banana Cavendish is extensively grown in Mindanao for export, generating around USD 300 million annually. However, FocTR4 became a limiting factor, forcing growers to abandon heavily infected fields and expand to non-infected areas. Through BPI Special Quarantine Administrative Order No. 1, series of 2012, Panama disease caused by FocTR4 was declared a dangerous and injurious disease of banana and abaca and placed provinces where the disease is present under quarantine. These provinces were Davao del Norte, Davao Oriental, Davao del Sur, Davao City, Compostella Valley, Bukidnon, and North and South Cotabato. The AO was made to safeguard the banana and abaca industries from diseases by establishing quarantine measures to arrest and prevent the spread of the diseases from infected to non-infected areas. Section 3 states the movement/transfer/carrying of plants and parts of *Musa* spp. varieties, soil, and plant parts used in packing materials for agricultural products from the infected areas to non-infected areas (including Visayas and Luzon) are strictly prohibited, except tissue-cultured banana and abaca plants and limited quantities intended for experimental purposes provided a prior permit for such purpose has been secured from the Director of the Bureau of Plant Industry. Additional quarantine and border control measures were implemented in the Special Quarantine Administrative Order. Quarantine checkpoints were established in municipalities and between provinces where the disease has been identified and verified as existing. Vehicles, farm equipment, and people passing the quarantine checkpoints were inspected, and disinfection was made using appropriate disinfectants (for shoes and wheels). Materials capable of spreading the disease will be confiscated. Foot baths were installed at every port/wharf airport, and entry and exit points were established in banana and abaca farms. In 2019, the BPI Special Quarantine Order No. 1 series 2019 was released to further strengthen the previous AO on the subject. The latest SPO was used to guide the transport and movement of *Musa* spp. and non-*Musa* spp. plants and planting materials from Mindanao to Luzon and Visayas. It outlines the responsibilities of the owner of the planting materials, including inspecting the quarantine officers to these plants and planting materials and granting Clearance for Domestic Transport (CDT).

## 9.5 PLANT BIOSECURITY MEASURES FOR QUARANTINED PLANT PATHOGENS

Presidential Decree No. 1433, also known as the Plant Quarantine Decree of 1978, signed by then President Ferdinand Marcos, provides the most important legal basis of plant biosecurity and quarantine rules and regulations in the country (Figure 9.2). Upon entry, clearance from the Plant Quarantine Officers assigned at the port is required (Section 5). Section 7 also indicates that plants, plant products, and materials should be inspected, and phytosanitary and quarantine measures should be implemented when necessary. The legal bases (Figure 9.2) used in the country are aligned with those of the International Plant Protection Convention, an international treaty under the United Nations Food and Agriculture Organization (FAO).

The first line of defense against the possible incursion of unwanted, foreign plant pathogens is to ensure that the allowable plants and planting materials (and any viable materials) for importation are free from any inoculum or pathogens before the materials leave the country of origin. This also includes 'any means that allows the entry or spread of a pest', as FAO's International Standards for Phytosanitary Measures (ISPM) states. This could be done by requiring a phytosanitary certificate and complying with the phytosanitary import conditions of the Philippines (e.g., inspection, testing, treatment, and certification prior to export by exporting country: ISPM 7) before importing the products before importing the products. Based on BPI Quarantine Admin Order 1 series of 1981 and DA DC 4 series of 2016, any 'person/company intending to import plants/plant products shall apply for Plant Quarantine Clearance (PQC) (BPI Q Form No. 1) (personal purposes) and Sanitary and Phytosanitary Import Clearance (SPSIC) (commercial purposes) with the NPQSD before importation. Importations which require PQC/SPSIC and are subject to inspection' (Based on DA DC 4 series of 2016). Specifically, the BPI-NPQSD outlined these materials that require PQC or SPSIC prior to entry. These are 'living plants, nursery stocks, including vegetative parts thereof used as propagating materials, seeds nut for planting, fresh fruits, vegetables, and other plant products, which have been declared as prohibited/restricted imports under Special Quarantine Orders by Virtue of their being known hosts of certain plant pests, or because they originate from restricted areas, and other plants cultures' (NPQSD 2023c). Aside from plant and planting materials, viable and live cultures of microorganisms and other organisms will also require a PQC or SPSIC, depending on purpose (e.g., pure cultures of fungi, bacteria, virus, nematodes, and other phytopathogenic materials, mushroom cultures including spawn, etc.).

Before the issuance of a PQC/SPSIC, there are two processes that the NPQSD will undertake. First is categorizing commodities in compliance with the Department of Agriculture (DA) Quarantine Administrative Circular No. 01 Series of 2014 or the 'Guidelines for Categorization of Commodities of Plant Origin'. This administrative order provides 'guidance in categorizing commodities according to their pest risk, which is important when establishing phytosanitary import requirements' (NPQSD 2023c). Category I is for commodities 'that have been processed to the point where they do not remain capable of being infested with quarantine pests and, therefore, should not be regulated' (NPQSD 2023c). Category II is for commodities 'that have been processed to the point where the commodity can be infested with quarantine pests. Intended use may be for consumption or further processing. PRA may be required for quarantine pests that the process may not eliminate' (NPQSD 2023c). Category III commodities are those 'that have not been processed. The intended use of commodity is for consumption or processing' (NPQSD 2023c), and Category IV commodities are those 'that have not been processed and intended use is for planting' (NPQSD 2023c). The Pest Risk Analysis (PRA), the other procedure, is necessary for commodities under Categories II, III, and IV.

The PRA is an 'international standard Plant Quarantine procedure conducted by the importing country in assessing the risk of entry of any plant/plant product capable of harboring insect pests, plant diseases, and other plant-associated organisms. The result of the PRA will be the basis for whether the importation of the commodity will be allowed or not' (NPQSD 2023c). Along with submitting the PQC/SPSIC form, the importer or its authorized representative should also submit a letter of intent addressed to the Director of the Bureau of Plant Industry. 'Upon approval by the BPI Director, the registration team will release the PQSC to the importer' (NPQSD 2023c). If the consignment or materials to be imported is for commercial purposes, the importer must register as a licensed importer. This procedure follows the DA Department Circular No. 04 series of 2016, the 'Guidelines on the Importation of Plants, Planting Materials and Plant Products for Commercial Purposes'.

The NPQSD has provided the outline and flow of the processes in a six-page document, BPI Operational Procedure Manual (BPI-QMS-OU-OP4, https://npqsd.bpi-npqsd.com.ph/wp-content/uploads/2023/05/Pest-Risk-Analysis-PRAR1.pdf). The document provides importer and authorized representatives to check the requirements and flow leading to the issuance of the PQC/SPSIC.

## 9.6 INSPECTION UPON ARRIVAL OF COMMODITIES

Once the imported plant, planting materials, or other viable materials have arrived in the country, the importer will apply a duly accomplished Application for Inspection form for inspection to NPQSD. Samples are also collected during the inspection of the consignment or imported materials for laboratory analysis. Four documents: (1) the Phytosanitary Certificate (original FAO/IPPC Model) issued by the Plant Quarantine (PQ) of the country of origin, (2) the PQC/SPSIC (original copy), the (3) bill of lading/airway bill, and (4) the inward cargo manifest and Bureau of Customs (BOC) entry declaration (photocopy), must be submitted to the Plant Quarantine Service (PQS)/NPQSD. The NPPO of the country of origin of the imported materials issue the Phytosanitary Certificate (PC). Any consignment that does not have this PC 'shall be held under BOC custody or transferred (upon importer's request) to PQS custody until such documents are presented' (NPQSD 2023c). If both the

PQC/SPSIC from the importing country and the PC (from the country of origin) are not provided or presented, 'the consignment shall either be returned to the country of origin or re-exported to another accepting country or destroyed (NPQSD 2023c)'.

## 9.7 PLANT BIOSECURITY MEASURES FOR DOMESTIC PLANT PATHOGENS

Because of the country's archipelagic geography, composed of more than 7,000 islands, particular outbreaks and epidemics of plant diseases are likely to occur and may be contained in specific islands. For instance, FocTR4 has been only detected so far on the island of Mindanao. However, with the frequency of travel of people and the movement of fresh produce between islands, there is a possibility that specific plants and planting materials could harbor the inoculum of regulated plant diseases. To prevent any further spread and increase in local incidence of regulated plant diseases and associated pathogens that already exist in the country, the Bureau of Plant Industry regulates the 'movement of plants, planting materials, plant products, and other articles capable of harboring plant pests via seaport, airport, and by land' (NPQSD 2023d) within the country. For domestic movement of plants and planting materials, a Clearance for Domestic Transport (CDT) of Plants/Planting Materials/Plant Products is needed from the NPQSD. The Plant Quarantine Officer (PQS) will inspect the commodity/consignment before transporting the materials to another region, island, or province. If necessary, phytosanitary measures will ensure the commodity/consignment is free from unwanted, regulated plant pathogens. The legal bases of the guidelines in the transport and movement of materials are outlined in Figure 9.2.

## 9.8 FURTHER STRENGTHENING PHILIPPINE PLANT BIOSECURITY PLANS

Plants provide food and are raw materials for products humans use for food, clothing, shelter, and survival. However, plant pathogens continue to pose threats to food supply, crop health, and plant biodiversity. They are worsened by globalization and climate change (Prasanna et al. 2022). Plant pathogens threaten crop production by reducing yield and may have a negative socioeconomic impact. Plant pathogens reported in other countries or within the country in different islands remain challenging and require coordination among regulating agencies, relevant industries/sectors, and the public. Preventing their entry into the country would be the best strategy.

A greater understanding of their biology, pathology, spread, and possible entry pathways of these TPPs could help the country prevent their entry and possible spread and contain their establishment. Better, state-of-the-art, and more sensitive diagnostic tools are needed to ensure effective monitoring and surveillance. For instance, by studying the population genetic structures of *Rhizoctonia solani* AG-1 1A, the sheath blight pathogen of rice, Cumagun et al. (2020) were able to provide strong evidence of the movement and dispersal of the pathogen from Southern China to the Philippines. Moreover, raising awareness of their impact on our food supply and economy and lobbying the country's legislators could help secure government and industry funding commitments that would help enhance research and development programs aiming toward preventive and integrated disease control measures. There is also a need to include emerging transboundary plant pathogens that are not on the list. For instance, *Xylella fastidiosa* and the *Tomato brown rugose fruit virus* are in the EPPO Global Database (2023). Another pathogen to consider is *Pseudocercospora ulei*, the causative agent of South American leaf blight of rubber (Portales et al. 2023). However, the regulated list needs to be updated.

Several countries, e.g., Australia and the United Kingdom, have plant biosecurity programs that help protect their borders from foreign and unwanted plant pathogens. Hence, developing countries, like the Philippines, could continue to benchmark with these countries and join the global efforts toward developing diagnostic and surveillance systems for transboundary plant pathogens. Alternatively, the country can refer to the IPPC framework to enhance phytosanitary capacity enhancement. Implementing phytosanitary measures may be very challenging for countries like the Philippines due to porous borders (many islands and population movements by boats and planes). It may need more officers employed to monitor and check incoming produce.

Additionally, the country could also benefit from implementing (or strengthening) these actions/practices through (a) continuous awareness campaign and involvement of various stakeholders, including those in the industry, (b) expand plant biosecurity campaigns, (c) continue the recruitment of plant quarantine officers, (d) continue to strengthen diagnostic capacity and building laboratories and diagnostic facilities in key ports, (e) regular monitoring, (f) continuous monitoring of re-emerging and new threats and promote early warning systems, (g) preparing for possible outbreaks, (h) collaboration with universities, local government units, and research institutions, (i) enforce compliance with phytosanitary standards, and (j) investment in plant health innovations. Some of these actions may have already been implemented (or continuously) by various organizations and regulatory bodies. For instance, the NPQSD has now allowed clients and stakeholders to lodge applications and process the SPSIC through an online system, speeding up the processing of phytosanitary certificates. The country has also participated in regional trainings and workshops to develop programs and reference materials against plant diseases (Portales et al. 2023).

## ACKNOWLEDGMENT

The authors thank Darwin Landicho for his technical assistance.

## DISCLAIMER

The authors' information, perspectives, and opinions expressed in this chapter do not necessarily reflect those of their institution, affiliation, and the government they represent. The contents of this chapter are not also the official position of the organization(s) mentioned unless otherwise indicated and supported with a published reference, e.g., administrative orders, information from official websites, and official reports.

## REFERENCES


Abiko, K., Kishi, K., & Yoshioka, A. (1977). Occurrence of oxycarboxin-tolerant isolates of Puccinia horiana P. Hennings in Japan. *Japanese Journal of Phytopathology*, 43(2), 145–150.

Altamirano, D. M., Gonzales, C. I., & Vinas, R. C. (1976). Analysis of the devastation of leaf mottling (Greening) disease of citrus and[MS1] its control program in the Philippines. In *International Organization of Citrus Virologists Conference Proceedings (1957–2010)* (Vol. 7, No. 7).

Amir, M. (1986). Udbatta disease in Indonesia. *Buletin Penelitian Balai Penelitian Tanaman Pangan Bogor*, 4, 23–34.

Aumentado, H. D., Aguilar, C. H., & Balendres, M. A. (2023a, in press). Natural infection of *Podosphaera xanthii* in *Solanum melongena* and tolerance of two eggplant genotypes to powdery mildew. *Indian Phytopathology.*

Aumentado, H. D., & Balendres, M. A. (2023b, in press). Identification of *Epicoccum poaceicola* causing eggplant leaf spot and its cross-infection potential to other solanaceous vegetable crops. *Archives of Phytopathology and Plant Protection.*

Aumentado, H. D., & Balendres, M. A. (2022). Molecular identification of Podosphaera xanthii and the susceptibility of Vigna species genotypes to natural infection of powdery mildew. *Malaysian Journal of Fundamental and Applied Sciences*, 18(6), 684–692.

Aumentado, H. D., & Balendres, M. A. (2022a). Pathogenicity of *Corynespora cassiicola* from eggplant to various plant species. *Journal of Phytopathology*, 170, 764–769.

Aumentado, H. D. R., & Balendres, M. A. O. (2022b). Identification of *Paramyrothecium foliicola* causing crater rot in eggplant and its potential hosts under controlled conditions. *Journal of Phytopathology*, 170(3), 148–151.

Baker, K. F., Dimock, A. W., & Davis, L. H. (1949). Life history and control of the Ascochyta ray blight of Chrysanthemum. *Phytopathology*, 39(10), 789–805.

Balendres, M. A. (2022). Plant diseases caused by fungi in the Philippines. In Guerrero, J., Dalisay, T., De Leon, M, Balendres, M. A., Notarte, K. I., Dela Cruz, T. (eds), *Mycology in the Tropics: Updates on Philippine Fungi*. Elsevier: San Diego, CA.

Balendres, M. A. O. (2023). Plant diseases caused[MS1] by fungi in the Philippines. *Mycology in the Tropics*, 163–188.

Balendres, M., Taguiam, J. D., Evallo, E., & Sison, M. L. (2021). Premature defoliation in Zinnia sp. and mycoparasitism of Ampelomyces Quisqualis against the powdery mildew pathogen Golovinomyces cichoracearum from the Philippines. *Indian Phytopathology*, 74(1), 283–284.

Bassimba, D. D. M., Mira, J. L., & Vicent, A. (2013). First report of Alternaria japonica causing black spot of turnip in Spain. *Plant Disease*, 97(11), 1505–1505.

Ben Othmen, S., Morán, F. E., Navarro, I., Barbé, S., Martínez, C., Marco-Noales, E., & López, M. M. (2018). 'Candidatus Liberibacter solanacearum' haplotypes D and E in carrot plants and seeds in Tunisia. *Journal of Plant Pathology*, 100(2), 197–207.

Bextine, B., Arp, A., Flores, E., Aguilar, E., Lastrea, L., Gomez, F. S.,....., & Rueda, A. (2013). First report of zebra chip and 'Candidatus Liberibacter solanacearum' on potatoes in Nicaragua. *Plant Disease*, 97(8), 1109–1109.

Bi, C., & Hu, Q. (2005). Integrated control technique of potato wart disease in Liupanshui of Guizhou Province. *Chinese Potato Journal*, 19(6), 369–370.

Blaustein, R. A., Lorca, G. L., & Teplitski, M. (2018). Challenges for managing Candidatus Liberibacter spp. (Huanglongbing disease). *Phytopathology*, 108(4), 424–435.

Booth, C. (1979). CMI Descriptions of Pathogenic Fungi and Bacteria no. 640. CAB International, Wallingford.

Bové, J. M., Danet, J. L., Bananej, K., Hassanzadeh, N., Taghizadeh, M., Salehi, M., & Garnier, M. (2000). Witches' broom disease of lime (WBDL) in Iran. In *International Organization of Citrus Virologists Conference Proceedings (1957–2010)*, 14(14).

Bowden, J., Gregory, P. H., & Johnson, C. G. (1971). Possible wind transport of coffee leaf rust across the Atlantic Ocean. *Nature*, 229, 500–501.

CAB, U. K. (2014). Stromatinia cepivora ((Berk.) Whetzel), white rot of onion and garlic [pest/pathogen]. Stromatinia cepivora ((Berk.) Whetzel), White Rot of Onion and Garlic.[pest/pathogen]. (AQB CPC record).

Caicedo, J., Crizón, M., Pozo, A., Cevallos, A., Simbaña, L., Rivera, L., & Arahana, V. (2015). First report of 'Candidatus Phytoplasma aurantifolia' (16SrII) associated with potato purple top in San Gabriel-Carchi, Ecuador. *New Disease Reports*, 32(1), 20.

Catley, A. (1987). Outbreaks and new records. Australia. Outbreak of chrysanthemum white rust in Australia. *FAO Plant Protection Bulletin*, 35(3), 99.

Chaerani, C. (2022). Plant parasitic nematodes in agricultural ecosystem of Indonesia. *Jurnal Perlindungan Tanaman Indonesia*, 26(1), 1–12.

Chang, S. C., & Lai, W. Y. (1970). Note on a Corn disease newly found in Taiwan. *Report of the Corn Research Center, Tainan Dais*, 8, 17–23.

Chen, C. H., Wang, T. C., & Seo, M. J. (2009). First report of soybean pod and stem blight caused by Diaporthe phaseolorum var. sojae in Taiwan. *Plant Disease*, 93(2), 202–202.

Cho, W. D., & Shin, H. D. (eds) (2004). *List of Plant Diseases in Korea*. Fourth ed. Korean Society of Plant Pathology, 779 pp.

Chu, X. L., Yang, B., Gao, L., Li, H. L., Hu, L., & Yang, N. (2010). Species diversity of cultivable bacteria isolated from the roots of Cymbidium faberi Rolfe. *Journal of Wuhan Botanical Research*, 28(2), 199–205.

Coles, R. B., & Wicks, T. J. (2003). The incidence of Alternaria radicina on carrot seeds, seedlings and roots in South Australia. *Australasian Plant Pathology*, 32(1), 99–104.

Cumagun, C. J. R., McDonald, B. A., Arakawa, M., Castroagudín, V. L., Sebbenn, A. M., & Ceresini, P. C. (2020). Population genetic structure of the sheath blight pathogen Rhizoctonia solani AG-1 IA from rice fields in China, Japan and the Philippines. *Acta Scientiarum:. Agronomy*, 42, 1–11.

Das, T. P. M., Anith, K. N., & Jayarajan, M. (2000). New weed host of Ephelis oryzae, the causal organism of Udbatta disease of rice from Kerala. *Indian Phytopathology*, 53(2), 234.

Department of Agriculture. (2023). 2022 DA annual report. Department of Agriculture, Quezon City, Philippines.

Department for Environment, Food, and Rural Affairs. (2023). Plant biosecurity strategy for Great Britain (2023 to 2028). Department for Environment, Food, and Rural Affairs.

Dimayacyac, D. A., & Balendres, M. A. (2022). First report of Colletotrichum nymphaeae causing post-harvest anthracnose of tomato in the Philippines. *New Disease Reports*, 46(2), e12125.

Dolores, L. M., Langres, J. A., Pinili, M. S., Caasi-Lit, M. T., Cortaga, C. Q., Retuta, Y. M., & Cueva, F. M. D. (2023). Incidence, distribution, and genetic diversity of 'Candidatus Phytoplasma luffae'-related strain (16SrVIII) associated with the cassava witches' broom (CWB) disease in the Philippines. *Crop Protection*, 169, 106244.

Eaton, M. J., Edwards, S., Inocencio, H. A., Machado, F. J., Nuckles, E. M., Farman, M., … Vaillancourt, L. J. (2021). Diversity and cross-infection potential of Colletotrichum causing fruit rots in mixed-fruit orchards in Kentucky. *Plant Disease*, 105(4), 1115–1128. Food and Agriculture Organization of the United Nations (2023) Transboundary Plant Pests and Diseases. https://www.fao.org/transboundary-plant-pests-diseases/en on 29 September 2023.

EPPO Global Database. (2023). European and Mediterranean Plant Protection Organization. https://gd.eppo.int, accessed on August 26, 2023.

Evallo, E., Taguiam, J. D., & Balendres, M. A. (2022). Colletotrichum fructicola associated with fruit anthracnose of persimmon. *Journal of Phytopathology*, 170(3), 194–201.

Evallo, E. S., Taguiam, J. D., Posada, I. B., & Balendres, M. A. (2023). Two additional *Colletotrichum* species causing leaf spot of Rambutan (*Nephelium lappaceum*). *Archives of Phytopathology and Plant Pathology*, 56(5), 349–362.

Farr, D. F., & Rossman, A. Y. (2023). *Fungal Databases; U.S.* Washington, DC: National Fungus Collections, USDA.

Firman, I. D., & Martin, P. H. (1968). White rust of chrysanthemums. *Annals of Applied Biology*, 62(3), 429–442.

Firrao, G., Carraro, L., Gobbi, E., & Locci, R. (1996). Molecular characterization of a Phytoplasma causing phyllody in clover and other herbaceous hosts in northern Italy. *European Journal of Plant Pathology*, 102(9), 817–822.

Freeman, S., & Shabi, E. (1996). Cross-infection of subtropical and temperate fruits byColletotrichumspecies from various hosts. *Physiological and Molecular Plant Pathology*, 49(6), 395–404.

Gilardi, G., Demarchi, S., Ortu, G., Gullino, M. L., & Garibaldi, A. (2015). Occurrence of Alternaria japonica on seeds of wild and cultivated rocket. *Journal of Phytopathology*, 163(5), 419–422.

Haapalainen, M., Latvala, S., Rastas, M., Wang, J., Hannukkala, A., Pirhonen, M., & Nissinen, A. I. (2018). Carrot pathogen 'Candidatus Liberibacter solanacearum' haplotype C detected in symptomless potato plants in Finland. *Potato Research*, 61(1), 31–50.

Hansen, A. K., Trumble, J. T., Stouthamer, R., & Paine, T. D. (2008). A new huanglongbing species,"Candidatus Liberibacter psyllaurous," found to infect tomato and potato, is vectored by the psyllid Bactericera cockerelli (Sulc). *Applied and Environmental Microbiology*, 74(18), 5862–5865.

Herradura, L. E., Magnaye, L. V., & Bajet, N. B. (2001). Occurrence of papaya ringspot potyvrirus in Mindanao. *Journal of Tropical Plant Pathology*, 37, 52–58.

Hickman, C. J. (1941). The red core root disease of the strawberry caused by Phytophthora fragariae n. sp. *Journal of Pomology and Horticultural Science*, 18(2), 89–118.

Hinutan E. (2023). Banana disease forces planters to shift to corn farming. *Newsline Philippines.* https://newsline.ph/top-stories/2023/02/28/banana-disease-forces-planters-to-shift-to-corn-farming/#:~:text=DAVAO%20CITY%20—–%20Some%20150%20banana,their%20farms%20since%20last%20year, accessed on March 8, 2024.

HonetŜlegrová, J. F., Vibio, M., & Bertaccinc, A. (1996). Electron microscopy and molecular identification of phytoplasmas associated with strawberry green petals in the Czech Republic. *European Journal of Plant Pathology*, 102(9), 831–835.

Hsu, S. T., Chang, R. J., & Tzeng, K. C. (1986). Bacterial ring rot of potato in Taiwan. *Plant Protection Bulletin(Taiwan)*, 28(2), 203–211.

Huan, P. E., Hui, L. I., Li, G. A., Jiang, R, Li, G. K., Gao, H. F., Wei, W. U., Jun, W. A., Zhang, Y., Huang, W. K., & Kong, L. A. (2022). Identification of *Heterodera schachtii* on sugar beet in Xinjiang Uygur Autonomous Region of China. *Journal of Integrative Agriculture*, 21(6), 1694–1702.

Huang, H. C., Hsieh, T. F., & Erickson, R. S. (2003). Biology and Epidemiology of Erwinia rhapontici, Causal Agent of Pink Seed and Crown Rot of Plants. *Plant Pathology Bulletin*, 12, 69–76.

IPPC. (2023). Aster yellows Phytoplasma on Grapevine. International Plant Protection Convention. https://www.ippc.int/en/countries/south-africa/pestreports/2008/05/aster-yellows-phytoplasma-on-grapevine/, accessed on August 26, 2023.

IPPC Secretariat. (2021). Scientific review of the impact of climate change on plant pests – A global challenge to prevent and mitigate plant pest risks in agriculture, forestry and ecosystems. FAO on behalf of the IPPC Secretariat. https://doi.org/10.4060/cb4769en.

Jee, H. J., Han, S. S., & Kweon, J. H. (2002). A simple method for sporangial formation of the rice downy mildew pathogen, Sclerophthora macrospora. *The Plant Pathology Journal*, 18(2), 77–80.

Kennedy, B. W., & King, T. H. (1962). Angular leafspot of Strawberry caused by Xanthomonas fragariae sp. nov. *Phytopathology*, 52, 873–875.

Khair, G. T. (1981). *List of Plant Parasitic Nematodes of Australia*. Australian Plant Quarantine Service, 46–47.Canberra, Australia.

Khan, A., & Tareen, J. K. (2012). Plant parasitic nematodes associated with cherry (Prunus avius L.) in Balochistan, Pakistan. *International Journal of Biology and Biotechnology*, 9(3), 293–294.

Kim, D. R., Gang, G. H., Jeon, C. W., Kang, N. J., Lee, S. W., & Kwak, Y. S. (2016). Epidemiology and control of strawberry bacterial angular leaf spot disease caused by Xanthomonas fragariae. *The Plant Pathology Journal*, 32(4), 290–299.

Kim, D. K., Shim, C. K., Bae, D. W., Lee, S. C., & Kim, H. K. (2001). Occurrence of blossom blight of Chrysanthemum boreale caused by Didymella chrysanthemi. *The Plant Pathology Journal*, 17(6), 347–349.

Kim, W. G., Lee, G. B., Shim, H. S., & Cho, W. D. (2023). White rot of Korean wild chive caused by Stromatinia cepivora. *Research in Plant Disease*, 29(2), 185.

Kmetz, K. T., Schmitthenner, A. F., & Ellett, C. W. (1978). Soybean seed decay: Prevalence of infection and symptom expression caused by Phomopsis sp., Diaporthe phaseolorum var. sojae, and D. phaseolorum var. caulivora. *Phytopathology*, 68(6), 836–840.

Lacorte, G., & Quiros, J. (2011). 1,000 ha of Mindanao banana farms ruined. Inquirer. https://newsinfo.inquirer.net/98657/1000-ha-of-mindanao-banana-farms-ruined, accessed on August 29, 2023.

Laun, N., & Zinkernagel, V. (1997). A comparison of the resistance against Phytophthora fragariae var. rubi, the causal agent of a raspberry root rot. *Journal of Phytopathology*, 145(5–6), 197–204.

Lebeda, A., & Mieslerová, B. (2011). Taxonomy, distribution and biology of lettuce powdery mildew (Golovinomyces cichoracearum sensu stricto). *Plant Pathology*, 60(3), 400–415.

Lebeda, A., Krístková, E., Sedláková, B., Coffey, M. D., & McCreight, J. D. (2011). Gaps and perspectives of pathotype and race determination in Golovinomyces cichoracearum and Podosphaera xanthii. *Mycoscience*, 52(3), 159–164.

Lee, H. A., & Medalla, M. G. (1921). Leaf stripe disease of sugar cane in the Philippines. *Science*, 54(1395), 274–275.

Lee, I. M., Gundersen-Rindal, D. E., Davis, R. E., Bottner, K. D., Marcone, C., & Seemüller, E. (2004). 'Candidatus Phytoplasma asteris', a novel Phytoplasma taxon associated with aster yellows and related diseases. *International Journal of Systematic and Evolutionary Microbiology*, 54(4), 1037–1048.

Leu, L. S., Kao, C. W., Yang, H. C., & Lin, H. S. (1982). Puccinia horiana: Occurrence in Taiwan, release and germination of sporidia, fungicide trial and screening for chrysanthemum resistance to white rust. *Plant Protection Bulletin, Taiwan*, 24(1), 9–18.

Liefting, L. W., Weir, B. S., Pennycook, S. R., & Clover, G. R. (2009). 'Candidatus Liberibacter solanacearum', associated with plants in the family Solanaceae. *International Journal of Systematic and Evolutionary Microbiology*, 59(9), 2274–2276.

Lipps, S. M., & Samac, D. A. (2022). Pseudomonas viridiflava: An internal outsider of the Pseudomonas syringae species complex. *Molecular Plant Pathology*, 23(1), 3–15.

Liu, H. L. (1990). Bacterial wilt of Gypsophila paniculata caused by Pseudomonas caryophylli. *Bulletin of Taichung District Agricultural Improvement Station*, 28, 33–42.

Lohsomboon, P., Manoch, L., Visarathanonth, N., Kakishima, M., Ono, Y., & Sato, S. (1986). Materials for the rust flora in Thailand II. *Transactions of the Mycological Society of Japan*, 27(3), 271–281.

Lumawag, R. J. (2019). DA: Panama disease-hit areas may double. Sunstar Davao. https://www.sunstar.com.ph/article/1823546/davao/business/da-panama-disease-hit-areas-may-double, accessed on August 29, 2023.

Magarey, R. C. (2006). Disease incidence: Regional variation in Queensland. In *Proceedings of the 2006 Conference of the Australian Society of Sugar Cane Technologists Held at Mackay, Queensland, Australia, 2-5 May 2006* (pp. 219–225). Australian Society of Sugar Cane Technologists.

Mahmodi, F., Kadir, J. B., Puteh, A., Wong, M. Y., & Nasehi, A. (2013). First report of pod and stem blight of lima bean caused by Diaporthe phaseolorum var. sojae in Malaysia. *Plant Disease*, 97(2), 287–287.

Marcuzzo, L. L., & Schmoeller, J. (2017). Survival and viability of sclerotia from Sclerotium cepivorum in soil. *Summa Phytopathologica*, 43(2), 161–163.

Meier, F. C., Drechsler, C., & Eddy, E. D. (1922). Black rot of carrots caused by Alternaria radicina n. sp. *Phytopathology*, 12(4), 157–166.

Molina, A., Fabregar, E., Sinohin, V., Fourie, G., & Viljoen, A. (2008). Tropical race 4 of *Fusarium oxysporum* f. sp *cubense* causing new Panama wilt epidemics in Cavendish varieties in the Philippines, Phytopathology. Amer. phytopath. Soc. 3340 Pilot Knob Road, St Paul, Mn 55121 USA p S108.

Molina, A. B., Fabregar, E. G., Soquita, R. O., & Sinohin, V. G. O. (2011). Comparison of host reaction to Fusarium oxysporum f. sp. cubense tropical Race 4 and agronomic performance of somaclonal variant 'GCTCV-119' (AAA, Cavendish) and 'grand naine' (AAA, Cavendish) in commercial farms in the Philippines, 897 ed. International Society for Horticultural Science (ISHS), Leuven, Belgium, 399–402.

Morita, H. (1965). Red stele root disease of strawberry caused by Phytophthora fragariae I. Identification of the causal fungus. *Japanese Journal of Phytopathology*, 30(5), 239–245.

Murty, V. S. T., & Alimeer, S. H. (1978). Occurrence of 'Udbatta' disease of rice in Chhattisgarh region. *Oryza*, 15(1), 92.

NPQSD. (2023a). List of allowable fresh fruits and vegetables. BPI-National Plant Quarantine Services Division. https://npqsd.bpi-npqsd.com.ph/import/, accessed on August 23, 2023.

NPQSD. (2023b). List of allowable plants/planting materials to import. BPI-National Plant Quarantine Services Division. https://npqsd.bpi-npqsd.com.ph/import/, accessed on August 23, 2023.

NPQSD. (2023c). How to import. BPI-National Plant Quarantine Services Division. https://npqsd.bpi-npqsd.com.ph/import/, accessed on August 25, 2023.

NPQSD. (2023d). How to import. BPI-National Plant Quarantine Services Division. https://npqsd.bpi-npqsd.com.ph/domestic/, accessed on August 25, 2023.

Onozaki, T., Ikeda, H., Yamaguchi, T., Himeno, M., Amano, M., & Shibata, M. (2002). 'Carnation Nou No. 1', a carnation breeding line resistant to bacterial wilt (Burkholderia caryophylli). *Horticultural Research (Japan)*, 1(1), 13–16.

Palicte, C. (2023). Davao Norte banana growers shift to corn due to Fusarium wilt. Philippine News Agency. https://www.pna.gov.ph/articles/1195852, accessed on August 29, 2023.

Pardo, J. M., Chittarath, K., Vongphachanh, P., Hang, L. T., Oeurn, S., Arinaitwe, W., … Cuellar, W. J. (2023). Cassava witches' broom disease in Southeast Asia: A review of its distribution and associated symptoms. *Plants*, 12(11), 2217.

Pattison, A., Limbaga, C., Gervacio, T., Notarte, A., Juruena, M., et al. (2020). Final report: Integrated Management of Fusarium Wilt of Bananas in the Philippines and Australia. Australian Centre for International Agricultural Research (ACIAR), 5–51.

Pedersen, P., & Grau, C. R. (2010). Effect of agronomic practices and soybean growth stage on the colonization of basal stems and taproots by Diaporthe phaseolorum var. sojae. *Crop Science*, 50(2), 718–722.

Peregrine, W. T. H., & Ahmad, K. B. (1982). *Phytopathological Papers*, 27, 1–87.

Phoulivong, S., McKenzie, E. H. C., & Hyde, K. D. (2012). Cross infection of *Colletotrichum* species; a case study with tropical fruits. *Current Research in Environmental & Applied Mycology*, 2(2), 99–11.

Pitman, A. R., Drayton, G. M., Kraberger, S. J., Genet, R. A., & Scott, I. A. (2011). Tuber transmission of 'Candidatus Liberibacter solanacearum'and its association with zebra chip on potato in New Zealand. *European Journal of Plant Pathology*, 129(3), 389–398.

Ploetz, R. C. (2021). Gone bananas? Current and future impact of fusarium wilt on production. In Scott, P., Strange, R., Korsten, L., & Gullino, M. L. (Eds.). *Plant Diseases and Food Security in the 21st Century* (pp. 21–32). Cham: Springer International Publishing.

Portales, L. A., Yago, J. I., & Dimayacyac, A. C. (2023). Fungal plant pathogens of quarantine importance in the Philippines. In *Mycology in the Tropics* (pp. 325–342). Academic Press.

Prasanna, B. M., Carvajal-Yepes, M., Kumar, P. L. et al. (2022). Sustainable management of transboundary pests requires holistic and inclusive solutions. *Food Security*, 14(6), 1449–1457.

Przetakiewicz, J. (2015). The viability of winter sporangia of Synchytrium endobioticum (Schilb.) Perc. from Poland. *American Journal of Potato Research*, 92(6), 704–708.

Putnam, M. L., & Miller, M. L. (2007). Rhodococcus fascians in herbaceous perennials. *Plant Disease*, 91(9), 1064–1076.

Quimio, T. H., & Nadayao, M. M. (1980). Powdery mildew of tomato caused by *Erysiphe cichoracearum* DC. Ex. Mecat. Philippine. *Phytopathology*, 16, 67–70.

Randles, J. W., Boccardo, G., & Imperial, J. S. (1980). Detection of the cadang-cadang RNA in African oil palm and buri palm. *Phytopathlogy*, 70, 185–189.

Ristaino, J. B. (2021). Potatoes, citrus and coffee under threat. In Scott, P., Strange, R., Korsten, L., & Gullino, M. L. (Eds.). *Plant Diseases and Food Security in the 21st Century* (pp. 3–19). Cham: Springer International Publishing.

Romero García, A., & Romero Cova, S. (1990). Chrysanthemum (Chrysanthemum morifolium Ram.) blight caused by Ascochyta chrysanthemi Stev. *Revista Chapingo*, 15(67–68), 38–43.

Ryley, M. J., Tan, Y. P., Kruse, J., Thines, M., & Shivas, R. G. (2022). More than meets the eye—Unexpected diversity in downy mildews (oomycetes) on grasses in Australia. *Mycological Progress*, 21(1), 297–310.

Salehi, M., Izadpanah, K., Siampour, M., Bagheri, A., & Faghihi, S. M. (2007). Transmission of 'Candidatus Phytoplasma aurantifolia' to Bakraee (Citrus reticulata hybrid) by feral Hishimonus phycitis leafhoppers in Iran. *Plant Disease*, 91(4), 466–466.

Schroeder, B. K., Lupien, S. L., & Dugan, F. M. (2002). First report of pink seed of pea caused by Erwinia rhapontici in the United States. *Plant Disease*, 86(2), 188–188.

Scott, D. J., & Wenham, H. T. (1973). Occurrence of two seed-borne pathogens, Alternaria radicina and Alternaria dauci, on imported carrot seed in New Zealand. *New Zealand Journal of Agricultural Research*, 16(2), 247–250.

Sharma, S., & Chakrabarti, S. K. (2020). Present status of wart disease of potato in Darjeeling and Kalimping districts of West Bengal. *Potato Journal*, 47(1), 41–45.

Soteros, I. J. (1979). Pathogenicity and control of Alternaria radicina and A. dauci in carrots. *New Zealand Journal of Agricultural Research*, 22(1), 191–196.

Sta Cruz, F. C., Tañada, J. M., Elvira, P. R. V., Dolores, L. M., Magdalita, P. M., Hautea, D. M., & Hautea, R. A. (2009). Detection of mixed virus infection with papaya ringspot virus (PRSV) in papaya (Carica papaya L.) grown in Luzon, Philippines. *Philippine Journal of Crop Science*, 34, 62–74.

Strange, R. N., & Scott, P. R. (2005). Plant disease: A threat to global food security. *Annual Review of Phytopathology*, 43, 83–116.

Tangonan, N. G. (1999). Host index of plant diseases in the Philippines (No. Ed. 3). Department of Agriculture-Philippine Rice Research Institute (DA-PHILRICE).

Tsuji, M., Nagasaka, A., & Kadota, I. (2019). Causal agent of bacterial rot of onion (Allium cepa) bulbs in the Tohoku region of Japan. *Japanese Journal of Phytopathology*, 85(3), 205–210.

van de Vossenberg, B. T., Prodhomme, C., Vossen, J. H., & van der Lee, T. A. (2022). Synchytrium endobioticum, the potato wart disease pathogen. *Molecular Plant Pathology*, 23(4), 461–474.

Whipps, J. M. (1993). A review of white rust (Puccinia horiana Henn.) disease on chrysanthemum and the potential for its biological control with Verticillium lecanii (Zimm.) Viégas. *Annals of Applied Biology*, 122(1), 173–187.

Wilkie, J. P., Dye, D. W., & Watson, D. R. W. (1973). Further hosts of Pseudomonas viridiflava. *New Zealand Journal of Agricultural Research*, 16(3), 315–323.

Zehra, R., Moin, S., Enamullah, S. M., & Ehteshamul-Haque, S. (2022). Detection and identification of quarantine bacteria and fungi associated with imported and local potato seed tubers. *Pakistan Journal of Botany*, 54(3), 1157–1161.

Zelanzy, B., RAndles, J. W., Boccardo, G., & Imperal, J. S. (1982). The viroid nature of the cadang-cadang disease of coconut palm. *Scientifia Filipinas*, 2, 45–63.

Zhang, J., He, Y., Ahmed, W., Wan, X., Wei, L., & Ji, G. (2022). First report of bacterial angular leaf spot of strawberry caused by Xanthomonas fragariae in Yunnan Province, China. *Plant Disease*, 106(7), 1978.

Zhang, B.-C., & Huang, Y.-C. (1990). A list of important plant diseases in China. *Review of Plant Pathology*, 69(3), 97–118.

Zhuang, W. Y. (ed) (2001). *Higher Fungi of Tropical China*. Ithaca: Mycotaxon, Ltd., 485.

# 10 Biosafety and Biosecurity in Laboratory Animal Facilities

*Rohani Cena-Navarro*
National Institutes of Health
University of the Philippines
Manila, Philippines

*Melissa Marie R. Rondina*
College of Veterinary Medicine
University of the Philippines
Los Baños

*Jan Irving A. Bibay*
Biological Resource Centre (BRC)
Agency for Science, Technology and Research (A*STAR)
Singapore

## 10.1 INTRODUCTION

### 10.1.1 Animal Biosafety

Animal biosafety is a set of preventive measures to minimize the risk involving laboratory animals. Low-middle-income countries (LMICs) may experience challenges implementing biorisk management in animal laboratories due to a lack of resources to comply with regulatory requirements. However, other mitigation strategies can be implemented to minimize the risk without compromising safety. This chapter provides an overview of Laboratory Animal Biosafety that may be applied to resource-limited facilities without compromising the safety of the personnel, laboratory animals, and the environment.

## 10.2 RISK ASSESSMENT

Working with laboratory animals poses different risks to humans, animal handlers, researchers, and the environment. People in the community and other animals may also be at risk if an animal escape occurs. Careful considerations must be made in order to maximize resources, without compromising safety. Therefore, a thorough risk assessment is necessary to make an informed decision on the mitigation controls

DOI: 10.1201/9781003426219-10

that will be employed to minimize the risk. Identification of hazards is an essential component when performing a risk assessment. When dealing with laboratory animals, hazards may be physical, zoonotic diseases, sharps, allergens, vectors, escape, or biosecurity. Being aware of the hazards is important to reduce the risk of injury or harm.

Physical hazards include injuries such as potential bites and scratches from laboratory animals. Bites should be treated as serious incidents because of the possible transmission of diseases from the contaminated flora such as *E.coli* and *Salmonella*, among others. Therefore, understanding animal behavior and proper training in animal handling are required before working with animals to prevent physical injury. Intense noise from equipment movement, cage washing, and animal noise may also pose a risk to the personnel. These hazards are often neglected but can pose a serious risk to humans if not mitigated.

Sharps are one of the most common hazards found in the laboratory. In particular, animals' claws and teeth may injure the handler during manipulation. Other common sharps include edges of animal cages and surgical or necropsy instruments. Animal facilities conduct necropsy on a regular basis, and appropriate measures to prevent cuts should be considered. Another source of hazard is the allergens such as dander, fur, dust, and bedding. Activities like bedding change and animal handling will expose the staff to potential allergens and can be dangerous to the handler. Some staff with pre-existing respiratory conditions must undergo health screening before working at the animal facility and inform the occupational health and safety committee.

The presence of vectors like insects and parasites is often found in any laboratory. The type of vectors present should be identified to determine the appropriate pest control measures. Examples of pest measures are the use of insecticides, pesticides, or manual removal. Standard operating procedures should include pest control management, and the facility should be aware of potential exposure of laboratory animals to chemicals when doing this activity. Animal escape is another important consideration in risk assessment. When an infected animal escapes from the laboratory, the people, animals, and the environment may be at risk of contamination by pathogens carried by the laboratory animal. Mitigation controls to prevent the escape of animals should be considered in the laboratory design and standard operating procedures.

Another consideration when performing a risk assessment is biosecurity. Animal research that will work on high-risk pathogens may require additional biosecurity measures before implementation. Access control, standard operating procedures, and combined containment design are considerations in biosecurity to prevent the accidental or deliberate release of pathogens. Activities in the laboratory must be thoroughly assessed before experimentation. Additional requirements, as required by local regulations, may be required prior approval. Researchers should work closely with the animal facility manager, biosafety officer, and other stakeholders to ensure safe and secure implementation of their research.

### 10.2.1 Risk Classification Factors

When performing risk assessment, several factors that would influence the risk assessment are important to select an appropriate mitigation control. One consideration is the type of animal species which would determine the level of risk that the animal will pose. For example, smaller laboratory animals like rodents and rabbits would pose a lower risk than larger animals like farm animals. Small-sized laboratory animals are relatively easier to handle as compared to medium-larger animals. The handling experience of the worker will also contribute to lowering the risk. For instance, workers with no prior experience in animal handling may be at increased risk of animal bites. Appropriate animal handling training should be required before handling animals.

The virulence and pathogenicity of the agent/s will also contribute to the severity of the risk. Organisms with increased virulence will increase risk as compared to the handling of low-virulent organisms. Information about specific details of some pathogens handled in the animal laboratory may be available in the Pathogen Safety Data Sheet and should be included in the risk assessment.

The route of transmission of the pathogen is one of the considerations in performing a risk assessment. Understanding the transmission will help determine the type of mitigation control. For example, agents that are transmitted via inhalation, such as influenza would require the wearing of personal protective equipment (PPE), such as respiratory protection. Transmission via percutaneous or via skin would require the wearing of a laboratory gown and gloves. Pathogens that would require a vector like dengue for transmission will need more information about the pathogen life cycle and should be included in the risk assessment.

The pathogen stability, infectious dose, and availability of data are also important considerations in risk assessment. Pathogens that are more stable would require additional steps for decontamination or longer contact time. Another consideration is the conditions that will allow the pathogen's prolonged survival such as environmental stability in a given temperature. The infectious dose, shedding patterns, and endemicity of an organism that would cause disease should be determined during the risk assessment, as some organisms may require a lower dose to become infectious. Local data on the morbidity and mortality associated with the pathogen should be collected if possible. Pathogens with high morbidity and mortality would require layers of risk mitigation as compared to low-pathogenic organisms. In effect, additional layers of protection would entail additional costs to reduce risk exposure. Some agents that are foreign to the region may have limited availability of vaccines or treatment. Agents that are endemic may or may not lower the risk depending on the type of activity. Active control or eradication programs for the disease should be considered as well as surveillance testing.

Another consideration in the risk assessment is the availability of effective prophylaxis, treatment, or vaccines against diseases. These measures will help reduce the consequence of risk in case of exposure. Diseases with no available treatment or vaccine may require an additional layer of protection. The type of activity or experiment also contributed to the decision-making during risk assessment. For example, experiments with short duration may have a lower risk compared to longer-duration experiments as the risk of exposure is higher for longer-duration activities. Activities involving aerosols, and needles, may require additional personal protective equipment and influence the assessment of risk. The type of housing of animals should also be considered when performing risk assessments. For example, filter-top cages provide protection for animal workers by preventing the dispersal of allergens as compared to conventional or open cages. However, filter-top cages are more expensive than conventional, therefore a thorough risk assessment is needed to determine the necessity of such specialized equipment.

## 10.3 DISEASES IN LAB ANIMALS

Working in the animal facility could potentially expose the workers to diseases that are transmitted from animals. Zoonoses are diseases of animals that may be secondarily transmitted to man and the pathogens are often asymptomatic in the animal. Therefore, the potential transmission of diseases should be considered when performing a risk assessment. Workers should become familiar with zoonotic organisms often associated with specific animal subjects. Table 10.1 shows some zoonotic agents associated with animal usage.

## 10.4 RISK MANAGEMENT

The primary goal of risk management is to minimize potential harm and enhance opportunities. This involves implementing strategies like containment for biohazards,

**TABLE 10.1**
**List of Zoonotic Organisms Often Associated with Specific Animal Subjects**

| Type | Example | Transmission |
|---|---|---|
| Bacteria | *Salmonella typhi (dogs)*<br>*Salmonella enteritidis*<br>*Coxiella burnetti*<br>*Leptospirosis* | Ingestion of contaminated feed ingredients and water and by contact with contaminated bedding and animal facility personnel |
| Virus | *Herpes B Virus (macaques)*<br>Hanta virus | Animal bites, scratches, exposures to the tissues or secretions of macaques, or by mucosal contact with monkey body fluid or tissue |
| Fungi | *Histoplasma capsulatum (birds)* | Inhalation of dust contaminated with fungus |
| Parasites | *Toxoplasma gondii (cats)* | Ingestion (feco-oral) |
| Rickettsia | *Coxiella burnetti (sheep)* | Inhalation of contaminated aerosol |

elimination or control of hazards, occupational health programs, and oversight from institutional committees. These measures collectively mitigate risks and promote safety.

In low-income countries, where resources might be limited, the importance of risk management in animal research facilities cannot be overstated. Implementing practical and cost-effective strategies to prevent disease spread and protect human and animal health can have far-reaching positive impacts on both local communities and the global research community.

### 10.4.1 Animal Biosafety Levels (ABSL)

Animal biosafety levels (ABSL) are a framework of containment measures to protect personnel, animals, and the environment from risks associated with working with animals in the laboratory. Table 10.2 defines the classifications of animal biosafety levels. ABSL classification is determined by assessing the potential hazards of specific pathogens, activities, and transmission risks, ranging from ABSL-1 (minimal risk) to ABSL-4 (highest risk), guiding appropriate safety protocols. The different requirements, recommended practices, and procedures for each animal biosafety level are defined in Table 10.3.

## 10.5 POLICIES AND PROCEDURES

Policies and procedures play an important role in risk management by providing guidelines, standards, and a systematic framework for detecting, assessing, and managing risks. The specific rules and procedures established will be determined by the nature of the organization, its industry, and the complexity of its activities. Listed below are some important policies and practices to consider when working with laboratory animals.

**TABLE 10.2**
**Four Classifications of Animal Biosafety Levels**

| ABSL 1 | ABSL 2 | ABSL 3 | ABSL 4 |
|---|---|---|---|
| Assigned for animal work involving well-characterized agents that are not known to cause infection in healthy humans | Assigned for animal work involving agents of moderate risk that can cause human diseases primarily through ingestion or mucosal exposure | Assigned to animal work involving agents that can cause serious and potentially lethal infections in humans and be transmitted through aerosols | Assigned to animal work involving non-indigenous or exotic agents that pose a high risk to humans, causing life-threatening disease, and for which no vaccine or treatment is available |

**TABLE 10.3**
**Requirements and Recommendations for the Different Animal Biosafety Levels**

| | ABSL 1 | ABSL 2 | ABSL 3 | ABSL 4 |
|---|---|---|---|---|
| | | **Personal Protective Equipment (PPE)** | | |
| i. Protective laboratory clothing | Minimum recommendations are coats, gowns, or uniforms (which should not be worn outside the animal facility) | Minimum recommendations are coats, gowns, or uniforms (which should not be worn outside the animal facility) | a. Disposable clothing includes non-woven, olefin cover-all suits, wrap-around, or solid-front gowns<br>b. Reusable clothing can be used but should be contained and decontaminated before laundering | a. Cabinet Laboratory:<br>- Solid-front gowns or coveralls<br>- Gloves<br>b. Suit Laboratory:<br>- Full-body, air-supplied, positive pressure suit |
| ii. Gloves | Required | Required | Required (two pairs are worn when appropriate) | |
| iii. Protective eyewear (e.g., laboratory glasses, goggles) | As needed (i.e., when there are anticipated splashes of infectious or hazardous agents or if the staff has contact with NHPs) | As needed (i.e., when there are anticipated splashes of infectious or hazardous agents or if the staff has contact with NHPs) | As needed (i.e., when there are anticipated splashes of infectious or hazardous agents, if contacting NHPs, or entering rooms containing infected animals) | |
| iv. Face protection (e.g., mask, face shield, or other splatter guard) | As needed (i.e., when there are anticipated splashes of infectious or hazardous agents or if the staff has contact with NHPs) | As needed (i.e., when there are anticipated splashes of infectious or hazardous agents, if contacting NHPs, or entering rooms containing infected animals) | As needed (i.e., when there are anticipated splashes of infectious or hazardous agents, if contacting NHPs, or entering rooms containing infected animals) | |
| v. Respiratory protection | Recommended but not required | As needed (i.e., when entering rooms containing infected animals) | Appropriate respiratory protection is required, and staff are enrolled in a respiratory protection program | |

(*Continued*)

**TABLE 10.3 (CONTINUED)**
**Requirements and Recommendations for the Different Animal Biosafety Levels**

| | ABSL 1 | ABSL 2 | ABSL 3 | ABSL 4 |
|---|---|---|---|---|
| | | **Primary Barriers (Safety Equipment)** | | |
| i. Containment equipment | Not generally required. Procedures can be performed on open benches | As needed (i.e., procedures involving infectious materials or if aerosols and splashes are expected)<br>* Containment should be appropriate for the animal's special procedure | ABSL 2 containment plus containment equipment for housing and cage dumping activities<br>*Biological materials that require containment are placed in a durable, leak-proof, sealed container and enclosed in a non-breakable, sealed container prior to removal from the laboratory | ABSL 3 containment plus maximum containment equipment is used for all procedures and activities |
| ii. Biological Safety Cabinets (BSC) | Not generally required | Class I, II, or III | Class I, II, or III<br>** All procedures must be conducted within BSC or physical containment devices. If not possible, a combination of PPE, administration, and engineering controls is used | Class III |

*(Continued)*

**TABLE 10.3 (CONTINUED)**
**Requirements and Recommendations for the Different Animal Biosafety Levels**

| | ABSL 1 | ABSL 2 | ABSL 3 | ABSL 4 |
|---|---|---|---|---|
| | **Secondary Barriers (Facility Design and Construction)** | | | |
| | Standard animal facility and should be separated from the general traffic patterns<br>- Access is restricted or controlled by laboratory doors<br>- Windows fitted with screens<br>- Lighting should be adequate for all activities<br>- Hand washing sink and eye wash station (recommended)<br>- Directional airflow (recommended)<br>- No recirculation of exhaust air | ABSL 1 plus<br>- Self-closing laboratory doors with locks<br>- Hand washing sink located near exit<br>- Autoclave is available to facilitate the decontamination of infectious materials and waste<br>- Vacuum lines are protected with disinfectant traps and in-line HEPA filters<br>- Mechanical cage washer (recommended)<br>- Negative airflow into animal and procedure rooms (recommended) | Has special engineering and design features ABSL 2 plus<br>- Physical separation from access corridors. Access is limited to those who need to enter<br>- Self-closing, double-door access<br>- Hands-free sink near the exit<br>- Sealed penetrations and windows<br>- Entry through the anteroom or airlock<br>- Negative airflow into animal and procedure rooms<br>- Ducted mechanical air ventilation system (required) | ABSL 3 plus<br>- Separate building or isolated zone<br>- Dedicated supply and exhaust, vacuum, and decontamination systems<br>- Dedicated non-circulating ventilation system (required)<br>- Double-door, pass-through autoclave (required)<br>- Clothing change before entry and shower on exit<br>- For the suit laboratory, entry is through an airlock with airtight doors |
| | **Other Practices** | | | |
| i. Medical Surveillance | Recommended appropriate to the species involved | Occupational medical services, including medical evaluation, surveillance, and treatment, as appropriate | Required | Required |

(*Continued*)

**TABLE 10.3 (CONTINUED)**
**Requirements and Recommendations for the Different Animal Biosafety Levels**

| | ABSL 1 | ABSL 2 | ABSL 3 | ABSL 4 |
|---|---|---|---|---|
| ii. Immunizations | - Recommended for ABSL 1, 2, and 3, and staff should be offered available immunizations for agents handled or present in the laboratory<br>- ABSL 4 agents have no vaccines or therapies available | | | |
| iii. Decontamination | Required for<br>- Wastes such as cultures, stocks, and other potentially infectious materials before disposal<br>- Work surfaces after use<br>- Any spills and splashes | ABSL 1 plus appropriate methods of decontamination such as autoclave, chemicals, incineration, etc. | ABSL 2 plus<br>- Routine decontamination of laboratory equipment after spills or splashes and before repairs, maintenance, or removal from the laboratory<br>- Decontamination of the entire laboratory when there is gross contamination of spaces, major changes in laboratory usage, renovations, or shutdown.<br>- Decontamination of clothes before laundering<br>- Decontamination of bedding before removal<br>- Provision of disinfectant foot baths as needed | ABSL 3 plus<br>- Daily inspections of essential containment and life support systems<br>- Decontamination of all wastes prior to removal from the laboratory |

*Reference: [CDC] Centers for Disease Control and Prevention. (2020). Biosafety in Microbiological and Biomedical Laboratories. 6th ed.

### 10.5.1 The Hierarchy of Controlling Hazards

#### 10.5.1.1 Elimination of the Hazard

This refers to the complete removal of the source of the hazards, such as sharp instruments, heavy equipment, toxic substances, or harmful pathogens. Since it involves completely removing the hazard from the work environment, it can lead to significant changes in the entire work process to make it safer and to improve overall work dynamics.

One common example in a laboratory animal facility is the manual handling of large animal cages and heavy equipment, which can result in ergonomic hazards and musculoskeletal injuries among laboratory animal workers. Implementing specialized equipment, such as automated cage washers, can eliminate the need for lifting and handling heavy loads, hence lowering the risk of staff injuries.

#### 10.5.1.2 Substitution of the Hazard

This refers to using safer alternatives to the source of the hazard. It is important to note that substitutes should reduce potentially harmful effects and not create new risks. An example is replacing the use of heavy equipment with lighter equipment. For instance, the traditional transport cages used to move animals within the room or between the animal holding room and procedure room can be bulky and challenging to move around. This can lead to a risk of musculoskeletal injuries. Replacing this with lightweight, maneuverable cages can reduce the risk of strain and injuries during animal transport.

#### 10.5.1.3 Engineering Control

This refers to reducing or preventing the hazards from coming into contact with the workers. This is achieved by creating protective barriers between an individual and a potential hazard or by modifying the equipment or the workplace. Also by setting the correct airflow in response to the function of a specific room (i.e., quarantine or isolation rooms) to protect adjacent rooms or the hallway from contamination.

For example, the use of heavy equipment in an animal facility, an engineering control that could be employed is to set up an automatic door locking system that can detect the presence of heavy equipment near doors and automatically lock it, preventing animals or even personnel from having accidental access to the room where the heavy equipment is operating. This system enhances animal and staff safety, ensures efficient operations, and provides flexibility where access is required.

#### 10.5.1.4 Administrative Control

This refers to employing work methods to reduce the length, frequency, or intensity of the workers' exposure to hazards. This is typically used when engineering controls appear to be impractical or unreasonably expensive. The most useful administrative controls in laboratory animal settings include creating and implementing rules and SOPs, providing personnel training, and conducting regular monitoring and maintenance.

With regard to the use of heavy equipment in the facility, creating an SOP on its proper use and providing instructions for when and where the equipment is used can help minimize interference with animal activities. Moreover, requiring training of technicians prior to handling the equipment can prevent injuries related to machine operation. Finally, ensuring regular maintenance of the equipment can prevent accidents that could arise from faulty machines.

#### 10.5.1.5 Personal Protective Equipment (PPE)

PPE should be used when hazards cannot be eliminated or reduced using other control methods. PPE is also utilized as a management strategy in laboratory animal facilities to protect the health of animals. Proper PPE should be chosen in response to the nature of the work done in the facility and the associated hazards. Refer to Table 10.2 for the required and recommended PPE for the different biosafety levels.

### 10.5.2 Occupational Health and Safety Program

The Guide for the Care and Use of Laboratory Animals (National Research Council, 2011) states that each institution must establish and maintain an occupational health and safety program (OHSP) as part of the overall program of animal care and use. The main goal of OHSP is to minimize the risk of injuries at the workplace. The nature of the OHSP will depend on the facility, research activities, hazards, and animal species involved. The following are necessary components of an effective OHSP.

#### 10.5.2.1 Hazard Identification and Risk Evaluation

A qualified person should evaluate potential dangers and select safeguards appropriate to the risks. The first step is to list the hazards related to the work and then identify the possible injuries for each hazard. After that, conduct a thorough risk evaluation.

Risk is the product of the severity of the hazard and the likelihood of its occurrence. A facility can develop its own matrix to rate both the severity of the identified injuries and the likelihood of their occurrence. The severity can be rated as negligible, minor, moderate, or significant. Meanwhile, the likelihood can be rated as rare, remote, occasional, or frequent (Clayton et al., 2019). After calculating the risk, it can then be categorized as low, medium, or high. If the risk is classified as high, no work should commence unless the risk is reduced to at least medium. Medium risks need careful evaluation and may require control measures such as administrative controls or PPE to reduce the level to low.

#### 10.5.2.2 Facilities, Equipment, and Monitoring

Washing facilities and supplies appropriate for the program's animal care and use should be made available. Facilities, machinery, and procedures should be designed, selected, and developed to ensure ergonomically sound operations.

#### 10.5.2.3 Personnel Education and Training Program

Personnel at risk should be provided with adequate information about the hazards associated with their work, and how these hazards are controlled. Continuous training and interactive mentoring should be made available to equip them with the necessary skills to carry out safe work practices proficiently. Refresher courses should also be provided to keep them updated. Therefore, good record-keeping is essential to keeping track of who needs to be trained and in what area.

#### 10.5.2.4 Personal Hygiene

All personnel must maintain a high level of personal cleanliness. Workers in animal facilities can contract diseases from animals, or they can develop allergies from animal exposure. Moreover, it is also important to protect the animals from potential pathogens coming from outside the facility. Hence, institutions must provide appropriate PPE to carry out their work. Appropriate policies must be established and enforced, like hand washing/disinfection, and changing of clothes before entering animal rooms; laundering or decontamination of soiled clothing; proper waste disposal; and the no-eating and drinking policy in animal rooms and laboratories.

#### 10.5.2.5 Animal Experimentation Involving Animals

An effective safety program should be established to ensure that the facility is adequate for the safe conduct of research. Animal protocols involving hazards should be carefully considered, including housing, waste, and carcass disposals, and the safe handling of the hazards involved. The program should have the capacity to accurately identify various hazards at the workplace, determine and impose safety and control measures, and provide necessary training and skills for staff.

#### 10.5.2.6 Medical Evaluation and Preventive Medicine for Personnel

Medical evaluation and preventative medicine programs should be developed and implemented. Implementation should include input from trained health professionals such as occupational health physicians and nurses. The preventive medicine program and evaluation should be tailored to each individual enrolled in the occupational health program.

### 10.5.3 Health Surveillance Program

To assess potential risks for individual employees, a pre-employment health evaluation prior to work assignment is recommended. Medical evaluations on a regular basis are strongly advised. A health evaluation is usually employed with the use of a detailed questionnaire about medical history. It is critical that employees disclose existing medical conditions and allergies that can affect their risk of being injured or acquiring illness at work. Personnel with existing conditions will be assigned to areas that do not pose significant health hazards.

### 10.5.4 Immunizations/Vaccinations

The kind of immunizations or vaccinations employed would depend on the type of research conducted in the facility and the species of animals involved. In general, it is essential to immunize animal care personnel against tetanus. Pre-exposure immunization should be provided to people who are at risk of infection or exposure to specific agents such as the rabies virus or the hepatitis B virus (e.g., if working with human blood or human tissues, cell lines, or stocks). Vaccination is advised if research on infectious diseases for which effective vaccines are available is to be conducted.

### 10.5.5 IACUC and IBC Management

Facilities working with infectious and toxic agents must ensure compliance with the regulations and guidelines on biosafety, biocontainment, and laboratory biosecurity. It is important that the potential hazards and risks are properly identified, assessed, and communicated to all staff involved in the experiment.

#### 10.5.5.1 Institutional Animal Care and Use Committee (IACUC)

One of the primary roles of the IACUC is to review and approve protocols utilizing animals in research and testing. During these processes, consideration is given to biohazards specific to the animal species and to the protocol in use. The IACUC should ensure proper handling, usage, and disposal of potentially hazardous materials such as genetically modified organisms, radioactive substances, chemicals, novel compounds, human-derived materials (i.e., human cell lines, stem cells, and tissues), non-human-derived materials (i.e., animal cell lines), and other waste products (i.e., biohazard and cytotoxic wastes).

#### 10.5.5.2 Institutional Biosafety Committee (IBC)

The IBC reviews projects involving recombinant DNA, RNAi, pathogens, potentially infectious materials (i.e., human-derived materials), and transgenic animals. In doing so, it takes into account both established local and institutional laws and regulations. The IBC provides recommendations in matters pertaining to the control of biohazards associated with the use of microbiological agents and their vectors. It also represents the interests of the surrounding community with respect to public health and the protection of the environment.

## 10.6 SANITATION, DECONTAMINATION, AND WASTE MANAGEMENT

Maintaining the well-being of laboratory animals and the personnel working in the laboratory animal facility involves effective sanitation and decontamination procedures and proper waste management. These are all essential in order to prevent or minimize contamination, control the spread of microorganisms between and to animals, people, and spaces. There is no single method that is applicable to all types of laboratory animal facilities at all times. It is important to regularly evaluate sanitation, decontamination, and waste disposal protocols if they remain effective and adhere to applicable guidelines and regulations.

## 10.6.1 Sanitation

Sanitation is the maintenance of the environmental conditions in a manner that will lower the number of microorganisms to a safe level. In an animal facility this usually involves changing the animal bedding (if applicable), cleaning, and disinfection. Cleaning is the physical removal of gross debris and organic materials that will allow for the disinfectant to penetrate. Disinfection, on the other hand, is the process to reduce, remove, inactivate, or destroy pathogenic microorganisms. Sanitation processes and frequency vary from facility to facility. It is usually determined by the species housed, the type of enclosures, the amount of debris generated by the animals (NRC, 2011), and the materials and equipment being used in the facility. No single method is adequate for all situations.

In the conduct of cleaning and disinfection, proper personal protective equipment must always be worn to reduce the likelihood of exposure from chemical hazards and any biological agents present. It is important to follow the PPE requirements stated in the safety data sheets (SDS) or material safety data sheets (MSDS) of these chemicals. Proper handling, use, storage, and disposal can also be found in these documents.

Ensure that all surfaces are thoroughly rinsed of detergents or disinfectants to prevent chemical burns or dermatitis because of residue exposure to footpads or skin. Perfumes or volatile chemicals may cause changes in the normal physiology, behavior, or metabolism of animals (Castelhano-Carlos & Baumans, 2009), so chemicals that are designed to cover odors and cleaning solutions that are scented should not be used in animal facilities (NRC, 2011). Those with residual odors should also be avoided.

### 10.6.1.1 Disinfectants

Disinfectants may be physical, chemical, or a combination of both. According to the World Health Organization, three important factors must be considered to attain optimum effectiveness against biological risks:

1. spectrum of activity in the animal facility, particularly on the type of agents that must be disinfected;
2. field of application, *e.g.*, for solid surfaces or liquids
3. application conditions such as contact time, disinfectant concentration, application temperature, presence of organic load, environmental conditions, water hardness, etc.

There are a number and variety of disinfectant products and formulations available in the market. It is important to select the appropriate disinfectant that addresses the specific needs of the animal facility based on the effectiveness of disinfection against known and unknown pathogens or organisms and compatibility with equipment and materials to be disinfected. Cost is also an important consideration as well as availability of the disinfectant product, since not all are available everywhere.

#### 10.6.1.2 UV Light

Ultraviolet light in the UV-C band (200–280 nm) is also known as ultraviolet germicidal radiation and is widely used for sterilizing equipment and creating sterile environments. UV disinfection is effective at inactivating most viruses, spores, and cysts, but some organisms can repair and reverse the damage caused by UV light. Most UV disinfection systems use germicidal lamps emitting UV radiation C (UVC) around 254 nm. However, it is well known that 254-nm UVC is harmful to the skin and eyes.

#### 10.6.1.3 Hot Water

Disinfection from the use of hot water alone is the result of the combined effect of the temperature and the length of time that a given temperature (cumulative heat factor) is applied to the surface of the item. Studies have shown that cleaning with cold water had no impact on bacteria count, raising the temperature to 60°C reduced the number of bacterial colonies by 90%, at 80°C bacterial colonies were reduced by 97%, and at 155°C they were eliminated entirely. The same cumulative heat factor can be obtained by exposing organisms either to very high temperatures for short periods or to lower temperatures for longer periods. Effective disinfection can be achieved with wash and rinse water at 62–82°C or more. Detergents and chemical disinfectants enhance the effectiveness of hot water but should be thoroughly rinsed from surfaces before reuse of the equipment. Proper PPE must be worn to protect against thermal burns.

#### 10.6.1.4 Iodine Compounds

Iodine compounds kill both enveloped and nonenveloped viruses by penetrating the cell wall of the organisms, causing a disruption of the cell metabolism and function. Secondarily, free iodine may bond in solution with the cell proteins to form salts. Iodine compounds are poor cleaners, requiring the use of another chemical to clean the surface. In addition, they are relatively unstable in terms of contact time and light exposure, and improper use can lead to permanent brownish stains on surfaces. They should not be used on aluminum or copper surfaces and materials. Iodine can be toxic, making it generally unsuitable for use as a disinfectant. However, iodophors and tincture of iodine are good antiseptics but unsuitable for use on medical/dental devices. Iodine should not be used on aluminum or copper. Iodine can be toxic. Organic iodine-based products must be stored at 4–10°C to avoid the growth of potentially harmful bacteria in them.

#### 10.6.1.5 Phenolic Compounds

Phenolic compounds are among the earliest germicides and are bactericidal, fungicidal, and virucidal. When properly formulated, they can be effective against mycobacteria. They work by penetrating cell walls and denaturing the cell's proteins. Some phenolic compounds are sensitive to and may be inactivated by water hardness and therefore must be diluted with distilled or deionized water. Recent safety concerns have restricted the use of phenols and phenolic compounds because they can be highly corrosive, absorbed by latex gloves, and can also penetrate the skin. They are extremely hazardous to cats, causing skin depigmentation.

#### 10.6.1.6 Chlorine Compounds

Widely used since the early 1800s, sodium hypochlorite (NaOCl), or more commonly known as bleach, is an intermediate-level fast-acting disinfectant and is among the most effective, most convenient, and least expensive germicides. Bleach exhibits sporicidal activity, is tuberculocidal, inactivates vegetative bacteria, and is fungicidal and virucidal. The effectiveness of bleach as a disinfectant relies on the concentration of free and available chlorine in the solution. It is believed that the free chlorine is able to denature proteins, inactivate sulfhydryl-containing enzymes, and damage nucleic acids (RNA and DNA).

Working bleach solutions must be between 0.5% and 2% to be effective, and a dilution of 1:10 of household bleach (which usually contains 5.25% sodium hypochlorite) is recommended. Lower or more dilute concentrations will not provide the proper level of disinfection. The disinfectant activity of bleach is considerably reduced by organic matter, thus a precleaning step is necessary before application. The temperature, pH, and hardness of water used to dilute bleach can also affect the level of available chlorine in the bleach solution. Exposure to sunlight, oxygen, and heat can significantly cause bleach solutions to rapidly lose potency. Undiluted household bleach stored at room temperature in the original container has a shelf-life of approximately one year. Working solutions of bleach should be prepared on a daily basis.

Chlorine is highly alkaline and can be corrosive to certain metals, concrete, and clothing. Bleach is reactive with acids and causes the release of chlorine gas, which is highly toxic. Its reaction to organic materials like feces can produce potentially carcinogenic compounds, one of which is trihalomethane, a toxic gas. Bleach must therefore be stored and used in well-ventilated areas only. Undiluted bleach must not be mixed with acids or other incompatible chemicals, such as ammonia-containing compounds.

#### 10.6.1.7 Alcohols

Alcohol products are active against vegetative bacteria, fungi, and lipid-containing viruses by denaturing cell proteins to cause cell death. They have no effect against spores, HBV, *Mycobacterium tuberculosis* (TB), and their action on non-lipid-containing viruses is variable. Ethanol (ethyl alcohol, C2H5OH) and 2-propanol (isopropyl alcohol, (CH3)2CHOH) are the most commonly used disinfectant alcohol derivatives. To be effective, they must be used at concentrations of approximately 70% (v/v) in water; higher or lower concentrations may not be as germicidal. A major advantage of aqueous solutions of alcohols is that they do not leave any residue on treated items. Mixtures with other agents are more effective than alcohol alone, for example, 70% (v/v) alcohol with 100 g/L formaldehyde and alcohol containing 2 g/L available chlorine. A 70% (v/v) aqueous solution of ethanol can be used to soak small pieces of surgical instruments. A contact time of ten minutes or more is necessary. Alcohol-based hand-rubs, alcohol mixed with emollients, are recommended for the decontamination of lightly soiled hands in situations where proper hand-washing is inconvenient or not possible. Note however that ethanol can dry hands. Alcohols are volatile and flammable and must not be used near open flames. Do not use 70% ethanol to clean a Class II, Type A recirculating biosafety cabinet. Working solutions should be stored in proper containers to avoid the evaporation of alcohol. Alcohols may harden rubber and dissolve certain types of glue. Proper inventory and storage of ethanol in the laboratory is very important to avoid its use for purposes other than disinfection. Bottles with alcohol-containing solutions must be clearly labeled to avoid autoclaving.

#### 10.6.1.8 Quaternary Ammonium Compounds

The quaternary ammonium compounds are the most widely used in animal facilities because they are relatively cost-effective, easy to use, and, if handled properly, safer for animals and people than the other currently available disinfectants. There are many types that are used as mixtures and often in combination with other germicides, such as alcohols, or detergents. They have good activity against some vegetative bacteria and lipid-containing viruses by working on denaturing cell proteins, influencing cell metabolic reactions, affecting cell permeability, stimulating cell glycolysis reaction, and affecting cell membrane enzyme activity. But they do not have tuberculocidal activity. The germicidal activity of certain types of quaternary ammonium compounds is considerably reduced by organic matter, water hardness, and anionic detergents. Care is therefore needed in selecting agents for pre-cleaning when quaternary ammonium compounds are to be used for disinfection. Potentially harmful bacteria can grow in quaternary ammonium compound solutions. Quaternary ammonium compounds, properly diluted, have low odor and are not irritating.

#### 10.6.1.9 Vaporized Disinfectants

When decontamination of an entire room or enclosure and complete sterilization are needed, such as after completion of an infectious disease study or in response to a pathogen outbreak, the use of vaporized hydrogen peroxide (VHP) or chlorine dioxide gas is extremely effective. Both VHP and chlorine dioxide have antimicrobial properties and are effective against spores. Neither is corrosive in the vapor or gaseous form. Chlorine dioxide can also be used in a liquid form for surface decontamination when sprayed on equipment such as animal change stations, biological safety cabinets, or countertops. Chlorine dioxide gas is different from chlorine gas.

### 10.6.2 Decontamination of Animal Carcasses

Decontamination is the process of removing dangerous substances present in an object or an area. Carcasses in a laboratory animal facility are considered hazardous wastes that contain or may potentially contain pathological or infectious substances. As such, prior to disposal, it is necessary to decontaminate the animal carcasses to prevent the release of microbial agents into the environment.

#### 10.6.2.1 Autoclave

Autoclaving is the use of a combination of high temperature (for example, 121 °C, 134 °C) as moist heat (steam) and pressure to destroy microorganisms, including spores. When used correctly, it is the most effective and reliable means to sterilize laboratory materials and decontaminate waste materials by destroying or inactivating biological agents. Different types of waste materials generally require different operating cycles to achieve appropriate inactivation temperatures. Therefore, laboratory autoclaves should be selected based on defined criteria such as intended use, and type and amount of waste to be decontaminated. Their effectiveness for the specific cycles that will be used should then be validated. Aside from monitoring temperature, pressure, and time, effective sterilization must be confirmed using biological indicators. Spores of *Geobacillus stearothermophilus* are most often used for efficiency testing.

#### 10.6.2.2 Incineration

A commonly used decontamination and disposal method, especially for animal carcasses is incineration. Incinerators must be appropriate for use with the material being incinerated, and a complete burn must be achieved; that is complete to ash.

Waste incineration must comply with applicable regulations in the locale, which may vary from place to place and time to time. Some areas may require permits to install and operate waste incinerators.

In the Philippines, the Republic Act 8749, otherwise known as the Philippine Clean Air Act of 1999, has banned the burning of municipal, biomedical, and hazardous wastes that emit poisonous and toxic fumes. The prohibition does not apply to traditional small-scale methods of community/neighborhood sanitation 'siga', traditional, agricultural, cultural, health, and food preparation and crematoria. Incinerators that function as crematoria should obtain a Permit to Operate from the Environmental Management Bureau (EMB).

### 10.6.3 Waste Management

An animal facility generates different types of wastes, and many of these are classified as hazardous wastes (wastes that present unreasonable risk and/or injury to health and safety and to the environment) such as syringe needles, animal beddings, animal carcasses, and contaminated media. Waste should be segregated and properly identified. Appropriate handling, treatment, and disposal of waste by type help reduce costs and protect public health.

Waste segregation must be done at the point of generation and kept isolated from each other. Hazardous waste should be placed in clearly marked containers that are appropriately labeled for the type and weight of waste. Except for sharps and fluids, hazardous wastes are placed in designated plastic bags, plastic lined cardboard boxes, or leakproof containers. Containers are color-coded depending on the type of waste as indicated in Table 10.4.

**TABLE 10.4**
**Classification of Garbage and Garbage Bags**

| Color of Container/Bag | Type of Waste |
|---|---|
| Black | Non-infectious dry waste |
| Green | Non-infectious wet waste (kitchen, dietary, etc.) |
| Yellow | Infectious and pathological wastes |
| Yellow with black band | Chemical waste including those with heavy metals |
| Orange | Radioactive waste |
| Red | Sharps and pressurized containers |

**Source:* Health Care Waste Management Manual, DOH

Animal carcasses as well as bedding and fecal material are classified as pathological wastes (yellow bag). Animals that have been infected with pathogens, including waste associated with them, require decontamination, usually by autoclaving prior to disposal.

### 10.6.3.1 Waste Storage

Until transport to a designated off-site treatment facility or collected by a third-party waste treatment company, all waste should be collected and temporarily stored in a designated area within the establishment. This area must be located away from animal rooms, laboratories, offices, food storage and preparation areas, or any public access areas. It should be marked with a warning sign: **'CAUTION: BIOHAZARDOUS WASTE STORAGE AREA—UNAUTHORIZED PERSONS KEEP OUT'**.

At a minimum, the temporary waste storage area should have an impermeable, hard-standing floor with good drainage and water supply to allow for easy cleaning and disinfection. There should be protection from sun, rain, strong winds, floods, etc. It should be designed to allow easy access for staff handling waste and for waste collection vehicles. However, it should be inaccessible to animals, insects, and birds.

### 10.6.3.2 Regulatory Requirements

Regulatory requirements regarding waste storage and disposal vary in each country. It is important to check the local and national guidelines and regulations in the facility's locale.

The Philippines has several policies and regulations on waste storage and disposal. One of which is the Department of Environment and Natural Resources (DENR) Administrative Order 36 series of 2004 (revision of the DENR Administrative Order No. 29 series of 1992, to further strengthen the Implementation of Republic Act 6969 or the Toxic Substances and Hazardous and Nuclear Wastes Control Act of 1990), which states that waste generators, like animal facilities, should have a Pollution Control Officer (PCO)/Environmental Officer who has been officially accredited by the DENR to perform responsibilities described in the regulation. In addition, waste generators are required to be registered with the EMB Regional Office having jurisdiction over their location, report on a regular basis the type and quantity of wastes, including hazardous wastes, generated and must have comprehensive emergency contingency plans to mitigate spills and accidents involving hazardous wastes. They must be responsible for the hazardous waste generated or produced on the premises until the hazardous waste has been certified by the waste treater as adequately treated, recycled, reprocessed, or disposed of.

If hazardous wastes will be disposed of by a third-party waste treatment company, this company has to likewise be accredited by DENR through the EMB.

## 10.7 EMERGENCY RESPONSE

It is imperative that animal facilities have a disaster plan that defines the actions necessary to ensure the safety of both personnel and animals in times of disaster and other emergencies. Contingency plans must address and prevent animal pain, distress, and deaths, as well as the safety of personnel. All personnel must be trained in the emergency response program of the facility, and a copy of the emergency response procedure must be accessible at all times.

As a general rule, a list of emergency contact numbers should be placed in conspicuous areas. This list should include emergency contact numbers of the laboratory animal facility officials such as (but not limited to) the director, the facility manager, attending veterinarian, biosafety officer, maintenance personnel, and community emergency hotline numbers such as the local police, nearest fire station, nearest medical institution, and other relevant emergency numbers applicable to the locale. Procedures for reporting emergencies should be clearly defined.

### 10.7.1 Internal Emergencies

Inside the animal facility, emergencies such as accidents, system malfunctions, and other problems may occur that require immediate attention and correction. Minor husbandry emergencies such as cage flooding, animal deaths, and wounds must be reported to the Attending Veterinarian and Principal Investigator if the animal/s affected belong to a study. The cause of cage flooding must be investigated, and the problem corrected. Cage flooding increases the humidity inside the cage, which can cause harm to the animals.

First aid kits should be available and accessible in all areas. Animal bites, scratches, punctures from contaminated needles, and other sharp objects may occur. Depending on the animal species, pathogen exposure of the animal, and the size of tissue damage, the injury may be considered a major medical emergency. In countries where rabies is still present, administration of post-exposure prophylaxis (PEP) consisting of a dose of human rabies immune globulin (HRIG) and rabies vaccine is necessary after a dog or cat bite. The biting animal must be placed in isolation for at least 10 days, depending on local regulations, for observation of rabies symptoms.

Accidental exposure to biological hazards, such as splashing of blood or body fluid onto the eyes or mouth, requires immediate flushing or rinsing for at least 15 minutes. Eyewash stations must be installed in strategic areas.

There are different types of chemicals used inside the animal facility, and some may be hazardous. For minor chemical burns, flooding the burned area with cool running water for at least 20 minutes may reduce pain and swelling. However, major chemical burns would require immediate medical attention. It is important to read the safety data sheets of all chemicals for proper use and storage.

Only personnel with proper training and wearing the appropriate PPE must work with radioactive materials. In cases of accidental radiation exposure, affected skin must be washed with soap and water, potentially contaminated PPE and clothing must be placed in disposable bags for radioactive wastes (usually orange in color), and immediate medical attention must be given.

Procedures for other medical emergencies such as heart attacks, poisoning, fainting, choking, shock, burns, fractures must also be established. It is recommended that there are personnel trained for medical emergencies and able to do cardiopulmonary resuscitation (CPR).

Plans for management of animals in cases of utility failures such as malfunction of the heating, ventilation, and air conditioning (HVAC) system and power outages should also be prepared. Special consideration must be given to animals that are physically dependent on electrical power for life support and well-being such as aquatic animals, animals in sealed chambers, cages, or enclosed rooms. Close monitoring may be necessary. Facilities with automatic locking doors must have a fail-safe system such as the existence of a physical key that can unlock doors in cases of power outages. For gas leaks, plumbing failures, fires, an evacuation plan for both humans and animals must be in place.

### 10.7.2 External Emergencies

In case of unprecedented events or unavoidable disasters like typhoons, fire, earthquakes, community quarantine, etc., necessary measures must be taken to ensure the safety of the personnel and the animals. Personnel safety must be prioritized at all times. A disaster plan should include actions that need to be taken in emergencies.

In case of natural disasters that require evacuation, a designated area for personnel must be identified where personnel can gather. In rare instances and upon risk assessment that would require animal evacuation, the location of a temporary animal facility must be determined and included in the disaster plan. Communication with temporary shelters should be coordinated for the proper transfer of animals.

Experiments should not be conducted during typhoons, storms, and floods if there is prior notice of a potential disaster. Personnel may have difficulty going to the facility during calamities. Therefore, a minimum number of staff is required in order to carry out tasks should be identified. Emergency supplies should keep enough feed and water for the animals during a disaster. Adequate supply must be available for temporary disasters. A list of emergency supplies and contact information should be available in the facility and these supplies should be accessible.

In case of fire, personnel should evacuate the building immediately, and if possible, animals should be evacuated as soon as possible. For earthquakes, all personnel should follow emergency procedures set by their institution. Check for building safety and evacuate if there is a fire or severe structural damage.

In the event of a pandemic, facilities should be prepared if personnel will not be able to work on-site. As animal facility workers are considered essential, the facility should ensure adequate staff are available to provide animal care.

It is recommended that end users be trained in disaster response and abide by their institution's policies. All personnel should participate in emergency drills and training exercises conducted regularly.

## 10.8 CHALLENGES AND MOVING FORWARD

Each animal facility is unique and may encounter different challenges in operation and biorisk management implementation. Therefore, a site-specific risk assessment is needed to tailor-fit the needs of the laboratory. Low-middle-income countries may encounter challenges in the lack of availability of safety equipment or access training. More often, at least for the Philippines, animal laboratories would typically rely on small research funding to sustain the operations of the animal laboratory. This becomes a barrier in terms of sustainability and ensuring availability of safety equipment.

Moving forward, institutional support is needed to fully operationalize an animal facility and ensure availability of resources. Investment to conduct capacity building of animal laboratories by creating a pool of trainers is essential that can sustain and ensure availability of training to workers. Biosafety officers together with stakeholders should work together in performing risk assessment. The formation of the Institutional Animal Care and Use Committee and Institutional Biosafety Committee is essential to ensure oversight of animal activities and safety in the institution.

## REFERENCES

Alderman, T. S., Carpenter, C. B., & McGirr, R. (2018). Animal research biosafety. *Applied Biosafety*, *23*(3), 130–142. https://doi.org/10.1177/1535676018776971.

Baker, D. G. (1998). Natural pathogens of laboratory mice, rats, and rabbits and their effects on research. *Clinical Microbiology Reviews*, *11*(2), 231–266. https://doi.org/10.1128/CMR.11.2.231.

Castelhano-Carlos, M. J., & Baumans, V. (2009). The impact of light, noise, cage cleaning and in-house transport on welfare and stress of laboratory rats. *Laboratory Animals*, *43*(4), 311–327. https://doi.org/10.1258/la.2009.0080098.

Centers for Disease Control and Prevention (U.S.), & National Institutes of Health (U.S.). (2020). *Biosafety in Microbiological and Biomedical Laboratories* (6th ed). (HHS Publication No. [CDC] 300859). Retrieved from https://stacks.cdc.gov/view/cdc/97733.

Clayton, A. M., Hayes, J., Lathrop, G. W., & Powell, N. (2019). Development of an occupational risk assessment tool for laboratory animal facilities. *Applied Biosafety*, *24*(2), 72–82. https://doi.org/10.1177/1535676019831915.

*DAO 2004–36, Procedural Manual Title III of DAO 92–29 "Hazardous Waste Management" DENR AO – Series of 2004*. DENR.

*Health Care Waste Management Manual* (4th ed.) (2020). Department of Health - Health Facility Development Bureau.

*Implementing Rules and Regulations for RA 8749.* DENR Administrative Order No. 2000-81. November 07, 2000.

Ingraham, A., & Marotta Fleischer, T. (2003). Disinfectants in laboratory animal science: What are they and who says they work? *Lab Animal*, *32*(1), 36–40. https://doi.org/10.1038/laban0103-36.

*Laboratory Biosafety Manual* (4th ed.). (2020). World Health Organization. Licence: CC BY-NC-SA 3.0 IGO.

Normand, J., Vlahov, D., Moses, L. E., National Research Council (U.S.), & Institute of Medicine (U.S.) (Eds.). (1995). *Preventing HIV Transmission: The Role of Sterile Needles and Bleach.* National Academy Press.

[NRC] National Research Council (US) Committee for the Update of the Guide for the Care and Use of Laboratory Animals. (2011). *Guide for the Care and Use of Laboratory Animals* (8th ed.). National Academies Press.

Philippine Clean Air Act of 1999, Rep. Act No. 8749, §1999.

Toxic substances, hazardous and nuclear wastes control act, Rep. Act No. 6969, 1990.

Weichbrod, R. H., Thompson, G. A., & Norton, J. N. (Eds.). (2018). *Management of Animal Care and Use Programs in Research, Education, and Testing* (2nd ed.). CRC Press/Taylor & Francis Group.

# 11 Mycological Biosafety and Biosecurity in the Philippines

*Jonathan Jaime G. Guerrero*
University of the Philippines
Manila College of Medicine and College of Public Health
Manila, Philippines

*Ric Ryan H. Regalado*
National Institute of Molecular Biology and Biotechnology
College of Science, University of the Philippines
Diliman, Quezon City, Philippines

*Gerard Lorenz M. Penecilla*
Department of Pharmacology, College of Medicine
West Visayas State University
Iloilo City, Philippines

*Charmaine A. Malonzo*
Department of Biology
College of Science, Bicol University
Legazpi City, Philippines

*Mheljor A. General*
Department of Biology College of Science
Bicol University
Legazpi City, Philippines

*Alleni T. Junsay*
Metrology in Chemistry Section
National Metrology Laboratory of the Philippines
Industrial Technology Development Institute
Department of Science and Technology
Bicutan, Taguig City, Philippines

 DOI: 10.1201/9781003426219-11

## 11.1 INTRODUCTION

Mycotoxin contamination is a recurring theme in the agriculture and food industries. The ubiquity of the fungi which produce them makes mycotoxins common contaminants in animal and human food supply (Murphy et al., 2006). The Food and Agriculture Organization (FAO) estimates that 25% of global food and feed outputs are contaminated with mycotoxins (Moretti et al., 2017), and major losses due to this, specifically financial, can be as diverse as the reduction in quality and production, and increased cost in finding alternative foods and better management (Marroquín-Cardona et al., 2014). Direct implications to humans have been recorded due to contaminants present in food such as in fruits and their processed products (Fernández-Cruz et al., 2010), spices such as chili and black pepper (Yogendrarajah et al., 2014), and infant foods (Asam & Rychlik, 2013), among others. Even as early as the middle ages, records show the frequency of ergotism throughout Europe, a mycotoxin-induced disease caused by contamination of rye bread by the fungus *Claviceps purpurea*, leading to burning sensations resulting in gangrene, neurological disease, and death (van Dongen & de Groot, 1995).

It is important to manage and control the contamination of food and feed products by mycotoxin-producing fungi to ensure a safe and secure food and agricultural environment. Mycotoxins may enter the food supply directly through the growth of molds on the actual food or indirectly through the use of contaminated raw materials in food processing and manufacture (Bullerman, 1979). FAO and the World Health Organization (WHO) declared high priority on mycotoxins due to their impact on human and animal health, and countries have set legal limits for specific mycotoxins (Fumagalli et al., 2021).

Fungi have already caused biosafety and biosecurity concerns in history. Disease outbreaks caused by the mycotoxins trichothecene and fumonisin were recorded in India in 1987 and 1995 due to contaminated wheat, sorghum, and maize (Raghavender & Reddy, 2009). Other cases link mycotoxin exposure to poor nutrition, the development of cancers, and even death (Bhat et al., 2010; Lukwago et al., 2019; Shephard, 2008). Mycotoxicosis in animals fed with contaminated feed also represents a rising concern in hog-raising (Magnoli et al., 2019), aquaculture (Gonçalves et al., 2020), and poultry (Murugesan et al., 2015) because of poor nutritional value, low performance, and increased susceptibility to disease.

Management of mycotoxin also has a serious economic impact. Data would show that losses in cereals, for example, are highest in years with the highest mycotoxin levels (Focker et al., 2021). For instance, the socioeconomic impact of mycotoxin contamination in Africa is among the rural poor, contributing more to threats to human health (Gbashi et al., 2019). Even in developed countries such as the United States, illnesses and death contribute to the cost associated with mycotoxins (Abbas, 2005). In Southeast Asia, mitigation strategies are recommended to control mycotoxins contamination in raw feed materials, specifically in rice and maize (Siri-anusornsak et al., 2022), a risk also reported to be high in the Philippines (Salvacion et al., 2015).

After the initial discovery of aflatoxin contamination in the Philippines in 1972, additional knowledge has been acquired regarding mycotoxins and the fungi that produce them. Despite the progress made in understanding mycotoxigenic fungi and mycotoxins in the country, there is still a lack of comprehensive knowledge regarding effective practices and measures to control fungi and toxins (Balendres et al., 2019). From the historical point of view, several mycotoxigenic fungi were discovered and isolated in the Philippines over the past two decades of scientific efforts (Figure 11.1).

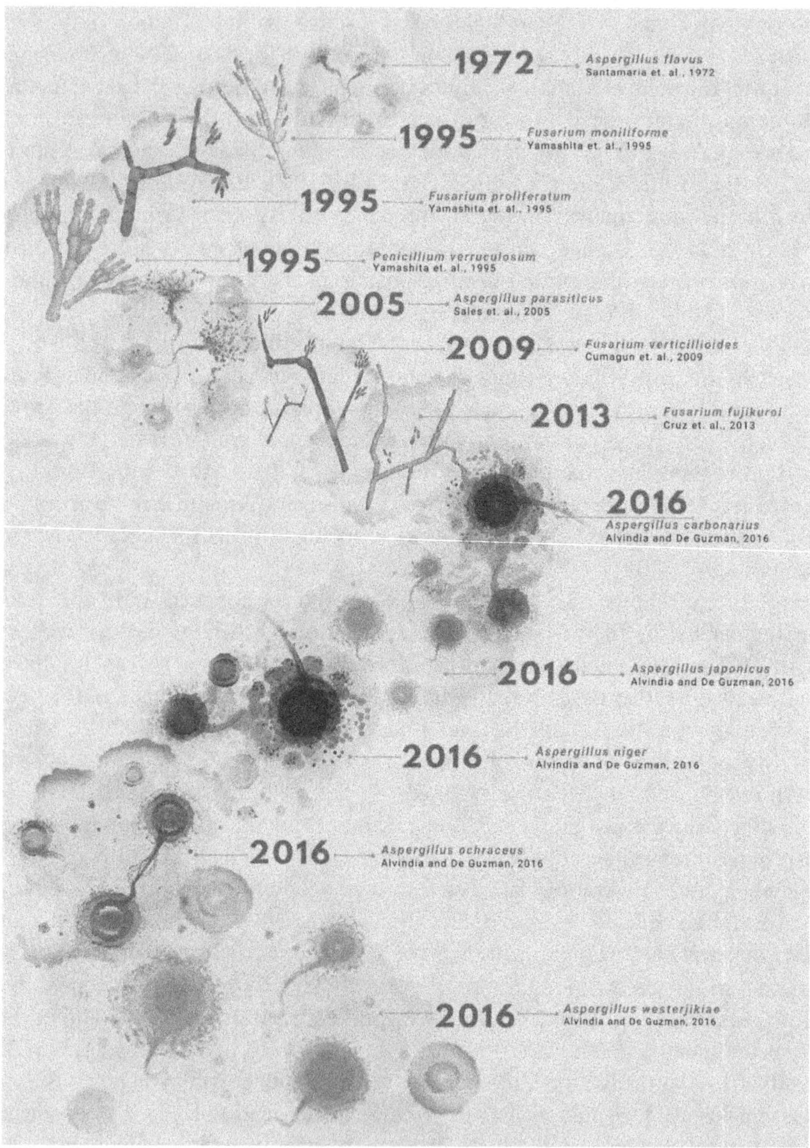


**FIGURE 11.1** The historical timeline of mycotoxigenic fungi isolated in the Philippines.

## 11.2 TRADITIONAL USES OF MACROFUNGI IN THE PHILIPPINES

Aside from mycotoxins, the wide use of mushrooms may also pose potential safety and security concerns. Studies of indigenous groups from Luzon Island in the Philippines, including the Aeta, Bugkalot, Gaddang, Kalanguya, and other Igorot groups, have revealed that gathering and utilization of wild edible mushrooms are influenced by the ecological environment and ethnicity where many of these mushroom species are utilized for medicinal and nutritional purposes (Dulay et al., 2023; dela Cruz and De Leon, 2023). In particular, many of the macrofungal species documented in these surveys were widely consumed as food, the most commonly eaten species being *Schizophyllum commune* (split-gill mushroom), *Auricularia polytricha* (wood ear mushroom), *Volvariella volvacea* (straw mushroom), *Pleurotus* (oyster mushroom), and *Termitomyces*. Beyond their culinary and medicinal uses, these fungi also serve diverse purposes: house decorations, utilized for body detoxification, base for a coffee-like beverage, natural insect repellents, and even as eye drop formulation from an ascomycete species (Table 11.1).

It was also noted that the different provinces have different species of mushrooms, with some areas having a more abundant fungal diversity than others (Dulay et al., 2023). For example, the coniferous type of forest in Benguet and Mt. Province areas are conducive to the growth of wild edible mushrooms, resulting in a stronger tradition of mushroom gathering in these areas than in Apayao, Kalinga, and Ifugao, where non-coniferous types of forests prevail (Licyayo, 2018). While there are numerous ethnomycological studies on Luzon, particularly in the Northern and Central regions, it is worth noting that there are limited studies on the Visayas and Mindanao regions, except for one in Northern Samar (Flores, Jr. et al., 2014). This is surprising given these regions' rich macrofungal diversity and robust cultural landscape. This vast knowledge gap, therefore, warrants more studies highlighting other ethnic and indigenous groups to expand further our understanding of macrofungi uses and associated risks among local communities.

Accordingly, the importance of management practices in the handling and storage of edible macrofungi species in the Philippines cannot be overstated, mainly because the Basidiomycota phylum, which includes many of these species, is the same group of fungi that produce mycotoxins. Improper handling and storage conditions can lead to mycotoxin contamination of various agriculturally-relevant commodities, putting consumers at risk. The risk of mycotoxin contamination is further exacerbated by the country's tropical, warm, and humid conditions, which are ideal for the growth and development of mycotoxigenic fungi. Expanding studies on the safe consumption of wild edible macrofungi in the Philippines is therefore crucial, especially considering their gaining commercial interest and recognition by the government as one of the priority crops in the recent road-mapping for agricultural development (Dulay et al., 2021). To date, edible macrofungi species are already commercially cultivated, including *Pleurotus* spp., *Volvariella volvacea*, and *Auricularia* spp., among other species. However, despite these species' commercialization success, research on their potential risks for mycotoxin contamination or production is still limited.

**TABLE 11.1**
**Species List and Distribution of Ethnomycologically Important Macrofungi Species in the Philippines**

| Family | Species Name[a] | Local Name | Sites | Uses | Reference/s |
|---|---|---|---|---|---|
| **Agaricaceae** | *Agaricus* sp. | *Chaminor* | Ifugao | Food | De Leon et al. (2019) |
| | *Coprinus cinereus* (Schaeff.) Gray | *Kulat pinkalan* | Nueva Vizcaya | Food | Torres et al. (2020a; 2020b) |
| | *Leucoagaricus cepaestipes* (Fr.) | *Gum-gumot* | Nueva Ecija | Food | De Leon et al. (2016) |
| | *Vascellum pratense* (Pers.) Kreisel | *Damurasin* | Ifugao | Food | De Leon et al. (2019) |
| **Albatrellaceae** | *Albatrellus* sp. | *Shapannan* | Benguet, Mt. Province | Food | Licyayo (2018) |
| **Amanitaceae** | *Amanita javanica* (Corner & Bas) T. Oda, C. Tanaka & Tsuda | *Bagel* | Benguet, Mt. Province | Food | Licyayo (2018) |
| **Auriculariaceae** | *Auricularia* spp. [4] | *Tengang-daga, lapa-lapayag, Ballutak, Gargarot* | Nueva Vizcaya, Kalinga, Apayao, Ifugao. Mt. Province | Food | Maslang et al. (2021); Licyayo (2018) |
| | *Auricularia auricula* (L. ex Hooker) Underwood | *Tengang-daga, Kuwat malabalugbog dagis, Kuwat kuling baki, Kuwat tangkiki* | Pampanga, Zambales, Nueva Vizcaya, Ifugao | Food | De Leon et al. (2012); Lazo et al. (2015); De Leon et al. (2019) |

*(Continued)*

**TABLE 11.1 (CONTINUED)**
**Species List and Distribution of Ethnomycologically Important Macrofungi Species in the Philippines**

| Family | Species Name[a] | Local Name | Sites | Uses | Reference/s |
|---|---|---|---|---|---|
| | *Auricularia auricula-judae* (Bull.) J. Schrott | *Tengang-daga, kulat kolang-kolang* | Bataan, Nueva Vizcaya | Food | Tantengco and Ragragio, (2018); Torres et al. (2020a; 2020b); Maslang et al. (2021) |
| | *Auricularia fusco-succinea* (Mont.) Henn. | *Tengang-daga* | Nueva Vizcaya | Food | Lazo et al. (2015) |
| | *Auricularia polytricha* (Mont.) Sacc. | *Kulat alenga baboy, Kuwat malabalugbog dagis, Kuwat kuling baki, Kuwat tangkiki* | Pampanga, Zambales, Northern Samar, Bataan, Nueva Vizcaya, Camarines Sur | Food | De Leon et al. (2012); Flores, Jr et al. (2014); Tantengco and Ragragio, (2018); Torres et al. (2020a; 2020b); Undan et al. (2021); Maslang et al. (2021) |
| **Bolbitaceae** | *Agrocybe* sp. | *Kuwat mayo* | Tarlac | Food | De Leon et al. (2012) |
| **Boletaceae** | *Boletus* spp.[2] | *Kulat pungkulan, Dangkiyan* | Nueva Vizcaya, Benguet, Mt. Province | Food | Torres et al. (2020a; 2020b); Licyayo (2018) |
| | *Boletus speciosus* Frost. | *Sinabog* | Benguet, Mt. Province | Food | Licyayo (2018) |
| **Cantharellaceae** | *Cantharellus* sp. | *Bagelni bishing* | Benguet, Mt. Province | Food | Licyayo (2018) |

*(Continued)*

**TABLE 11.1 (CONTINUED)**
**Species List and Distribution of Ethnomycologically Important Macrofungi Species in the Philippines**

| Family | Species Name[a] | Local Name | Sites | Uses | Reference/s |
|---|---|---|---|---|---|
| | *Cantharellus cibarius* Fr. | *Banay* | Northern Samar | Food | Flores, Jr et al. (2014) |
| **Fomitopsidaceae** | *Fomitopsis* sp. | *Kulat bungkog* | Nueva Vizcaya | Medicine | Torres et al. (2020a; 2020b) |
| **Ganodermataceae** | *Ganoderma* spp. [3] | *Kulat baklag/Kulat bungkog* | Nueva Vizcaya | Medicine | Torres et al. (2020a; 2020b) |
| | *Ganoderma applanatum* (Pers.) Pat | *Kulat bungkog* | Nueva Vizcaya | Medicine | Torres et al. (2020a; 2020b) |
| | *Ganoderma australe* (Fr.) Pat. | *Kulat bungkog* | Nueva Vizcaya | Medicine | Torres et al. (2020a; 2020b) |
| | *Ganoderma lucidum* (Curtis) Karst. | *Kulat baklag, Kulat betang, Kuwat kahoy, Kabuteng-kahoy, Tibig* | Zambales, Bataan, Nueva Vizcaya | Medicine, food (as coffee), decoration | De Leon et al. (2012); Tantengco and Ragragio, (2018); Torres et al. (2020a; 2020b) |
| | *Ganoderma tsugae* Murrill | *Kulat betang* | Nueva Vizcaya | Medicine | Torres et al. (2020a; 2020b) |
| **Gomphaceae** | *Ramaria botrytis* (Pers.) Ricken | *Pansit-pansitan* | Benguet, Mt. Province | Food | Licyayo (2018) |
| **Hygrophoraceae** | *Hygrophorus russula* (Schaeff.) Kauffman | *Lumsek* | Benguet, Mt. Province | Food | Licyayo (2018) |
| **Hymenochaetaceae** | *Phellinus* sp. | *Gorgor* | Ifugao | Food, medicine | De Leon et al. (2019) |
| **Inocybaceae** | *Inocybe rimosa* (Bull.) P. Kumm. | *Ligbos* | Northern Samar | Food | Flores, Jr et al. (2014) |

*(Continued)*

**TABLE 11.1 (CONTINUED)**
**Species List and Distribution of Ethnomycologically Important Macrofungi Species in the Philippines**

| Family | Species Name[a] | Local Name | Sites | Uses | Reference/s |
|---|---|---|---|---|---|
| **Meripilaceae** | *Meripilus giganteus* (Pers.) P. Karst. | *Bang-ugan* | Nueva Ecija | Food | De Leon et al. (2016) |
| **Mycenaceae** | *Mycena* spp. [3] | *Uong ginikan, Kulat kalansepay* | Pampanga, Ifugao, Nueva Vizcaya | Food, medicine | De Leon et al. (2012); De Leon et al. (2019); Torres et al. (2020a; 2020b) |
| **Lycoperdaceae** | *Calvatia* sp. | *Kuwat Bola/Duldul* | Tarlac | Food | De Leon et al. (2012) |
| **Lyophyllaceae** | *Lyophyllum fumosum* (Pers.) P.D. Orton | *Bagel* | Benguet | Food | Licyayo (2018) |
| | *Termitomyces* spp. [9] | *Kuwat mayo, Kuwat yabot, Kuwat malakamawey, Kuwat malakamay, Kwatkuyog/kidlat/ susongbuyok, Kabuteng-mamarang, O-ong bunton* | Pampanga, Tarlac, Zambales, Bataan, Nueva Vizcaya, Apayao, Ifugao, Kalinga, Mt. Province | Food, medicine | De Leon et al. (2012); Tantengco and Ragragio, (2018); Licyayo (2018); Maslang et al. (2021) |
| | *Termitomyces albuminosus* (Berk.) R. Heim | *Tariktik* | Nueva Vizcaya | Food | Maslang et al. (2021) |

*(Continued)*

**TABLE 11.1 (CONTINUED)**
**Species List and Distribution of Ethnomycologically Important Macrofungi Species in the Philippines**

| Family | Species Name[a] | Local Name | Sites | Uses | Reference/s |
|---|---|---|---|---|---|
| | *Termitomyces clypeatus* R. Heim. | *Kuwat gilatgilatan, Kuwat kuyog, Kuwat lupa/Uong* | Pampanga, Tarlac, Zambales, Nueva Vizcaya | Food | De Leon et al. (2012); Maslang et al. (2021) |
| | *Termitomyces robustus* (Beeli) R. Heim. | *Kuwat malakamawey, Kuwat malakamay* | Pampanga, Zambales, Nueva Vizcaya | Food | De Leon et al. (2012); Maslang et al. (2021) |
| **Physalacriaceae** | *Oudemansiella canarii* (Jungh.) Hohn | *Balutak* | Ifugao | Food | De Leon et al. (2019) |
| **Pleurotaceae** | *Pleurotus* spp. [8] | *Kuwat kasoy, Tarulok/ Uong* | Zambales, Nueva Vizcaya, Ifugao, Apayao, Kalinga | Food | De Leon et al. (2012); Lazo et al. (2015); Licyayo (2018) |
| | *Pleurotus djamor* (Rumph. ex Fr.) Boedijn | *Kabuteng kawayan* | Camarines Sur | Food | Undan et al. (2021) |
| | *Pleurotus dryinus* (Pers.) P. Kumm. | *Kulat paangan* | Nueva Vizcaya | Food | Torres et al. (2020a; 2020b) |
| | *Pleurotus ostreatus* (Jacq. ex Fr.) Kumm. | *Gek-gek* | Ifugao | Food | De Leon et al. (2019) |
| | *Pleurotus pulmunarius* (Fr.) Quél. | *Kabuteng kawayan* | Pampanga, Tarlac, Zambales | Food | De Leon et al. (2012) |

(*Continued*)

**TABLE 11.1 (CONTINUED)**
**Species List and Distribution of Ethnomycologically Important Macrofungi Species in the Philippines**

| Family | Species Name[a] | Local Name | Sites | Uses | Reference/s |
|---|---|---|---|---|---|
| **Pluteaceae** | *Volvariella volvacea* (Bull.) Singer | *Uong ti saba, Kwat-saging, Kuwat ginikan, Kuwat amucao* | Pampanga, Tarlac, Zambales, Nueva Vizcaya, Bataan, Ifugao, Camarines Sur | Food | De Leon et al. (2012); Lazo et al. (2015); Tantengco and Ragragio, (2018); De Leon et al. (2019); Undan et al. (2021) |
| **Podoscyphaceae** | *Podoscypha brasiliensis* D.A. Reid | *Kuyupan* | Nueva Ecija | Food | De Leon et al. (2016) |
| **Polyporaceae** | *Lentinus* spp. [3] | *Kulat bitkalan anoy/ lukong, Tarulok* | Nueva Vizcaya | Food | Lazo et al. (2015); Torres et al. (2020a; 2020b) |
| | *Lentinus sajor-caju* (Fr.) Fr. | *Ulat, Kuwat kawayan* | Tarlac, Ifugao | Food | De Leon et al. (2019) |
| | *Lentinus squarrosulus* Mont. | - | Tarlac | Food | De Leon et al. (2012) |
| | *Lentinus tigrinus* (Bull.) Fr. | *Kuwat kikitban, Kuwat miyapol, Kulat bitkalan sipsip* | Pampanga, Tarlac, Zambales, Nueva Vizcaya | Food | De Leon et al. (2012); Torres et al. (2020a; 2020b) |
| | *Lenzites elegans* (Spreng.) Pat. | *Kiki* | Ifugao | Food, medicine | De Leon et al. (2019) |
| | *Microporus* sp. | *But-taytay* | Nueva Ecija | Food | De Leon et al. (2016) |
| | *Polyporus* spp. [4] | *Tarulok/Uong, Kulat kaneg, Kulat kuyong, Kulat simbed* | Nueva Vizcaya | Food, medicine | Lazo et al. (2015); Torres et al. (2020a; 2020b) |

(*Continued*)

**TABLE 11.1 (CONTINUED)**
**Species List and Distribution of Ethnomycologically Important Macrofungi Species in the Philippines**

| Family | Species Name[a] | Local Name | Sites | Uses | Reference/s |
|---|---|---|---|---|---|
| | *Polyporus grammocephalus* Berk. | - | Zambales | Food | De Leon et al. (2012) |
| | *Polyporus picipes* Fr. | *Kulat kaneg* | Nueva Vizcaya | Medicine | Torres et al. (2020a; 2020b) |
| | *Polyporus squamosus* (Huds.) Fr. | *Bannog* | Ifugao | Food | Licyayo (2018) |
| | *Trametes* sp. | *Fatang* | Ifugao | Remedy for stomach ache and headache, and used for body detoxification | De Leon et al. (2019) |
| | *Trametes elegans* (Spreng.) Fr. | *Fatang* | Ifugao | Medicine | De Leon et al. (2019) |
| **Psathyrellaceae** | *Coprinellus disseminatus* (Pers.) J.E. Lange | *Uong kawayan* | Ifugao | Food | De Leon et al. (2019) |
| | *Coprinopsis atramentaria* (Bull.) Redhead, Vilgalys & Moncalvo | *Kulat guko-guko* | Nueva Vizcaya | Food | Torres et al. (2020a; 2020b) |
| | *Coprinopsis cinerea* (Schaeff.) Redhead, Vilgalys & Moncalvo | *Kabuteng mais* | Camarines Sur | Food | Undan et al. (2021) |

(*Continued*)

**TABLE 11.1 (CONTINUED)**
**Species List and Distribution of Ethnomycologically Important Macrofungi Species in the Philippines**

| Family | Species Name[a] | Local Name | Sites | Uses | Reference/s |
|---|---|---|---|---|---|
| | *Coprinopsis lagopus* (Fr.) Redhead, Vilgalys & Moncalvo | *Kulat guko-guko* | Nueva Vizcaya | Food | Torres et al. (2020a; 2020b) |
| | *Panaeolus* sp. | *Kulat awang* | Nueva Vizcaya | Food | Torres et al. (2020a; 2020b) |
| | *Psathyrella* sp. | *Kuwat kunyayabi* | Zambales | Food | De Leon et al. (2012) |
| **Pyronemataceae** | *Trichaleurina celebica* (Hennings) M. Carbone, Agnello & P. Alvarado | *Lateg-lateg, Lateg kabayo, Lusi lusi* | Nueva Vizcaya | Food, eye drop | Maslang et al. (2021) |
| **Russulaceae** | *Lactarius volemus* (Fr.) Fr. | *Gatas-gatasan* | Benguet, Mt. Province | Food | Licyayo (2018) |
| | *Russula virescens* (Schaeff.) Fr. | *Kapputan, Ofot/Upot* | Nueva Ecija, Benguet, Mt. Province | Food | Licyayo (2018); De Leon et al. (2016) |
| **Sarcosomataceae** | *Galiella celebica* (Henn.) Nannf. | *Lusi-lusi, Fofotoy* | Kalinga, Apayao | Food | Licyayo (2018) |

*(Continued)*

**TABLE 11.1 (CONTINUED)**
**Species List and Distribution of Ethnomycologically Important Macrofungi Species in the Philippines**

| Family | Species Name[a] | Local Name | Sites | Uses | Reference/s |
|---|---|---|---|---|---|
| **Schizophyllaceae** | *Schizophyllum commune* Fr. | *Utchipe, Kuwat kuritdit, Kwat-kawayan, Kurakding/Kuratding* | Pampanga, Zambales, Northern Samar, Nueva Vizcaya, Bataan, Apayao, Ifugao, Kalinga, Camarines Sur | Food | De Leon et al. (2012); Lazo et al. (2015); Flores, Jr et al. (2014); Tantengco and Ragragio, (2018); Licyayo (2018); De Leon et al. (2019); Torres et al. (2020a; 2020b); Undan et al. (2021); Maslang et al. (2021) |
| **Sclerodermataceae** | *Scleroderma citrinum* Pers. | *Buo* | Nueva Ecija | Food | De Leon et al. (2016) |
| **Sparassidaceae** | *Sparassis crispa* (Wulfen) Fr. | *Mala-koli* | Benguet, Mt. Province | Medicine | Licyayo (2018) |
| **Stereaceae** | *Stereum* sp. | *Kwat-kawayan* | Bataan | Food | Tantengco and Ragragio, (2018) |
| | *Stereum lobatum* (Kunze ex Fr.) Fr. | *Kulat kagkagen* | Nueva Vizcaya | Food | Torres et al. (2020a; 2020b) |
| **Tricholomataceae** | *Clitocybe* sp. | *Kulat tegatan* | Nueva Vizcaya | Food | Torres et al. (2020a; 2020b) |
| | *Collybia reineckeana* Henn. | *O-ong dir-an* | Ifugao | Food | Licyayo (2018) |

*(Continued)*

**TABLE 11.1 (CONTINUED)**
**Species List and Distribution of Ethnomycologically Important Macrofungi Species in the Philippines**

| Family | Species Name[a] | Local Name | Sites | Uses | Reference/s |
|---|---|---|---|---|---|
| | *Tricholoma robustum* (Alb. & Schwein.) Ricken | *Panggayya* | Benguet, Mt. Province | Food | Licyayo (2018) |

[a]Numbers enclosed in brackets [ ] indicate the number of different morphotaxa bearing the same genera reported in the literature but not yet identified to species level

Additionally, climate change scenarios pose an additional challenge to managing edible macrofungi, complicating efforts to ensure their safety along the food supply chain. Changes in temperature and precipitation patterns favor some mycotoxigenic fungal species over others (Mannaa & Kim, 2017), requiring continuous evaluation and monitoring of handling and storage practices. The lack of clear regulation for mycotoxins in macrofungi adds to the concern. While some mycotoxins have been found in various agricultural products, well-established national standards for mycotoxins per commodity are still poorly defined, considering the diverse range of mycotoxins reported in the country (Table 11.4). Climate change, inadequate regulations, and limited research call for a proactive approach to mycotoxins in the country. A strengthened food safety system backed by science is key, and if properly managed, macrofungi will continue to play an important role in food security, health, and cultural heritage in the Philippines.

## 11.3 Current Mycotoxin Research in the Philippines

Much of the current research in mycotoxins and fungi-associated risks in the Philippines are focused on agricultural commodities, specifically during post-harvest period. Due to conditions during storage such as humidity, light, and temperature, mycotoxigenic fungi thrive among grains and other stored agricultural products. *Aspergillus* and *Fusarium* species are the most commonly encountered, with *Penicillium* species also reported (Table 11.2). A list of mycotoxins and their associated mycotoxicogenic fungi in agricultural commodities can be found in the works of Balendres et al. (2019).

## 11.4 MYCOTOXICOSES

It is important to differentiate mycotoxicoses from other conditions, such as mycoses and mushroom poisoning, which are excluded from this discussion. Mycotoxicosis results from exposure to or poisoning from mycotoxins which are toxic secondary metabolites produced by molds or microfungi (Bennett et al., 2003), generally a condition affecting vertebrates. In contrast, mycoses are the diseases that result from the growth of fungi on their hosts, while mushroom poisoning, the result of exposure to macrofungi, is arbitrarily excluded among the mycotoxicoses (Bennett et al., 2003). While there are several hundred mycotoxins that have been identified (Alshannaq et al., 2017) with a wide variety of chemical composition and properties, only a few dozen are currently considered of medical importance, and from these, the most relevant and clinically significant are highlighted (Table 11.3).

Poisoning from these mycotoxins can be acute, that is, their effects occur with a rapid onset from the time of exposure, or they can be chronic (e.g., occurring several months or years after repeated low-level exposure) (Ostry et al., 2017), which is mostly the case as a result of ingestion of contaminated food (Peraica et al., 2014). However, there is a paucity of information about how many people are affected by mycotoxicoses (Bennett et al., 2003).

**TABLE 11.2**
**Mycotoxin Research in the Philippines in the Last 20 Years**

| Mycotoxin | Fungi | Substrate | Location | Focus of Research | Reference |
|---|---|---|---|---|---|
| Fumonisin | *Fusarium* spp. | Corn | | Fuzzy modeling of mycotoxin risk under current and projected climate conditions | (Salvacion et al., 2015) |
| Ochratoxin A Sterigmatocystin | *Aspergillus* is the dominant species | Coffee | Benguet | Post-harvest assessment of mycotoxin contamination | (Culliao & Barcelo, 2015) |
| Ochratoxin A | *Aspergillus ochraceus, A. westerdijkiae, A. japonicus, A. carbonarius, Penicillium verruculosum* | Coffee | Benguet, Ifugao, Abra, Cavite | Assessment of the distribution of mycotoxigenic fungi and associated mycotoxin | (Alvindia & de Guzman, 2016) |
| Fumonisin | *Fusarium* spp. | Corn | Cauayan, Isabela Kibawe, Bukidnon | Modeling using insect damage to ears and weather variables as predictor | (Campa et al., 2005) |
| Fumonisin | *Fusarium verticillioides* | Corn Soil | Sampling areas in Visayas and Mindanao | Identification and quantification of fumonisin-producing fungi in corn and soil samples | (Hussien et al., 2017) |
| Fumonisin | *Fusarium verticillioides, F. proliferatum, F. graminearum, Fusarium spp.* | Corn | Corn-producing regions in the Philippines | Identification of *Fusarium* species in corn with Fusarium ear-rot disease of corn contaminated with fumonisin | Pascual et al., 2016 |

*(Continued)*

**TABLE 11.2 (CONTINUED)**
**Mycotoxin Research in the Philippines in the Last 20 Years**

| Mycotoxin | Fungi | Substrate | Location | Focus of Research | Reference |
|---|---|---|---|---|---|
| Fumonisin | *Fusarium verticillioides* | Corn | Laguna<br>Isabela | Genetic characterization of *F. verticillioides* Detection of fumonisin was one of the analysis | (Cumagun et al., 2009) |
| Aflatoxin | *Aspergillus nomius, A. flavus* | Peanut<br>Soil | Various locations in Visayas and Mindanao | Classify isolated aflatoxigenic fungi into chemo and genotypes | AboDalam et al., 2020 |
| Aflatoxin | | Corn and corn products | Ilocos, Isabela, Iloilo, South Cotabato, Davao | Detection of aflatoxin using the minicolumn method designed initially for copra (coconut) meal | (Arim et al., 1999) |
| Fumonisin | *Fusarium verticillioides* | Corn | Isabela<br>Laguna | Genetic diversity of fumonisin-producing *F. verticillioides* | (Magculia & Cumagun, 2011) |
| Aflatoxin | *Aspergillus* spp. | Peanut | | Risk profiling of aflatoxin in peanut to consumers in the Philippines | Rustia et al., 2022 |
| Aflatoxin | *Aspergillus flavus* | Cavendish banana | Davao del Norte | Evaluation of natural and in vitro aflatoxin production | (Sales et al., 2003) |
| Fumonisin | *Fusarium fujikoroi* | Rice | Muñoz, Nueva Ecija<br>Victoria, Laguna | Phylogenetic analysis, fumonisin production, and pathogenicity of *F. fujikoroi* | (Cruz et al., 2013) |
| Ochratoxin<br>Zearalenone | | Pig feed | | Evaluation of the effect of the mycotoxin-deactivating agent on the growth performance of pigs | (Acda et al., 2008) |

## TABLE 11.3
## Medically Important Mycotoxins

| Toxin | Mechanism of Toxic Action | Targets Organs/ Associated Toxic effects |
|---|---|---|
| Aflatoxins | Binding to cellular macromolecules (proteins and nucleic acids) leading to cytotoxic effects<br>(Alshannaq et al., 2017) | Acute aflatoxicosis: vomiting, abdominal pain, pulmonary & cerebral edema, coma, convulsions, death<br>(Alshannaq et al., 2017; Choudhuri et al., 2019)<br>Chronic:<br>Hepatotoxicity<br>Hepatocellular carcinoma<br>Carcinogen (IARC group I)<br>Immune suppression<br>(Choudhuri et al. 2019; Klaassen, 2019; Alshannaq et al., 2017) |
| Ergot alkaloids | Mimic vasoactive monoamines | Ergotism<br>Blood Vessels: Vasospasm<br>Gangrene<br>Central Nervous System Effects: Convulsions<br>Abortion<br>(Klaassen, 2019; Choudhuri et al., 2019) |
| Fumonisins | Interference in sphingolipid metabolism<br>(Alshannaq et al., 2017) | Esophageal cancer<br>Liver & renal cancer<br>(Klaassen, 2019; Choudhuri et al., 2019)<br>IARC Group 2B (possibly carcinogenic)<br>Gastrointestinal: Abdominal pain, borborygmus, diarrhea<br>Neural tube defects in experimental animals |
| Ochratoxins | Inhibition of phenylalanine metabolism<br>Inhibition of mitochondrial ATP production<br>Stimulation of lipid peroxidation<br>(Alshannaq et al., 2017) | Nephrotoxicity (Choudhuri et al., 2019)<br>Immunosuppression<br>Hepatotoxicity<br>Teratogenicity in animal studies<br>IARC Group 2B (possibly carcinogenic)<br>(Alshannaq et al., 2017). |

(*Continued*)

**TABLE 11.3 (CONTINUED)**
**Medically Important Mycotoxins**

| Toxin | Mechanism of Toxic Action | Targets Organs/ Associated Toxic effects |
|---|---|---|
| Patulins | | Nausea, vomiting, ulceration, and hemorrhage<br>Hepatotoxicity<br>IARC Group C<br>(Alshannaq et al., 2017; Klaassen, 2019) |
| Trichothecenes | Inhibition of protein synthesis (Alshannaq et al., 2017) | Gastrointestinal: GI hemorrhage, vomiting<br>Dermatitis on direct contact<br>Alimentary toxic aleukia<br>(Alshannaq et al., 2017; Choudhuri et al. 2019) |
| Zearalenone | Mimics 17β-estradiol binds to estrogen receptors | Estrogenic effects<br>Endocrine disruption<br>IARC Group C<br>(Alshannaq et al., 2017; Klaassen, 2019) |

The clinical manifestations of mycotoxicoses are diverse and depend on the properties of the specific mycotoxin, host factors, and the characteristics of the exposure (dose, frequency, and duration of exposure). The mechanisms of the toxic action of mycotoxins are varied and specific organs may be more affected than others leading to a variety of possible clinical effects. For example, the liver is the primary target of aflatoxin B1, considered the most toxic and significant (Bennett et al., 2003; Alshannaq et al., 2017). On the other hand, ochratoxin A is known for its nephrotoxic effects. Evidence for the epidemiologic link between mycotoxin exposure and human disease may be abundant. For example, aflatoxin B1 is known to be the most potent natural carcinogen (Bennett et al., 2003; Joint FAO/WHO Expert Committee on Food Additives et al., 2002) and is associated with hepatocellular carcinoma (Peraica et al., 2014; Alshannaq et al., 2017). In contrast, ochratoxins are considered severely toxic in laboratory animals, but cases are scarce in humans (Peraica et al., 2014).

Patient characteristics, including age, nutrition, genetic background, and comorbidities, may contribute to increased susceptibility to the toxic effects of mycotoxins. For example, children may be more susceptible to the toxic effects of mycotoxins because of their lower body mass, faster metabolic rate, immature organ functions, and detoxification mechanisms (Peraica et al., 2014). In addition, patients with hepatitis B infection with concomitant exposure to aflatoxin are at increased risk for hepatocellular carcinoma (Bennett et al., 2003). Genetic variations in metabolism, as in aflatoxicosis, would have different effects as aflatoxins are metabolized to reactive metabolites by cytochrome P450 enzymes (Bennett et al., 2003).

The route of exposure to mycotoxins is mostly through ingestion from contaminated food (Peraica et al., 2014). However, depending on the toxin, dermal and inhalation exposures can also happen. Generally, it is not communicable from person to person (Bennett et al., 2003). While it is impossible to eliminate mycotoxins from food, regulatory agencies have set regulatory standards. The Joint Food and Agriculture Organization/World Health Organization Expert Committee on Food Additives (JECFA) is responsible for evaluating the risks of mycotoxins to human health. It sets tolerable intake limits, while the Codex Alimentarius Commission establishes the standards to limit exposures based on the JECFA evaluation (World Health Organization, 2018; Habschied et al., 2021).

Diagnosis of mycotoxicosis is difficult. Human biological monitoring, while needing validation, can be a helpful tool and provide information for exposure assessment. Biological samples that can be used include urine, plasma, serum, and breast milk (Escrivá et al., 2017; Habschied et al., 2021). Some of the most commonly used methods for detecting mycotoxins in biological samples include high-performance liquid chromatography coupled with fluorescence detection (HPLC-FLD), enzyme-linked immunosorbent assays (ELISA), liquid chromatography-tandem mass spectrometry (LC-MS/MS), and gas chromatography-tandem mass spectrometry (GCMS) (Habschied et al., 2021; Escrivá et al., 2017).

In general, there is no specific treatment for mycotoxicosis. Management consists of supportive therapy (e.g., adequate hydration) (Bennett et al., 2003) and management of the sequelae (e.g., referral to a specialist for those with hepatocellular carcinoma associated with chronic exposure to aflatoxin B1). Since most of the associated chronic effects are nearly irreversible (carcinogenesis), emphasis is placed on exposure prevention strategies.

## 11.5 MYCOTOXIN CHEMISTRY AND DETECTION

The commonly tested mycotoxins in the Philippines are aflatoxin B1 (AFB1), aflatoxin B2 (AFB2), aflatoxin G1 (AFG1), aflatoxin G2 (AFG2), fumonisin B1 (FB1), ochratoxin A (OTA), and zearalenone (ZEN) (Figure 11.2). The analytical methods used to analyze mycotoxins are immunochemical-based methods, biosensors (Zhang et al., 2023), and chromatographic techniques. Enzyme-linked immunosorbent assay (ELISA) is an example of an immunochemical-based method that is simple, cheap, and portable. This method is used for first-level screening and survey studies (Anfossi et al., 2016). Chromatographic techniques such as high-performance liquid chromatography (HPLC) with an ultraviolet (UV) or fluorescence (FLD) detector and gas chromatography (GC) are for the quantitative determination of single or few related mycotoxin compounds for confirmatory analysis. In comparison, liquid chromatography-tandem with triple quadrupole mass spectrometry (LCMS/MS) and ultrahigh performance LC with high-resolution mass spectrometry (UHPLC-HRMS) provides a more sensitive, accurate, and selective method for the determination of mycotoxin.

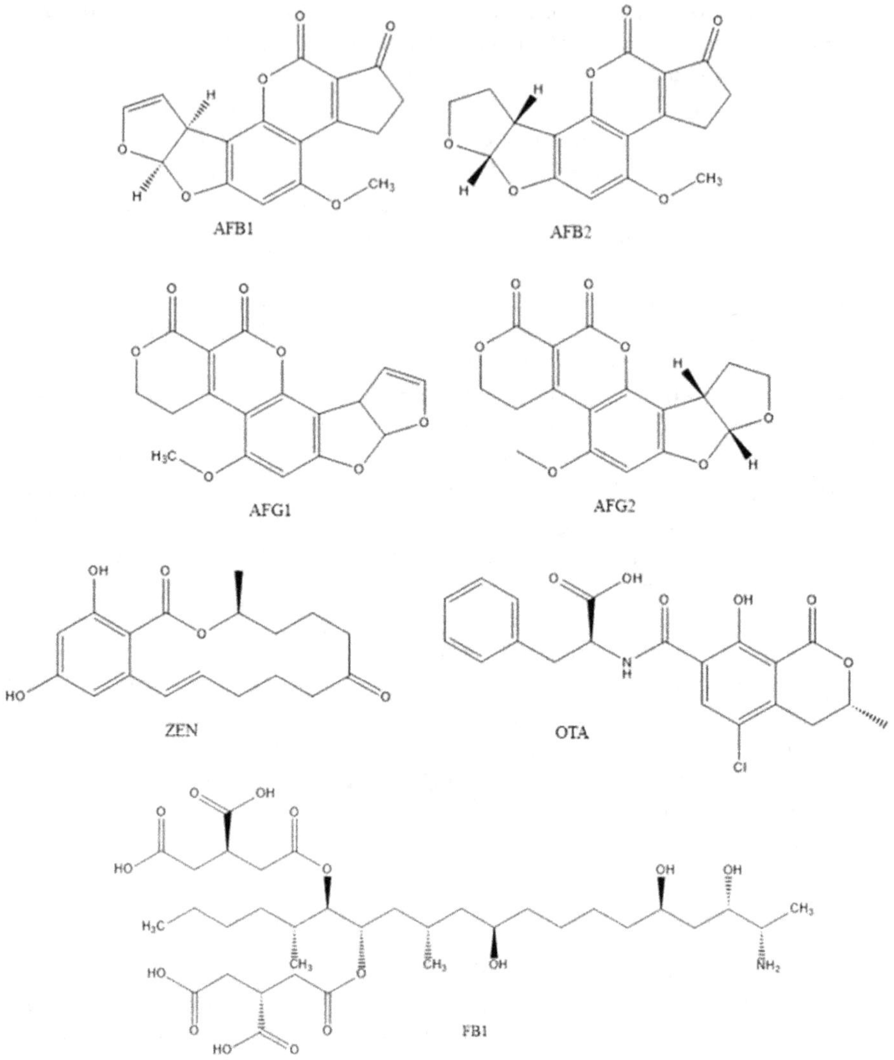

**FIGURE 11.2** Structure of commonly tested mycotoxins in the Philippines. Aflatoxin B1 (AFB1), Aflatoxin B2 (AFB2), Aflatoxin G1 (AFG1), Aflatoxin G2 (AFG2), Zearalenone (ZEN), fumonisin B1 (FB1), ochratoxin A (OTA)

These instruments are used for multi-residue analysis and structure elucidation of unknown compounds (Leite et al., 2023).

Before instrumentation, the samples are subjected to extraction methodologies to isolate and concentrate the compound. Examples of extraction methodologies are liquid-liquid extraction (LLE), solid-liquid extraction (SLE), solid phase extraction (SPE), immunoaffinity column (IAC) (Kim et al., 2017), and QuEChERS (Quick, Easy, Cheap, Effective, Rugged, and Safe) (Rodríguez-Cañás et al., 2023). After the extraction, clean-up can be employed to remove other interferences. The chosen method and clean-up method will vary depending on the matrix compatibility (Castell et al., 2023).

Mycotoxins are naturally occurring contaminants, and therefore, imposing a ban on them is impossible. In order to assure food safety and protection of consumers, maximum admissible levels of mycotoxins are regulated by various national and international agencies depending on the commodity or compound and its intended use (Anfossi et al., 2016). These organizations are the European Union (EU) Commission, United States Food and Drug Administration (US-FDA), World Health Organization (WHO), Codex Alimentarius Commission (CODEX), and Food and Agriculture Organization (FAO) (Abdolmaleki et al., 2021). Most of the set regulatory limits of different countries are based on the EU Commission regulation (EC) No. 1881/2006, while the criteria for sample preparation and confirmatory methods of analysis used for food control purposes are in the EU Commission regulation (EC) No 401/2006. In the Philippines, the maximum levels for mycotoxin are published in the Philippine National Standard/Bureau of Agriculture and Fisheries Standard (PNS/BAFS) 194:2022 (Table 11.4).

The strategies to control mycotoxin contamination in commercially important food and feed products are outlined in the standards set in the Philippine National Standards (PNS) developed by the Bureau of Agriculture and Fisheries Products Standards (BAFPS) under the Department of Agriculture (Table 11.5). So far, the available PNS dealing with mycotoxins are only for agricultural products and none for fishery products. While PNS 194:2002 also includes aquatic food commodities, the standards for these products only apply to maximum limits for cadmium, lead, and methylmercury. Mycotoxin management generally relies on controlling the crucial level of moisture contents, water activities, relative humidity, and temperature storage, depending on the product.

## 11.6 WAY FORWARD

Biosafety and biosecurity concerns surrounding fungi depend largely on sound policies and practices to avoid adverse effects. Risk reduction principles are similar to how laboratories manage risks from bacteria, viruses, and other hazards and threats. Health implications and agricultural losses can be detrimental, even catastrophic, when the risks remain uncontrolled. Thus, it is important to continuously monitor and evaluate existing practices and policies to respond to the dynamic challenges of mycological safety and security. Research, both in the Philippines and neighboring ASEAN countries, may be done to better understand the network of factors at play that shape mycotoxin vulnerability, and institute mitigation measures to reduce risk to the minimum.

## ACKNOWLEDGMENT

Authors express their gratitude to the researchers in mycology and biorisk who have contributed to the growth in this field.

**TABLE 11.4**
**Maximum Level (ML) of Mycotoxin per Commodity**

| Mycotoxin | Commodity | Maximum Level (µg/kg) PNS/BAFS 194:2022 | Maximum Level (µg/kg) EU EC No. 1881/2006 |
|---|---|---|---|
| Total Aflatoxin | Nuts, almonds, dried figs, maize grain, rice, spices | 10–15 | 15 |
| | Cereals | - | 4 |
| Aflatoxin M1 | Milk | 0.50 | |
| DON | Meal derived from wheat or maize | 1,000 | 1,750 |
| | Cereal grains for further processing | 2,000 | 1,250 |
| | Pasta | - | 750 |
| | Bread | - | 500 |
| | Processed cereal-based foods and baby foods for infants and young children | - | 200 |
| Fumonisins | Raw maize grain | 4,000 | 4,000 |
| | Maize meal | 2,000 | 1,000 |
| Ochratoxin A | Wheat, barley, rye | 5.0 | |
| | Cereals | - | 3–5 |
| | Dried vine fruit | - | 10 |
| | Coffee | - | 5–10 |
| | Wine, juice | - | 2 |
| | Processed cereal-based and dietary foods for infants and young children | - | 0.5 |
| | Spices | - | 15–30 |
| Patulin | Fruit juices, spirit drinks, cider, and other fermented drinks from apple | - | 50 |
| | Apple products | - | 25 |
| | Baby foods, apple juice, and apple products for infants and young children | - | 10 |
| Zearalenone | Unprocessed cereals | - | 100 |
| | Unprocessed maize | - | 350 |
| | Cereals | - | 75 |
| | Refined maize oil | - | 400 |
| | Bread, pastries, biscuits, and cereals | - | 50 |
| | Processed cereal-based foods, processed maize-based food, and baby foods for infants and young children | - | 20 |

**TABLE 11.5**
**Philippine National Standards for Preventing and Controlling Mycotoxins in Specific Agricultural Commodities**

| PNS/BAFPS Number | Title | Salient Recommendation to Prevent Mycotoxin Contamination |
|---|---|---|
| 44:2009 | Code of Practice for the Prevention and Reduction of Aflatoxin Contamination in Copra | Copra moisture content (MC) of 6% to 7% |
| 130:2014 | Code of Practice for the Prevention and Reduction of Ochratoxin A Contamination in Philippine Cacao Beans | Cacao beans maximum MC at 7.5% |
| 131:2014 | Code of Practice for the Prevention and Reduction of Ochratoxin A Contamination in Philippine Tablea | Removal of shell or husk after roasting |
| 173:2015 | Code of Practice for the Prevention and Reduction of Aflatoxin Contamination in Tree Nuts | Storage conditions:<br>water activities less than 0.7;<br>relative humidity below 70%;<br>temperatures below 10°C<br>Maximum allowable levels for total aflatoxins in tree nuts (almonds, Brazil nuts, hazelnuts, pistachios: 10 μg/kg for ready-to-eat nuts; 15 μg/kg for further processing) |
| 175:2015 | Code of Practice for the Prevention and Reduction of Aflatoxin Contamination in Peanuts | Storage conditions:<br>water activities less than 0.7;<br>relative humidity below 70%;<br>temperatures between 0 and 10°C |
| 170:2016 | Code of Practice for the Prevention and Reduction of Ochratoxin A Contamination in Coffee | Maximum aw of 0.67 to 0.70;<br>MC of beans not to exceed 12% |
| 27:2018 | Code of Practice for the Prevention and Reduction of Aflatoxin Contamination in Corn | MC of corn not greater than 14% |
| 194:2022 | General Standard for Contaminants and Toxins in Food and Feed-Product Standard | Maximum allowable limits (μg/kg) for aflatoxin M1, total aflatoxins, ochratoxin A, deoxynivalenol, fumonisins (B1 + B2) |
| 146:2019 | Code of Practice for the Prevention and Reduction of Mycotoxin Contamination in Cereals | Corn MC at 14% before storage;<br>Rice and sorghum grains MC at 14% before storage |

## REFERENCES


Abbas, H. K. (2005). The costs of mycotoxin management in the United States. In H. K. Abbas (Ed.). *Aflatoxin and Food Safety*. CRC Press (pp. 26–37).

Abdolmaleki, K., Khedri, S., Javanmardi, F., Oliveira, C. A. F., and Khaneghah, A. M. (2021). The mycotoxin in edible oils: An overview of prevalence, concentration, toxicity, detection, and decontamination techniques. *Trends in Food Science and Technology, 115*, 500–511.

AboDalam, T. H., Amra, H., Sultan, Y., Magan, N., Carlobos-Lopez, A. L., Cumagun, C. J. R., & Yli-Mattila, T. (2020). New genotypes of aflatoxigenic fungi from Egypt and the Philippines. *Current Research in Environmental & Applied Mycology (Journal of Fungal Biology), 10*(1), 142–155.

Acda, S. E., Batungbacal, M. R., Centeno, J. R., & Carandang, N. F. (2008). Effects of mycotoxin-deactivating agent on the growth performance of pigs fed ochratoxin and zearalenone-contaminated diets. *Philippine Journal of Veterinary Medicine, 45*(1), 14–21.

Alshannaq, A., & Yu, J. H. (2017). Occurrence, toxicity, and analysis of major mycotoxins in food. *International Journal of Environmental Research and Public Health, 14*(6), 632.

Alvindia, D. G., & de Guzman, M. F. (2016). Survey of Philippine coffee beans for the presence of ochratoxigenic fungi. *Mycotoxin Research, 32*(2), 61–67.

Anfossi, L., Giovannoli, C., & Baggiani, C. (2016). Mycotoxin detection. *Current Opinion in Biotechnology, 37*, 120–126.

Arim, R. H., Ferolin, C. A., Ramirez, R. P., Aguinaldo, A. R., & Yoshizawa, T. (1999). Determination of aflatoxins in corn and its processed products in the Philippines by minicolumn method. マイコトキシン, *1999*(Suppl2), 193–196.

Asam, S., & Rychlik, M. (2013). Potential health hazards due to the occurrence of the mycotoxin tenuazonic acid in infant food. *European Food Research and Technology, 236*(3), 491–497.

Balendres, M. A. O., Karlovsky, P., & Cumagun, C. J. R. (2019). Mycotoxigenic fungi and mycotoxins in agricultural crop commodities in the Philippines: A review. *Foods, 8*(7), 249.

Bennett, J. W., & Klich, M. (2003). Mycotoxins. *Clinical Microbiology Reviews, 16*(3), 497–516.

Bhat, R., Rai, R. V., & Karim, A. (2010). Mycotoxins in food and feed: Present status and future concerns. *Comprehensive Reviews in Food Science and Food Safety, 9*(1), 57–81.

Bullerman, L. B. (1979). Significance of mycotoxins to food safety and human Health1,2. *Journal of Food Protection, 42*(1), 65–86.

Bureau of Agriculture and Fisheries Standards. (2022). General standards for contaminants and toxins in food and feed — Product standard PNS/BAFS 194:2022.

Campa, R. de la, Hooker, D. C., Miller, J. D., Schaafsma, A. W., & Hammond, B. G. (2005). Modeling effects of environment, insect damage, and Bt genotypes on fumonisin accumulation in maize in Argentina and the Philippines. *Mycopathologia, 159*(4), 539–552.

Castell, A., Arroyo-Manzanares, N., Campillo, N., Torres, C., Fenoll, J., & Viñas, P. (2023). Bioaccumulation of mycotoxins in human forensic liver and animal liver samples using a green sample treatment. *Microchemical Journal, 185*, 108192.

Cruz, A., Marín, P., González-Jaén, M. T., Aguilar, K. G. I., & Cumagun, C. J. R. (2013). Phylogenetic analysis, fumonisin production and pathogenicity of *Fusarium fujikuroi* strains isolated from rice in the Philippines. *Journal of the Science of Food and Agriculture, 93*(12), 3032–3039.

Culliao, A. G. L., & Barcelo, J. M. (2015). Fungal and mycotoxin contamination of coffee beans in Benguet Province, Philippines. *Food Additives and Contaminants: Part A, 32*(2), 250–260.

Cumagun, C. J. R., Ramos, J. S., Dimaano, A. O., Munaut, F., & Van Hove, F. (2009). Genetic characteristics of *Fusarium verticillioides* from corn in the Philippines. *Journal of General Plant Pathology, 75*(6), 405.

De Leon, A. M., Cruz, A. S., Evangelista, A. B. B., Miguel, C. M., Pagoso, E. J. A., Dela Cruz, T. E. E., Nelsen, D. J., & Stephenson, S. L. (2019). Species listing of macrofungi found in the Ifugao indigenous community in Ifugao Province, Philippines. *Philippine Agricultural Scientist, 102*, 118–131.

De Leon, A. M., Reyes, R. G., & dela Cruz, T. E. E. (2012). An ethnomycological survey of macrofungi utilized by Aeta communities in Central Luzon, Philippines. *Mycosphere, 3*(2), 251–259.

De Leon, A. M., Kalaw, S. P., Dulay, R. M. R., Undan, J. R., Alfonzo, D. O., Undan, J. Q., & Reyes, R. G. (2016). Ethnomycological survey of the Kalanguya indigenous community in Caranglan, Nueva Ecija, Philippines. *Current Research in Environmental & Applied Mycology, 6*(2), 61–66.

dela Cruz, T. E. E., & De Leon, A. M. (2023). Edible mushrooms of the Philippines: Traditional knowledge, bioactivities, mycochemicals, and in vitro cultivation. In J. J. G. Guerrero, T. U. Dalisay, M. P. De Leon, M. A. O. Balendres, K. I. R. Notarte, & T. E. E. Dela Cruz (Eds.), *Mycology in the Tropics: Updates on Philippine Fungi* (pp. 271–292). Academic Press.

Dulay, R. M. R., Batangan, J. N., Kalaw, S. P., Leon, A. M. D., Cabrera, E. C., Kimura, K., Eguchi, F., & Reyes, R. G. (2023). Records of wild mushrooms in the Philippines: A review. *Journal of Applied Biology and Biotechnology, 11*(2), 11–32.

Dulay, R. M. R., Cabrera, E. C., Kalaw, S. P., & Reyes, R. G. (2021). Optimization of submerged culture conditions for mycelial biomass production of fourteen *Lentinus* isolates from Luzon Island, Philippines. *Biocatalysis and Agricultural Biotechnology, 38*, 102226.

Choudhuri, S., Chanderbhan, R. F., & Mattia, A. (2019). Food toxicology: Fundamental and regulatory aspects. In C.D. Klaassen (Ed.), *Casarett and Doull's Toxicology the Basic Science of Poisons* (9th edn., pp. 1315–1359). McGraw-Hill Education.

Escrivá, L., Font, G., Manyes, L., & Berrada, H. (2017). Studies on the presence of mycotoxins in biological samples: An overview. *Toxins, 9*(8), 251.

Fernández-Cruz, M. L., Mansilla, M. L., & Tadeo, J. L. (2010). Mycotoxins in fruits and their processed products: Analysis, occurrence and health implications. *Journal of Advanced Research, 1*(2), 113–122.

Flores, Jr, Alvarez, & Cortez (2014). Inventory and utilization of macrofungi species for food and medicine. *International Conference on Biological, Chemical and Environmental Sciences (BCES-2014) June 14-15, 2014 Penang (Malaysia)*. International Conference on Biological, Chemical and Environmental Sciences, Penang, Malaysia.

Focker, M., van der Fels-Klerx, H. J., Magan, N., Edwards, S. G., Grahovac, M., Bagi, F., Budakov, D., Suman, M., Schatzmayr, G., Krska, R., & de Nijs, M. (2021). The impact of management practices to prevent and control mycotoxins in the European food supply chain: My ToolBox project results. *World Mycotoxin Journal, 14*(2), 139–154.

Fumagalli, F., Ottoboni, M., Pinotti, L., & Cheli, F. (2021). Integrated mycotoxin management system in the feed supply chain: Innovative approaches. *Toxins, 13*(8), 572.

Gbashi, S., Edwin Madala, N., De Saeger, S., De Boevre, M., Adekoya, I., Ayodeji Adebo, O., & Berka Njobeh, P. (2019). The socio-economic impact of mycotoxin contamination in Africa. In P. Berka Njobeh & F. Stepman (Eds.), *Mycotoxins—Impact and Management Strategies*. IntechOpen.

Gonçalves, R. A., Schatzmayr, D., Albalat, A., & Mackenzie, S. (2020). Mycotoxins in aquaculture: Feed and food. *Reviews in Aquaculture, 12*(1), 145–175.

Habschied, K., Kanižai Šarić, G., Krstanović, V., & Mastanjević, K. (2021). Mycotoxins-biomonitoring and human exposure. *Toxins*, *13*(2), 113.

Hussien, T., Carlobos-Lopez, A. L., Cumagun, C. J. R., & Yli-Mattila, T. (2017). Identification and quantification of fumonisin-producing *Fusarium* species in grain and soil samples from Egypt and the Philippines. *Phytopathologia Mediterranea*, *56*(1), 146–153.

Joint FAO/WHO Expert Committee on Food Additives, World Health Organization & International Programme on Chemical Safety. (2002). *Evaluation of Certain Mycotoxins in Food: Fifty-Sixth Report of the Joint FAO/WHO Expert Committee on Food Additives*. World Health Organization. https://apps.who.int/iris/handle/10665/42448

Kim, H. J., Lee, M. J., Kim, H. J., Cho, S. K., Park, H. J., & Jeong, M. H. (2017). Analytical method development and monitoring of aflatoxin b1, B2, G1, G2 and ochratoxin A in animal feed using HPLC with Fluorescence detector and photochemical reaction device. *Cogent Food and Agriculture*, *3*(1), 1419788.

Klaassen, C. D., ed. (2019). *Casarett and Doull's Toxicology: The Basic Science of Poisons* (9th edn.). New York: McGraw-Hill Education.

Lazo, C. R. M., Kalaw, S. P., & De Leon, A. M. (2015). Ethnomycological survey of macrofungi utilized by gaddang communities in Nueva Vizcaya, Philippines. *Applied and Environmental Microbiology*, *5*(3), 256–262.

Leite, M., Freitas, A., Barbosa, J., & Ramos, F. (2023). Comprehensive assessment of different extraction methodologies for optimization and validation of an analytical multi-method for determination of emerging and regulated mycotoxins in maize by UHPLC-MS/MS. *Food Chemistry Advances*, *2*, 100145.

Licyayo, D. C. M. (2018). Gathering practices and actual use of wild edible mushrooms among ethnic groups in the Cordilleras, Philippines. In *Diversity and Change in Food Wellbeing: Cases from Southeast Asia and Nepal* (pp. 71–86). Wageningen Academic Publishers.

Lukwago, F. B., Mukisa, I. M., Atukwase, A., Kaaya, A. N., & Tumwebaze, S. (2019). Mycotoxins contamination in foods consumed in Uganda: A 12-year review (2006–18). *Scientific African*, *3*, e00054.

Magculia, N. J. F., & Cumagun, C. J. R. (2011). Genetic diversity and PCR-based identification of potential fumonisin-producing *Fusarium verticillioides* isolates infecting corn in the Philippines. *Tropical Plant Pathology*, *36*, 225–232.

Magnoli, A. P., Poloni, V. L., & Cavaglieri, L. (2019). Impact of mycotoxin contamination in the animal feed industry. *Current Opinion in Food Science*, *29*, 99–108.

Mannaa, M., & Kim, K. D. (2017). Influence of temperature and water activity on deleterious fungi and mycotoxin production during grain storage. *Mycobiology*, *45*(4), 240–254.

Marroquín-Cardona, A. G., Johnson, N. M., Phillips, T. D., & Hayes, A. W. (2014). Mycotoxins in a changing global environment – A review. *Food and Chemical Toxicology*, *69*, 220–230.

Maslang, J. A. L., Asuncion, C., Jubay, A. Z., Saludarez, M. U., Liday, D. M., Villanueva, H. T., & Bolonquita, M. C. (2021). A survey on the ethnomycology and laboratory analyses of wild mushrooms utilized as food among multicultural groups in selected municipalities of Nueva Vizcaya, Philippines. *The PASCHR Journal*, *4*(1), 62–80.

Moretti, A., Logrieco, A. F., & Susca, A. (2017). Mycotoxins: An underhand food problem. In A. Moretti & A. Susca (Eds.), *Mycotoxigenic Fungi: Methods and Protocols* (pp. 3–12). Springer.

Murphy, P. A., Hendrich, S., Landgren, C., & Bryant, C. M. (2006). Food mycotoxins: An update. *Journal of Food Science*, *71*(5), R51–R65.

Murugesan, G. R., Ledoux, D. R., Naehrer, K., Berthiller, F., Applegate, T. J., Grenier, B., Phillips, T. D., & Schatzmayr, G. (2015). Prevalence and effects of mycotoxins on poultry health and performance, and recent development in mycotoxin counteracting strategies1. *Poultry Science*, *94*(6), 1298–1315.

Ostry, V., Malir, F., Toman, J., & Grosse, Y. (2017). Mycotoxins as human carcinogens-the IARC Monographs classification. *Mycotoxin Research*, *33*(1), 65–73.

Pascual, C. B., Barcos, A. K. S., Mandap, J. A. L., & Ocampo, T. M. (2016). Fumonisin-producing Fusarium species causing ear rot of corn in the Philippines. *Philippine Journal of Crop Science (PJCS)*, *41*(1), 12–21.

Peraica, M., Richter, D., & Rašić, D. (2014). Mycotoxicoses in children. *Archives of Industrial Hygiene and Toxicology*, *65*(4), 347–363.

Pitt, J. I., & Miller, J. D. (2017). A concise history of mycotoxin research. *Journal of Agricultural and Food Chemistry*, *65*(33), 7021, 7033.

Raghavender, C., & Reddy, B. (2009). Human and animal disease outbreaks in India due to mycotoxins other than aflatoxins. *World Mycotoxin Journal*, *2*(1), 23–30.

Rodríguez-Cañás, I., González-Jartín, J. M., Alvariño, R., Alfonso, A. Vieytes, M. R., & Botana, L. M. (2023). Detection of mycotoxins in cheese using an optimized analytical method based on a QuEChERS extraction and UHPLC-MS/MS quantification. *Food Chemistry*, *408*, 135182.

Rustia, A., Mariano, C., Bautista, K., Mahoney, D., Barrios, E., Villarino, C., Limon, M., & Capanzana, M. (2022). Risk profiling of aflatoxin in peanut (*Arachis hypogaea* L.) to the Filipino consuming population. *Philippine Journal of Science*, *151*(5), 1557–1577.

Sales, A. C., & Yoshizawa, T. (2005). Mold counts and *Aspergillus* section *Flavi* populations in rice and its by-products from the Philippines. *Journal of Food Protection*, *68*(1), 120–125.

Sales, A. C., Azanza, P. V., & Yoshizawa, T. (2003). Evaluation of *aspergillus* Section *flavi* populations, natural and in vitro aflatoxin production in dried Cavendish banana (*Musa cavendishii*) chips from Southern Philippines. *マイコトキシン*, *2003*(Suppl3), 325–329.

Salvacion, A. R., Pangga, I. B., & Cumagun, C. J. R. (2015). Assessment of mycotoxin risk on corn in the Philippines under current and future climate change conditions. *Reviews on Environmental Health*, *30*(3), 135–142.

Shephard, G. S. (2008). Impact of mycotoxins on human health in developing countries. *Food Additives and Contaminants: Part A*, *25*(2), 146–151.

Siri-anusornsak, W., Kolawole, O., Mahakarnchanakul, W., Greer, B., Petchkongkaew, A., Meneely, J., Elliott, C., & Vangnai, K. (2022). The Occurrence and Co-occurrence of Regulated, Emerging, and Masked Mycotoxins in Rice Bran and Maize from Southeast. *Toxins, 14*(8), 567.

Tantengco, O. A., & Ragragio, E. (2018). Ethnomycological survey of macrofungi utilized by Ayta communities in Bataan, Philippines. *Current Research in Environmental & Applied Mycology*, *8*(1), 104–108.

Torres, M. L. S., Ontengco, D. C., Tadiosa, E. R., & Reyes, R. G. (2020a). Ethnomycological studies on the Bugkalot indigenous community in Alfonso Castañeda, Nueva Vizcaya, Philippines. *International Journal of Pharmaceutical Research and Allied Sciences*, *9*(4), 43–54.

Torres, M., Tadiosa, E. R., & Reyes, R. G. (2020b). Species listing of macrofungi on the Bugkalot Tribal community in Alfonso Castañeda, Nueva Vizcaya, Philippines. *Current Research in Environmental & Applied Mycology*, *10*(1), 475–493.

Undan, J. R., Fermin, S. M. C., Pajarillaga, L. M. A., Malonzo, M. A. C., Kalaw, S. P., & Reyes, R. G. (2021, November). Ethnomycological survey and molecular identification of macrofungi utilized by Bicolano community in Camarines Sur, Southern Luzon, Philippines. *Ecology, Environment and Conservation*, *27*, S1–S7.

U.S. Food & Drug. (2022). *Mycotoxins: Toxins Found in Food Infected by Certain Molds or Fungi.* https://www.fda.gov/food/natural-toxins-food/mycotoxins#:~:text=As%20an%20individual%20consumer%2C%20you,to%20buy%20is%20not%20contaminated

van Dongen, P. W. J., & de Groot, A. N. J. A. (1995). History of ergot alkaloids from ergotism to ergometrine. *European Journal of Obstetrics and Gynecology and Reproductive Biology, 60*(2), 109–116.

World Health Organization International Agency for Research on Cancer. (2002). *IARC Monographs on the Evaluation of Carcinogenic Risks to Humans.* Lyon: International Agency for Research on Cancer.

World Health Organization. (2018). Mycotoxins. https://www.who.int/news-room/fact-sheets/detail/mycotoxins

Yogendrarajah, P., Jacxsens, L., Lachat, C., Walpita, C. N., Kolsteren, P., De Saeger, S., & De Meulenaer, B. (2014). Public health risk associated with the co-occurrence of mycotoxins in spices consumed in Sri Lanka. *Food and Chemical Toxicology, 74*, 240–248. https://doi.org/10.1016/j.fct.2014.10.007

Zhang, M., Guo, X., & Wang, J. (2023). Advanced biosensors for mycotoxin detection incorporating miniaturized meters. *Biosensors and Bioelectronics, 224*, 115077.

# 12 Genetically Modified Organisms in Research and Biotechnology

*Ma. Socorro Edden P. Subejano*
Department of Microbiology Anatomy
Physiology and Pharmacology, School of Agriculture
Biomedicine and Environment La Trobe University
Melbourne, Victoria, Australia

*Jann Eldy L. Daquioag*
Institute of Biology, National Science
Complex, College of Science
University of the Philippines
Diliman, Quezon City, Philippines

*Gil M. Penuliar*
Institute of Biology, National Science
Complex, College of Science
University of the Philippines
Diliman, Quezon City, Philippines

## 12.1 INTRODUCTION

Imagine a world where crops could defy the limits of nature, where they possess the power to resist pests and conquer harsh environmental conditions. This is not science fiction, but a possibility through the creation of genetically modified organisms (GMOs). As we stand at the crossroads of genetic engineering and agriculture, the promise of GMOs beckons us with wonder, but also raises profound questions about food and the environment. The history of GMOs is as captivating as their potential. In 1973, Herbert Boyer and Stanley Cohen developed genetic engineering by inserting DNA from one bacteria into another; and in 1982, the Food and Drug Administration approved human insulin to treat diabetes, as the first consumer GMO product developed through genetic engineering (Baeshen et al., 2014).

Ingenious engineering has endowed crops like soybeans and maize with herbicide tolerance, empowering farmers to weed out unwanted plants without harming their crops. The introduction of *Bt* crops, armed with proteins from *Bacillus thuringiensis* Berliner, has allowed plants to repel certain insects naturally (Hautea et al., 2016).

DOI: 10.1201/9781003426219-12

The potential of GMOs even extends to environmental remediation, as genetically engineered organisms show promise in detoxifying polluted environments (Rafeeq et al., 2023).

The creation of GMOs, however, has not been without controversies. The safety of these organisms, their environmental impact, and the ethical considerations surrounding their creation have sparked debates within the scientific community. To address these concerns, regulations have been established to ensure the responsible use and safety of GMOs in various countries. It is crucial to recognize that biosafety issues are multifaceted, encompassing scientific, economic, political, sociological, and public health factors. A comprehensive understanding of risk assessment and management procedures will facilitate responsible decision-making and harness the full potentials of GMOs.

## 12.2 WHAT ARE GENETICALLY MODIFIED ORGANISMS?

A GMO is an animal, plant, or microbe whose DNA has been altered using genetic engineering techniques. Throughout history, humans have employed selective breeding methods to modify organisms, intentionally cultivating desirable traits in crops, cattle, and even domesticated animals over generations (WHO, 2020). These traditional breeding practices involve choosing organisms with desired traits and allowing them to reproduce, gradually accumulating the desired characteristics. For instance, early agriculturalists selectively bred wild grasses to develop modern-day wheat varieties (Kanchiswamy et al., 2015).

However, recent advances in biotechnology have ushered in a revolution in genetic manipulation, enabling scientists to directly modify the DNA of microorganisms, crops, and animals. This transformation is primarily attributed to the development of recombinant DNA technology in the 1970s, which allowed for the transfer of genes between organisms that would not naturally interbreed (Cohen et al., 1972). Genetic engineering techniques, such as CRISPR-Cas9, have further enhanced the precision and efficiency of modifying specific genes within an organism's genome (Doudna & Charpentier, 2014).

The implications of these advances are profound, as they provide opportunities to enhance crop yields, develop disease-resistant plants, and even potentially address global food security challenges. However, GMOs also raise concerns about their impact on ecosystems, human health, and ethical considerations, sparking ongoing debates and regulatory discussions on their usage and labeling (Pellegrino et al., 2018).

## 12.3 IMPORTANCE OF GENETICALLY MODIFIED ORGANISMS

GMOs have heralded a transformative era in agriculture and biotechnology by enabling precise incorporation of desired traits into crop genetics. This precision has accelerated the development of crops with heightened resistance to pests, diseases, and environmental stressors, effectively addressing global food production challenges (Brookes & Barfoot, 2020). As the world's population continues to grow and arable land diminishes, GMOs play a pivotal role in fostering sustainable food production, offering improved yields, reduced food wastage, and enhanced nutritional content (Qaim, 2010).

GMOs have not only revolutionized staple crops like corn and soybeans in the United States, boosting agricultural efficiency and environmental sustainability, but they have also expanded into specialty crops and fruits, meeting consumer demands for higher quality and safety (Fernandez-Cornejo et al., 2014). By reducing the reliance on chemical pesticides and promoting eco-friendly farming practices, GMOs help safeguard ecosystems and conserve valuable land resources (Raman, 2017). However, their responsible use must always consider ethical, environmental, and social implications to fully harness their potential for a sustainable future. In essence, GMOs have become an indispensable tool in modern agriculture, addressing the challenges posed by a burgeoning global population while advancing the cause of sustainable and efficient food production.

*Advantages of using genetically modified organisms*: GMOs offer significant advantages in agriculture and biotechnology. They enhance crop yields by introducing genetic traits that improve nutrient utilization and resist environmental challenges, which is critical for addressing global food demand and regions with limited arable land (Qaim & de Janvry, 2003). GMOs also reduce the need for chemical pesticides by providing built-in pest resistance, thereby promoting environmental sustainability and human health (Fernandez-Cornejo et al., 2014).

Moreover, GMOs have the potential to boost the nutritional content of crops, thus contributing to the fight against malnutrition. As the global population continues to grow, GMOs play a vital role in ensuring food security by increasing production and reducing losses from pests, diseases, and environmental stressors (Klümper & Qaim, 2014). They contribute significantly to sustainable agriculture by decreasing chemical inputs, optimizing resource use, minimizing water consumption, and supporting responsible land management, which ultimately helps in reducing deforestation.

## 12.4 ETHICAL CONCERNS RELATED TO GENETICALLY MODIFIED ORGANISMS

GMOs are at the forefront of global discussions concerning food production, biotechnology, and ethics. Advocates emphasize their potential to address pressing issues like food security and malnutrition, while critics raise ethical concerns and controversies (Krimsky, 2015). These encompass safety concerns regarding potential unintended consequences for human health and the environment, including environmental impacts such as harm to non-target species, disruptions to ecosystems, and biodiversity loss (Domingo & Bordonaba, 2011). Additionally, concerns arise about corporate consolidation in the seed industry, which may reduce seed diversity and farmer autonomy. Ethical dilemmas stem from altering the genetic makeup of organisms, challenging natural evolution, and the impact on vulnerable populations due to patents and access limitations (Qaim & Kouser, 2013). Transparency in GMO labeling is another ethical issue, impacting the right to informed food choices. The introduction of genetically modified animals adds complexities related to animal welfare. Despite these challenges, GMO research significantly contributes to biotechnology, spanning medicine, agriculture, and environmental fields. Ethical concerns necessitate thoughtful regulation and public education for responsible GMO integration, balancing potential benefits with these multifaceted ethical considerations to ensure a sustainable and equitable future.

## 12.5 REGULATORY FRAMEWORK FOR GENETICALLY MODIFIED ORGANISMS

Regulatory frameworks for GMOs encompass a set of rules, guidelines, and policies established to govern the research, development, and commercialization of genetically engineered organisms. This framework aims to ensure the safety of GMOs for human health, the environment, and agriculture while promoting scientific innovation. In the United States, GMO regulation is a complex process coordinated by three key federal agencies: the United States Department of Agriculture (USDA), the Environmental Protection Agency (EPA), and the Food and Drug Administration (FDA). These agencies collectively ensure the safe development, deployment, and consumption of genetically modified organisms (GMOs) (US FDA, 2021).

The USDA assumes a pivotal role in overseeing GMOs by managing field testing and the release of genetically modified crops into the environment. This includes evaluating the potential environmental impacts and ensuring compliance with regulations and safety protocols. The agency collaborates closely with developers and researchers to assess the safety of GMOs in agricultural settings, aiming to strike a balance between innovation and environmental stewardship (US FDA, 2021). Concurrently, the EPA focuses its regulatory efforts on genetically engineered plants endowed with pest-resistant traits. This includes GMOs designed to produce insecticides, thereby reducing the need for chemical pesticide application. The EPA rigorously evaluates these traits to ensure they pose minimal harm to non-target species and the broader environment. Their oversight helps safeguard ecosystems and human health, offering a sustainable alternative to conventional pest control methods. The FDA concentrates on assessing the safety of genetically modified foods intended for human consumption. They conduct comprehensive evaluation to determine potential risks to human health and scrutinize the genetic modifications and the safety of the end products. Additionally, the FDA mandates labeling requirements for genetically modified foods, allowing consumers to make informed choices about their dietary selections (US FDA, 2021).

In the European Union (EU), the regulatory framework for genetically modified organisms (GMOs) is characterized by stringent guidelines and a steadfast commitment to the precautionary principle, prioritizing the safety of human health and the environment (European Commission, 2019a). The EU has implemented a comprehensive regulatory framework for GMOs and their products. This framework encompasses a series of regulations adopted in the early 2000s, such as Regulation (EC) 1829/2003 and Regulation (EC) No 1830/2003, which establish criteria for evaluating potential risks, harmonized procedures for risk assessment and authorization, provisions for labeling feed and food products containing GMOs, and requirements for traceability and labeling. In addition, Commission Regulation (EC) No 65/2004 introduces a system for the development and assignment of unique identifiers for GMOs, facilitating tracking and identification. Commission Regulation (EC) No 641/2004 provides detailed rules for the application process for authorizing new genetically modified food and feed products and addressing the presence of genetically modified material. This comprehensive regulatory framework prioritizes safety, transparency, and traceability in the handling and marketing of GMOs and their derivatives within the European market, aligning with the EU's commitment to safeguarding human health and the environment while enabling informed consumer choices (European Commission, 2019b).

In the Philippines, the use of recombinant DNA (rDNA) technology in biotechnology research in the early 1970s necessitated the establishment of the National Committee on Biosafety of the Philippines (NCBP), which oversees the regulatory framework for GMOs. An Executive Order No. 430 (EO 430), enacted in 1990, created the NCBP, with the task of regulating genetic engineering research and development activities and their products in the Philippines (Mendoza, Garcia, Sahagun, & Laurena, 2009). On May 24, 2000, the Philippines became a signatory to the Cartagena Protocol on Biosafety to the Convention on Biological Diversity, an international treaty governing the movements of living modified organisms (LMOs) resulting from modern biotechnology from one country to another (Convention on Biological Diversity). In 2021, the Philippines introduced DOST-DA-DENR-DOH-DILG Joint Department Circular No. 1, which outlines the rules and regulations for the research and development, handling and use, transboundary movement, release into the environment, and management of genetically modified plant and plant products derived from the use of biotechnology (National Committee on Biosafety of the Philippines, 2021).

## 12.6 FUTURE APPLICATIONS OF GENETICALLY MODIFIED ORGANISMS

Future applications of GMOs are poised to revolutionize various domains, driven by the precision in manipulating genetic material. These advancements offer innovative solutions to critical challenges. For example, engineered crops like *Bt* cotton and *Bt* corn produce insect-toxic proteins, reducing the reliance on chemical pesticides and contributing to sustainability in agriculture. Additionally, GMOs can be tailored to enhance nutritional content, extend shelf life, and increase crop yields, vital for ensuring global food security amid a growing population (Giudice et al., 2021). In the medical sphere, GMOs play a pivotal role, particularly in pharmaceutical production. Engineered bacteria and yeast serve as cost-effective platforms for synthesizing complex proteins and antigens, revolutionizing vaccine manufacturing (Webb & Hong, 2021). Furthermore, gene therapy, involving the introduction of therapeutic genes into patients' cells, holds transformative potential in treating genetic disorders and various diseases, marking a significant milestone in healthcare innovation.

Beyond agriculture and medicine, GMOs promise to impact environmental conservation and restoration significantly. Engineered microorganisms are leveraged for bioremediation, effectively breaking down or sequestering environmental contaminants (Rafeeq et al., 2023). This technology holds promise for cleaning polluted environments and potentially conserving ecosystems by enhancing the resilience of endangered species and facilitating the restoration of habitats affected by human activities. In the industrial sector, genetically modified microorganisms are actively explored for biofuel, bioplastic, and specialty chemical production. These bio-based alternatives not only reduce reliance on fossil fuels but also contribute to mitigating the environmental impact associated with conventional manufacturing processes. This sustainable approach aligns with broader goals of environmental responsibility and resource conservation (Peralta-Yahya et al., 2012).

## 12.7 BIOLOGICAL, ECOLOGICAL, SAFETY, AND SECURITY CONCERNS

GMO research in the Philippines has been ongoing for over two decades and was officially monitored after the establishment of the NBCP in 1990 (Mendoza et al., 2009). This was in response to the significant progress in the development of pest- and disease-resistant transgenic crops in the international community. The gradual introduction of these transgenic crops evoked concerns about establishing a governing body that will monitor and deal with issues concerning the entry of GMOs and other exotic materials that may pose hazards not only to human health but also to the endemic flora and fauna of the country (Mendoza et al., 2009).

The establishment of NCBP and the introduction of modern biotechnology in the country generated considerable interest among scientists and the public. The introduction of powerful tools capable of analysis and modification of genes advanced research in modern biotechnology in the Philippines. From 2010 to 2022, a total of 91 GMO research projects were approved by the Department of Science and Technology Biosafety Committee (DOST BC) from the following private and public institutions: Bayer CropScience, Inc., Cotton Development Authority, International Rice Research Institute (IRRI), Nippon Steel / Roxas Holdings Inc., Pioneer Hi-Bred Philippines, Inc., Philippine Rice Research Institute (PhilRice), Sumikin Engineering CO., LTD., Syngenta Philippines, Inc., University of the Philippines Diliman (UPD), University of the Philippines Los Baños (UPLB), University of the Philippines Mindanao (UPMin), University of the Philippines Genome Center and Visayas State University Leyte (DOST BC, 2011, 2012, 2013, 2014, 2015, 2015, 2016, 2017, 2018, 2019, 2020, 2021, 2022; Panopio & Navarro, 2011). IRRI, a key player in GMO research, has been significantly involved, contributing 45% of the approved projects. The research from the different agencies focused on a variety of topics including development of bacterial strains for fermentation, effects of consumption of GMO in human health, and developing new GMO crops that are resistant to pests and diseases. The focus of these research was to develop products that are beneficial to mankind, help improve food security, and reduce poverty. However, the use of GMO crops in the Philippines has been controversial with some expressing concerns about the safety of GMOs in biodiversity, human health, and the ethical considerations related to social justice, human rights, and environmental stability (Funk, 2020).

## 12.8 BIOLOGICAL CONCERNS: POTENTIAL EFFECTS OF GMO ON BIODIVERSITY AND ECOSYSTEMS

GMO has emerged as a significant topic of debate in the realm of agriculture and food production. While they offer promising solutions to enhance crop yields, improve resistance to pests, and address global food shortages, their introduction into natural ecosystems has raised valid biological concerns. One of the foremost concerns is the potential impact of GMOs on biodiversity and ecosystems. This complex issue delves into the intricate relationships between genetically modified crops, native species, and the delicate balance of ecosystems. This section explores the potential effects of GMOs on biodiversity and ecosystems, shedding light on the ecological implications of these novel organisms and the importance of responsible practices to safeguard the environment for future generations.

The development of genetically modified (GM) crops with enhanced resistance to biotic stresses, such as insect pests, diseases, and improved nutritional quality, has contributed to a significant increase in GM crop cultivation. This rise in adoption led to a record-breaking 191.7 million hectares of GM crops being cultivated by 2018 (Verma et al., 2021). GM crops have the potential to deliver greater yields compared to conventional crops and offer additional advantages, such as affordability and resistance to pesticides and weed killers. By using lower amounts of pesticides, GM crops can contribute to a decrease in harmful gas emissions into the environment. Moreover, in developing countries, GM crops have demonstrated proven health benefits for farm workers due to reduced spraying of chemical pesticides, particularly benefiting small and economically disadvantaged farmers (Azadi et al., 2022). In spite of numerous studies affirming the safety of GM crops and food for both consumers and the environment, public skepticism still persists, and the debate continues, often fueled by unsubstantiated and misleading claims, generally from nongovernmental organizations. The complex subject of GMOs' potential impact on biodiversity has garnered significant attention, particularly concerning biological conservation and the resilience of ecosystems.

## 12.9 GMOS AND BIODIVERSITY AND ECOSYSTEM SERVICES

The intricate subject of GMOs' potential influence on biodiversity has garnered significant attention, especially concerning biological conservation and the resilience of ecosystems. Preservation of biodiversity is universally acknowledged as a top priority by both the scientific community and society as a whole and has been a prominent focus in the field of genetic engineering. The literature on this topic is diverse and covers various aspects, as the potential impact of genetically engineered crops on biodiversity can be studied at the crop, farm, and landscape levels, while also considering different organisms or microorganisms, both target and non-target (Nicolia et al., 2013).

GM crops available in the market primarily possess resistance to herbicides and/or pests. They contribute to the crop's genetic diversity, especially when applied to enhance underutilized crops, thus making them viable for large-scale cultivation.

Nonetheless, the development of GM crop plants frequently relies on a restricted set of high-performing breeding lines, which may lead to a decrease in the variety of cultivars being grown on farmlands (Sneller, 2003). There are also concerns about the potential adverse effects of horizontal gene transfer (HGT) from genetically modified organisms, although its impact continues to be a subject of controversy. There have been claims that HGT of an introduced gene in a GMO has the potential to confer a new trait in another organism, posing a potential risk to human health or the environment (Keese, 2008). Gene flow to wild and related species is a possibility (Andersson & de Vicente, 2010) and in the worst-case scenario, it could result in the total genetic extinction of wild populations. Nevertheless, specific examples of such gene flow in GM crops are limited, whereas this phenomenon has been observed more frequently in conventional crops (Nawaz et al., 2020).

At the farm level, the reliance on just four major crops, namely wheat, maize, rice, and potato, to fulfill global food demands can directly impact the biota present on farmlands, encompassing plants, insects, pests, farm animals, birds, fungi, and microorganisms. Because of this, the genetic and varietal diversity of these dominant plant species can play a role in limiting, maintaining, or enhancing biodiversity within the local landscape. However, this phenomenon is specific to crop plants and species, as in some countries, approved GM varieties are crossbred with high-yielding cultivars/lines tolerant to other stresses (Nawaz et al., 2020).

At the individual crop species level, the gene transfer that can occur through pollination, seed-mediated transfer, and vegetative propagule-mediated transfer may take place between crops, from crops to wild species, or to microorganisms. Each type of gene flow can potentially impact biodiversity. Rare hybridization events between a GM crop and its wild relatives may produce hybrids with enhanced fitness (Nawaz et al., 2020). While this could be the case, understanding the ecological significance of transgenic plants is crucial, as a single inserted trait could lead to the creation of a crop plant that can impact the existing ecological network. For instance, the unintended spread of transgenes to wild species may have unforeseen consequences, such as the potential propagation of more invasive weeds (Vrbnicanin et al., 2017).

As for the effects of GM crops on biodiversity and non-target species, such as birds, non-target arthropods, soil micro and macrofauna, there is limited to no evidence of negative effects on biodiversity. Specifically concerning GM crops that produce *Bt* toxin, there have been occasional claims that non-target species might be at risk of endangerment. The monarch butterfly case in North America serves as an illustrative example. It was initially claimed, based on laboratory experiments, that *Bt* toxin engineered into maize could be expressed in such large quantities in the maize pollen that it might poison monarch caterpillars when it coats milkweed plants (their food). However, such a thick pollen coating rarely occurs naturally, and all current maize strains used have been engineered to prevent *Bt* toxin production in the pollen, thus resolving the concern (Raven, 2010). Studies have reported that agricultural fields with GM crops were found to support a more abundant and diverse population of invertebrates compared to fields subjected to frequent pesticide applications. A meta-analysis of 42 field experiments comparing *Bt* maize and *Bt* cotton to non-*Bt* crops revealed a greater abundance of non-target invertebrates in *Bt* crops

(Marvier et al., 2007). Over a 15-year period, another study found little to no significant effects of *Bt* on soil biodiversity, attributing reported differences to variations in temperature, soil type, and crop varieties (Carpenter, 2011). Overall, the adoption of GM crops has contributed to more environmentally friendly farming practices, particularly through the enhancement of soil biodiversity.

In the context of GMOs, it is crucial to regard farmland as an ecosystem that sustains essential aspects of human life, such as food, animal feed, pharmaceuticals, bioenergy, and shelter. Similar to any other ecosystem, farmland ecosystems comprise an intricate network of organisms interacting with each other under diverse biotic and abiotic conditions. An imbalance within this intricate food web can lead to the disruption of numerous symbiotic relationships (Nawaz et al., 2020). There are studies reporting the effects of GM crops on soil biological activity and soil health. For example, GM soybean roots exude soluble carbohydrates and amino acids that alter the rhizosphere microbial community structure compared to non-GM cultivars (Kremer et al., 2005). Another study reported that the resistance of two *Bt* cotton lines to the *Fusarium oxysporum* root phytopathogen was inferior compared to the parental lines (Li et al., 2009). Considering the vital roles fungi play in the soil environment, these reports are concerning, but Turrini et al. (2015) highlighted the need for monitoring key sensitive microbial groups that mediate crucial soil health processes in future environmental impact assessments of GM crops.

## 12.10 SAFETY CONCERNS: POTENTIAL HEALTH RISKS ASSOCIATED WITH THE CONSUMPTION OF GMO PRODUCTS

Another common concern regarding GMO and products derived from them is the potential health risks associated with their consumption. The first genetically engineered products were introduced into food products in 1990, which included chymosin for cheese production, yeast for baking, and even bacteria for food fermentation. Hundreds of products containing ingredients derived through genetic engineering were sold in the early 1990s. However, they were not widely publicized and did not elicit much public reaction mainly because the products were not considered to be novel or controversial at the time (Shelton et al., 2018)

The controversy regarding the potential health risks associated with the consumption of GMO products in the world began in 1996 with the introduction of new products including Flavr Savr tomato, recombinant bovine somatotropin (rBST) for cow milk production, and hormones for cattle raising . These new products led to several protests and boycotts in the international community accompanied by the establishment of numerous anti-GMO non-governmental organizations (NGOs) which led to raise awareness of the issue of genetically modified food. These NGOs influence the policies regarding application of GMOs on food and challenge scientific claims that GMOs are safe.

In the Philippine context, consumed GMO and GMO derivatives were introduced to agriculture by researchers to combat nutrient deficiency and malnutrition.

However, the acceptance of GMO and GMO derivatives in food met strong resistance from anti-GMO organizations, particularly about the possible health hazards from GMO crops and their derivatives that are directly and indirectly consumed by people as food and beverage. Local ordinances banning the planting of GMOs have been enacted after the local government saw the possible negative effects of GMOs on the environment and on the local health of people (Maghari &Ardekani, 2011; Bawa &Anilakumar, 2013).

The country has a regulatory framework in place for the assessment and approval of GMOs, but there is still some public uncertainty about the safety of these products. In 2013, the Philippine Food and Drug Administration (FDA) issued an advisory stating that all GM food products available in the market have passed food safety assessment based on international standards. The advisory also stressed that GM food products are as safe as food from conventional crops.

However, there are still some concerns about the potential risks of GMOs, such as the possibility of allergic reactions, the transfer of genes to other organisms, and the impact on the environment. These concerns have led to protests against the use of GMOs in the Philippines, and some people have chosen to avoid GM food altogether.

## 12.11 ETHICAL AND LEGAL CONCERNS ON GMOS

The use and regulation of GMOs in the Philippines have significant implications for the country's agricultural sector, its environmental impact, biodiversity, public health, and the corporatization of agriculture. In the Philippines, concerns about GMOs are multifaceted. Economically, there are fears about the monopolization of seed supply by large multinational corporations, which could marginalize small-scale farmers whose agricultural margins are already meager. The Philippine agricultural sector is often dominated by a few large plantations responsible for the highest yields and the majority of output, with middlemen often coordinating agricultural output to markets (Boquet &Boquet, 2017). The vast majority of small farmers results to be at the whims of middlemen and the agricultural market when it comes to the success or failure of their harvest. Further complicating this issue is a patchwork system of government support for agriculture that often leaves farmers stuck in debt traps and hostage to predatory financing for new agricultural initiatives (Boquet & Boquet, 2017).

Health concerns include potential allergies and antibiotic resistance, as raised by prior GMO scares, which often drive resistance against the proliferation of GMOs in the agricultural sector. Environmental concerns encompass potential harm to non-target organisms in the form of genetic drift, loss of biodiversity with the rise of monocultures, and the introduction of 'genetic pollution' into the ecosystem. Despite these concerns, there is also support for the use of GMOs in the Philippines as a way to combat nutrient deficiency and malnutrition and to boost lagging agricultural yields in the sector. The Philippines ranked last among ASEAN countries in Total Factor Productivity, the most informative measure of agricultural productivity (Dy, 2017). In 2021, the Philippines became the first country to approve GM golden rice that could fight vitamin A deficiency and blindness, which is a concern for the country owing to its high poverty rate and high rate of food insecurity (Quisumbing, 2003; Garg et al., 2018).

The use and experimentation of GMOs also raise ethical issues. The concept of genetic pollution refers to the uncontrolled spread of genetic information, which could disrupt ecosystems and lead to a loss of biodiversity. There are also concerns about the safety of GMOs for human consumption, with long-term health effects relatively backed as safe for human consumption by the scientific community but struggles to gain traction with the general public. While the Philippines has no laws requiring GMO labeling in ingredients, no less than three bills have been proposed in its Senate to require disclosure of GMOs in food ingredients.

The Philippines has a complex history of legal frameworks and challenges surrounding GMOs. The country has ratified the Cartagena Protocol on Biosafety (CPB), which requires the regulation of GMOs before they are subjected to transboundary movements. A science-based regulation permitted the Philippines to be the first in Asia to have a GM crop commercialized in 2002—the *Bt* Corn, which led to the country being the leader in cultivation of GM crops in the Southeast Asian region (Vistan, 2018; Polinag, 2020).

Despite the Department of Agriculture being responsible for regulating GMOs in the Philippines, GMO-free Zones in the Philippines have been enacted and implemented via GMO Ban Ordinances, which are the legal bases for such zones at the local level (Aruelo, 2017). The Department of Agriculture also released Order No.8s. 2002, *Rules and Regulations on the Importation and Release of Plant and Plant Products Derived from the Use of Modern Biotechnology*, which covers only plant and plant products derived from modern biotechnology (Kuiper et al., 2001; Nap et al., 2003).

In 2015, the Philippine Supreme Court issued the first legal decision on GMOs in the Philippines using the Writ of *Kalikasan*, a legal remedy to protect one's right to a balanced and healthy ecology when it is violated or threatened. Greenpeace filed a suit to stop ongoing trials of *Bt* eggplant which the court ruled in their favor with a permanent ban on field trials of GM eggplant and a temporary halt on approving applications for the contained use, import, commercialization, and propagation of GMO crops, including the import of GMO products (Aruelo, 2017; Vistan, 2018).

More controversially, it also invalidated Administrative Order No. 08-2002 and prevented the Department of Agriculture from carrying any further approval of GMO trials as the court adopted the precautionary principle, which holds that it is best to err on the side of caution in the absence of scientific consensus regarding GM products in its decision. This decision would ultimately be reversed in a 2016 ruling for mootness as the said trials have been completed with no release of products. While it reversed the invalidation of the GMO regulations, it still endorsed the precautionary principle approach for GMO analysis (Domingo, 2016).

In 2023, the Court followed up with a further Writ of Kalikasan that built on the previous ruling to block commercial release and planting of *Bt* eggplant and Golden rice. Petitioners before the court sought to block commercial propagation of both GMO crops and have biosafety permits voided. They also asked the court to require the Department of Agriculture to perform independent risks and impact assessments, obtain the prior and informed consent of farmers and indigenous peoples, and implement liability mechanisms in case of damage, as required by law (Ponce et al., 2019).

The legal concerns surrounding GMOs in the Philippines are complex and multifaceted, involving a range of economic, health, environmental, and ethical issues. The country's legal frameworks, legislations, and court decisions have both facilitated and impeded the use of GMOs. As the debate continues, it is crucial to strike a balance between harnessing the potential benefits of GMOs and using scientific research to inform policy debates. Future research should continue to explore this delicate balance, informing policy decisions that serve the best interests of the Filipino people and their environment.

## 12.12 APPLICATION FOR CONTAINED USE AND TEST OF GMO IN THE PHILIPPINES

The policies and procedures specified in the Philippines Biosafety Guidelines for Contained Use of Genetically Modified Organisms (2014) provide a comprehensive framework for institutions involved in or planning to conduct GMO-related work. Specifically, any institution engaging in such activities must first establish an Institutional Biosafety Committee (IBC), recognized by the Department of Science and Technology Biosafety Committee (DOST BC). The primary function of the IBC is multifaceted. Initially, it involves evaluating project proposals and ensuring that research facilities—like laboratories, screenhouses, greenhouses, or glasshouses—meet the required biosafety and biosecurity standards for the intended study. Subsequently, approved project proposals are endorsed to the DOST-BC for final approval. Once a project is approved, the IBC takes on a supervisory role, responsible for monitoring its progress and providing regular reports to the DOST BC. A critical aspect of its role is prioritizing the protection of the environment and human health throughout the institution's GMO-related activities. Equally important is the aspect of transparency and community engagement. The IBC is crucial in informing nearby communities about planned confined tests, ensuring that any potential risks are communicated effectively. This proactive approach aims to cultivate understanding and awareness among community members regarding potential risks associated with the GMO activities, enabling informed decision-making and risk mitigation strategies.

## 12.13 CASE STUDIES ON GMOS IN THE PHILIPPINES

GMOs have become a significant aspect of modern agriculture in the Philippines. These organisms, with their genetically engineered desirable traits, such as increased resistance to pests, diseases, and environmental stressors, make them a promising tool for improving agricultural productivity and ensuring food security in the country. The adoption of GMOs has generated both enthusiasm and controversy in the Philippines. While proponents highlight their benefits such as boosting crop yields and reducing pesticide use, critics raise issues concerning environmental impacts and potential health risks. In this section, the history, benefits, concerns, and/or ongoing dialogue surrounding the use of GMOs in Philippine agriculture will be discussed. Understanding the multifaceted nature of GMO adoption in the Philippines is essential for fostering informed discussions and making well-balanced decisions about the future of food production in the country.

### 12.13.1 Bt Eggplant

Eggplant (*Solanum melongena* L.) is extensively grown in many Asian countries, particularly in the South and Southeast Asian regions (Srinivasan, 2008). *Talong* (eggplant), as it is locally called in the Philippines, is considered one of the most economically important agricultural commodities, with production in 2021 estimated at 244,7034.56 metric tons and area planted at 21,828.70 hectares (Philippine Statistics Authority, 2023). It is among the top 30 commonly consumed food items in Filipino households (Department of Science and Technology - Food and Nutrition Research Institute, 2022).

One of the major problems in eggplant production is the prevalence of pests and diseases. The eggplant fruit and shoot borer (*Leucinodes orbonalis* Guenée) (EFSB) is an insect pest known to cause extensive damage and is considered a significant problem in eggplant production in the Philippines. The EFSB is a moth species that lays eggs primarily on the leaves, and in their larval form, feed on plant tissues by boring into the shoots, leaves, and fruits. This results in wilting of young shoots, drying, and drop off, ultimately affecting plant growth. Moreover, the presence of holes and feeding tunnels that get filled with larval excrement render the fruit unfit for consumption and marketing. This infestation leads to yield reduction that can go as high as 80%, and poor produce quality (Hautea et al., 2016; Srinivasan, 2008, 2009).

Many farmers in the Philippines use chemical insecticides in the control of ESB and prefer this approach over other measures such as the manual removal of damaged fruits and the use of biological control arthropods, as they are impractical, ineffective, and expensive. To avoid massive losses, farmers resort to intensive broad-spectrum insecticide applications (20–72 times for 5–6 months/season), which consequently increases production costs. Furthermore, this control strategy can pose a significant risk to human health and the environment. A 2015 study reported insecticide exposure-related health issues described by eggplant farmers, such as skin irritation, muscle pain, and headaches, in addition to the detection of insecticide residues in soil samples from eggplant farms in Sta. Maria, Pangasinan (Del Prado-Lu, 2015). While there have been attempts to develop eggplant cultivars that are highly resistant to EFSB via conventional breeding, there has been no success in producing commercial cultivars with adequate levels of resistance (Srinivasan, 2008).

To curb heavy dependence on insecticide use, the India-based Maharashtra Hybrid Seed Company (Mahyco), in 2000, incorporated the *cry1Ac* gene from *Bacillus thuringiensis* (*Bt*) Berliner into eggplant to make it EFSB-resistant. This genetically engineered *Bt* eggplant (referred to as Event EE-1) can express the Cry1Ac toxin that is toxic to certain insect pests, particularly EFSB. The *cry1Ac* gene is used in *Bt* cotton and has a long history of safe use. In late 2003, a partnership involving Mahyco, Cornell University, USAID, and public sector partners in India, Bangladesh, and the Philippines was established under the Agricultural Biotechnology Support Project II cooperative agreement. All three countries utilized the resistant EE-1 event and introgressed into local eggplant varieties (Hautea et al., 2016).

In 2010, a two-year Biosafety Permit was issued by the Department of Agriculture-Bureau of Plant Industry (DA-BPI) to evaluate the field performance of *Bt* eggplants in Sta. Maria, Pangasinan, Philippines. Hautea et al. (2016) reported outstanding results in the control of EFSB shoot and fruit damage, in addition to decreased larval infestation. Furthermore, moths that developed from larvae collected and reared in their *Bt* eggplant hosts were found to be sterile. This translates to increased yields and improved quality of eggplant crops, as well as reduced reliance on conventional chemical insecticides, making it an environmentally friendly option for farmers (Shelton et al., 2018). Another study documented that *Bt* eggplants can selectively control EFSB and do not adversely impact non-target arthropods that perform important ecosystem services (Navasero et al., 2016).

While the Sta. Maria, Pangasinan field trials for *Bt* eggplant were given the green light, anti-biotech activists, in 2011, reportedly destroyed confined field trials by forcibly entering the premises and illegally uprooting the plants (Shelton, 2021). Legal disputes spanning four years ensued in response to these actions. In December 2015, the successful campaign against the technology by Greenpeace Southeast Asia, along with the Filipino farming development network MASIPAG and 14 individuals, resulted in the December 2015 Philippine Supreme Court decision that ordered the halting of field testing, which had already been completed when the ruling came out. The existing biotechnology regulations were also deemed null and void by the same ruling. The non-governmental organizations and citizens argued that citizen consultation was not included in the existing biosafety protocols and that the GMOs pose potential risks to human health and the environment (Laursen, 2013). On July 26, 2016, the Supreme Court reversed its 2015 ruling, and by 18 October 2022, the University of the Philippines Los Baños was granted the 'Biosafety Permit for Commercial Propagation (No. 22-001 Propa)' by the DA-BPI, allowing the commercial cultivation of *Bt* eggplant and making the Philippines the second country to commercialize *Bt* eggplant after Bangladesh (Caraan, 2022; Feed the Future Insect-Resistant Eggplant Partnership, 2023). However, on April 18, 2023, the Supreme Court issued the Writ of Kalikasan that effectively stops the commercial propagation and release of GM eggplant (Supreme Court of the Philippines, 2023). The Writ of Kalikasan is a judicial instrument that offers safeguarding measures against human-induced activities that can result in natural disasters.

### 12.13.2 Golden Rice

Vitamin A deficiency (VAD) is a significant global public health issue that is particularly prevalent in low-income and middle-income countries. VAD exposes children to an array of health problems, such as diarrhea, measles, vision impairments, compromised immune functions, and anemia (Xu et al., 2021). Pregnant and lactating women are also vulnerable. The poor are most affected as their diets primarily consist of staple foods that offer limited nutritional value due to their limited purchasing power and lack of awareness of healthier options. In the Philippines, the 2018–2019 Expanded National Nutrition Survey (ENNS) of the Department of Science and Technology's Food and Nutrition Research Institute (DOST-FNRI) reported that one out of six or 15.5% of Filipino infants and children aged six months to five years had VAD, especially those in the rural areas (Luci-Atienza, 2021). Over the past decade, extensive efforts have been made to address VAD in developing countries. These include initiatives such as food fortification, supplementation, and educational programs focusing on dietary improvements. Another viable approach involves enhancing the β-carotene content of major staple crops, such as rice, through plant breeding. Rice (*Oryza sativa* L.), unfortunately, does not have naturally occurring provitamin A (that the body converts to Vitamin A) in its endosperm, and this necessitates a genetic engineering approach (Mallikarjuna Swamy et al., 2021).

In the late 1990s, plant scientist Ingo Potrykus and biochemist Peter Beyer initiated the development of Golden Rice (GR) to address VAD in the developing world (Owens, 2018). GR was developed by introducing two gene constructs into the US *japonica* rice variety Kaybonnet, which completed the carotenoid pathway needed for β-carotene production in the endosperm. The genes included phytoene synthase (*zmpsy1*) from *Zea mays* and the carotene desaturase (*crtI*) gene derived from the common soil bacterium *Pantoea ananatis* (syn. *Erwinia uredovora*) (Mallikarjuna Swamy et al., 2021). The grain acquires a yellow-orange or golden color due to the β-carotene, hence its name. For the successful implementation and widespread adoption of GR in Asia, it is crucial to transfer this golden rice trait from Kaybonnet into locally adapted and commonly cultivated rice varieties.

In the Philippines, the development and testing of GR, or *Malusog* (Filipino term for healthy) Rice, have been in progress since 2004. The GR event GR2E has undergone crossbreeding with locally preferred rice varieties in the Philippines (Mallikarjuna Swamy et al., 2021), catering to the preferences of both farmers and consumers. The research and development phase of GR in the Philippines also proved to be contentious. In August 2013, small GR test plots in the Bicol region were targeted by protestors from two anti-GMO organizations, namely KMB and Sikwal-GMO. Protesters tore down fences and uprooted and trampled the plants, arguing that the promotion of GR is influenced by international agrochemical corporations and the US and that the rice trial poses risks to human health and biodiversity (Kupferschmidt, 2013; McGrath, 2013).

In 2017, the Philippine Rice Research Institute (PhilRice) and the International Rice Research Institute (IRRI) submitted an application to the DA-BPI for approval for Food, Feed, and/or Processing (FFP) purposes. Additionally, PhilRice separately submitted an application for a biosafety permit to conduct a field trial. On May 20, 2019, the DA-BPI issued the biosafety permit for the field trial that took place at the DA-PhilRice stations in Munoz, Nueva Ecija, and San Mateo, Isabela. It was successfully concluded in October 2019. On December 18, 2019, the DA-BPI granted the FFP permit, authorizing the direct utilization of GR as food and feed, as well as for processing purposes (International Rice Research Institute (IRRI), 2019). The FFP approval is important before GR can be approved for commercial propagation and subsequent deployment to the public. In July 2021, the milestone approval finally came and the Philippines became the first country in the world to approve the commercial cultivation of nutrient-rich GR that will help reduce childhood malnutrition (IRRI, 2021). In 2022, the pilot deployment of GR seeds was spearheaded by DA-PhilRice. The rice variety has been officially registered with the National Seed

Industry Council under the names NSIC 2022 Rc682GR2E or Malusog 1 (Tababa, 2023). Despite all these positive developments, GR will remain off market, as with *Bt* eggplant, because of the Writ of Kalikasan issued by the Philippine Supreme Court on April 18, 2023 in response to a petition aiming to prevent its commercial release (Supreme Court of the Philippines, 2023).

### 12.13.3 Papaya Ringspot Virus-Resistant Papaya

Papaya (*Carica papaya* L.) is a fruit crop that has significant economic value in numerous tropical and subtropical countries, including the Philippines. It is renowned for its nutritional and medicinal properties as it is a rich source of vitamins A, B, and C. The fruit is particularly valued for its high β-carotene content, which has been associated with potential health benefits such as preventing cancer, diabetes, and heart disease. Papaya also finds applications in the pharmaceutical and cosmetics industries (Azad et al., 2014). Papaya cultivation is widespread in the Philippines, with production volume estimated at 165,911.50 metric tons and area planted at 7,747.53 hectares in 2021 (Philippine Statistics Authority, 2023).

Papaya ringspot virus (PRSV) is responsible for a devastating disease that significantly impacts the cultivation of papaya on a global scale (Tripathi et al., 2008). It was initially reported in Hawaii in 1945, as well as in India, Bangladesh, and Sri Lanka. PRSV emerged as a significant problem in the Philippines in the early 1980s as it led to the decline in papaya production in Southern Luzon, where papaya is an integral part of the multiple cropping system. The initial observation of PRSV symptoms was in Silang, Cavite, in 1982, and in 1984, the presence of PRSV was confirmed. Since then, the disease has spread throughout Luzon and has been identified in other papaya-growing islands such as Mindoro, Palawan, Marinduque, Leyte, and Negros Oriental (Villegas, 2001).

PRSV can be transmitted through both mechanical means and by various aphid species acting as vectors. Papaya infected with PRSV exhibits leaf mottling, reduced leaf size, tapering of the stem, presence of oil streaks on petioles, and ringspots on the fruits. When infected during the seedling stage, papaya leaves develop mottling and the plant loses vigor, and eventually succumbs to the disease before reaching the flowering stage. If the infection occurs at a later stage, the stem tapers and the plant exhibits all the aforementioned symptoms (Villegas, 2001). Disease management strategies include implementing quarantine measures, eradicating infected plants, planting papaya in isolated areas to avoid infection, regularly removing infected plants, utilizing tolerant varieties to minimize economic losses, employing cross-protection techniques, and utilizing transgenic resistance (Tripathi et al., 2008).

Traditional strategies have shown limited effectiveness in managing PRSV. The virus spreads easily through several aphid species, with transmission occurring rapidly and aphids requiring only a brief feeding period on an infected plant to acquire the virus. This makes it extremely challenging to control the virus through insect control or by discarding infected plants. Furthermore, the cultivated species *C. papaya* has not exhibited any natural resistance to PRSV. Certain wild *Carica* species like *C. cauliflora*, *C. pubescens*, and *C. quercifolia* have shown resistance to PRSV but are sexually incompatible with the cultivated species (Azad et al., 2014). Cross-protection techniques have been employed in Hawaii, Thailand, and Taiwan but with limited success because the protection provided was strain-specific and was never complete. Additionally, the control measures mentioned only offered temporary solutions (Tripathi et al., 2008). Enhancing population resistance is considered the most efficient approach for controlling plant viruses.

Pathogen-derived resistance (PDR) has sparked research efforts aimed at achieving virus resistance in papaya through gene technology. PDR refers to the ability of transgenic plants that contain genes or sequences from a parasite, such as the coat protein (*cp*) gene of a virus, to exhibit protection against the harmful effects of the same or related pathogens. Gonsalves' laboratory at Cornell University, in collaboration with Richard Manshardt and Maureen Fitch from the University of Hawaii and Jerry Slightom of The Upjohn Company, embarked on utilizing the PDR concept in 1986. They initiated a project to genetically engineer papaya by incorporating the *cp* gene of the mild PRSV strain HA5-1. When this gene is expressed in papaya, the plant becomes resistant to the virus. This collaboration resulted in the commercialization of the transgenic cultivars SunUp and Rainbow in 1998 (Gonsalves, 1998).

In 1998, the Papaya Biotechnology Network of Southeast Asia was established with the support of the International Service for the Acquisition of Agri-biotech Applications (ISAAA) and brought together national laboratories from Indonesia, Malaysia, the Philippines, Thailand, and Vietnam. The primary approach involved *cp*-mediated protection, based on the successful development of PRSV-resistant papaya in Hawaii. The initial effort to develop transgenic papaya with PRSV resistance was undertaken by the Institute of Plant Breeding (IPB) under the College of Agriculture (now College of Agriculture and Food Science) at the University of the Philippines Los Baños (UPLB). The genetic construct included the *cp* gene obtained from a virulent strain of PRSV isolated in Cavite (Tecson Mendoza et al., 2008).

Confined field trials have been conducted by the UPLB IPB for candidate lines of GM papaya with PRSV resistance. This confined field trial serves as an intermediate step between greenhouse testing and open field trial. The National Committee on Biosafety of the Philippines (NCBP) oversees and regulates the first two stages, while the DA-BPI supervises and regulates the open field trial (Tecson Mendoza et al., 2008). The UPLB IPB is also working on the development of biotech papaya with delayed ripening and resistance to papaya ringspot virus. In 2012, this GM papaya underwent a contained test, followed by a confined field trial in 2014. The technical advisory team of the Department of Agriculture Biotech Program Office suggested backcrossing the F1 hybrid with the transgenic line instead of conducting a second field trial in 2017. Currently, the dossiers are being prepared for the contained trial and eventual registration of the variety (ISAAA, 2017).

Concerns such as heteroencapsidation, the process in which the genetic material of one virus is encapsidated by the coat protein of another virus, have only been documented in transgenic herbaceous plants and not in transgenic vegetable plants carrying the viral *cp* gene constructs. While changes in vector specificity and host range may occur within a single generation, they are not permanent since they do not affect the viral genome and will not be passed on to the viral progeny. Therefore, the phenomenon of heteroencapsidation in transgenic plants that express virus *cp* genes has limited significance and is expected to have negligible adverse effects on the environment (Fuchs & Gonsalves, 2007). Furthermore, the perceived risk of transgenic crops influencing diversity and dynamics of non-target organisms have been examined and demonstrated to have minimal or no negative consequences. PRSV-resistant transgenic papaya does not exert a significant influence on the overall abundance and variety of actinomycetes across various soil layers, proving that GM crops that are resistant to viruses have not demonstrated any adverse effects on non-target organisms (Hsieh & Pan, 2006).

### 12.13.4 Bt Cotton

Cotton (*Gossypium* spp. L.) is responsible for approximately 40% of the global production of natural fibers and is grown commercially in 78 countries across temperate, subtropical, and tropical regions worldwide (Naranjo, 2011). It holds significant importance as a global commodity crop, offering valuable resources for various industries, including fiber, feed, food, and potentially, fuel production (Thorp et al., 2014). Numerous surveys have identified over 1300 species of herbivorous insects inhabiting cotton fields but only a small fraction of these insects are considered economically significant pests. Despite this, cotton has historically been one of the largest consumers of insecticides on a global scale (Naranjo, 2011). Cotton growers have faced significant pressure because of the resistance developed by pests against synthetic insecticides, which resulted in substantial cotton losses caused by *Pectinophora gossypiella* (pink bollworm) and *Heliothis virescens* (tobacco budworm) (Abbas, 2018). Although the Philippines plays a minor role in global cotton production, ensuring an adequate supply of local cotton to meet domestic demand is crucial (Rola et al., 2010).

In 1996, Bollgard cotton, developed by Monsanto Company, became the first commercially available *Bt* cotton in the US. It produced the Cry1Ac toxin that effectively targeted tobacco budworm and pink bollworm. Bollgard II, introduced in 2003, represented the next generation of *Bt* cotton that can produce the Cry2Ab toxin. In 2004, US-based Dow Agro-sciences released WideStrike cotton, containing Cry1Ac and Cry1F, which demonstrated better efficacy against a broader range of caterpillar insects compared to the original Bollgard. The most recent advancement in *Bt* cotton is the third generation, which incorporates three genes: Bollgard 3 (Cry1Ac + Cry2Ab + Vip3A), Twin Link Plus (Cry1Ab + Cry2Ac + Vip3Aa19), and Wide Strike 3 (Cry1Ac + Cry1F + Vip3A) (Abbas, 2018).

The global adoption of *Bt* cotton has increased significantly, expanding from 800,000 hectares in 1996 to 5.7 million hectares by 2003, including both standalone and herbicide-tolerant cotton varieties. By 2002, *Bt* cotton was commercially cultivated in the United States, Mexico, Argentina, South Africa, China, India, Australia, and Indonesia, while pre-commercial plantings were grown in Colombia. While large-scale farmers in industrialized nations like the United States and Australia derive significant benefits from *Bt* cotton, the majority of growers are from developing countries (Purcell & Perlak, 2004).

The Philippine Fiber Development Administration (PhilFIDA) is currently engaged in the development of *Bt* cotton, which will tackle pest-related challenges while simultaneously boosting crop yields and generating greater profits for farmers. This technology, provided by Nath Biogene Ltd. and Global Transgene Ltd. from India, underwent initial testing in a confined field trial in 2010, followed by multi-location field trials in 2012 and 2013. In 2015, the necessary data to complete the required regulatory dossiers were obtained, alongside some related laboratory experiments conducted in 2017. Through evaluation, the bioefficacy of the *Bt* cotton hybrids against the cotton bollworm was further confirmed (The International Service for the Acquisition of Agri-biotech Applications (ISAAA), 2017). The multilocational field trials for *Bt* cotton conducted by PhilFIDA in Luzon and Mindanao last year reported outcomes that were highly encouraging in terms of revitalizing the country's cotton industry (Rivera, 2023). On March 17, 2023, PhilFIDA officially submitted its application for a biosafety permit to push for the commercial propagation of *Bt* cotton GFM cry1A to the BPI. PhilFIDA has high hopes for the commercial propagation of *Bt* cotton and sees its potential to revive the Philippine cotton industry (PhilFIDA, 2023). As of May 2023, the Joint Assessment Group, consisting of representatives from the Department of Science and Technology, Department of Agriculture, Department of Environment and Natural Resources, and Department of Health Biosafety Committees, has requested further information from PhilFIDA (Rivera, 2023).

## REFERENCES


Abbas, M. S. T. (2018). Genetically engineered (modified) crops (*Bacillus thuringiensis* crops) and the world controversy on their safety. *Egyptian Journal of Biological Pest Control*, 28(1), 52.

Andersson, M. S., & de Vicente, C. (2010). Gene flow between crops and their wild relatives. *Evolutionary Applications*, 3(4), 402–403.

Aruelo, L. (2017). Philippines GMO-free zones: Successful roots in organic policy and law. https://www.gmo-free-regions.org/index.php?id=3959. Retrieved on September 18, 2023.

Azad, Md. A. K., Amin, L., & Sidik, N. M. (2014). Gene technology for papaya ringspot virus disease management. *The Scientific World Journal*, 2014, 768038.

Azadi, H., Taheri, F., Ghazali, S., Moghaddam, S. M., Siamian, N., Goli, I., Choobchian, S., Pour, M., Özgüven, A. I., Janečková, K., Sklenička, P., & Witlox, F. (2022). Genetically modified crops in developing countries: Savior or traitor? *Journal of Cleaner Production*, 371, 133296.

Baeshen, N. A., Baeshen, M. N., Sheikh, A., Bora, R. S., Ahmed, M. M., Ramadan, H. A., Saini, K. S., & Redwan, E. M. (2014). Cell factories for insulin production. *Microbial Cell Factories*, 13, 141.

Bawa, A. S., & Anilakumar, K. R. (2013). Genetically modified foods: Safety, risks and public concerns—A review. *Journal of Food Science and Technology*, 50(6), 1035–1046.

Boquet, Y. (2017). The Philippine agriculture: Weaknesses and controversies. In Y. Boquet (ED.), *The Philippine Archipelago*. (pp. 259–299). Springer International Publishing.

Brookes, G., & Barfoot, P. (2020). Environmental impacts of genetically modified (GM) crop use 1996–2018: Impacts on pesticide use and carbon emissions. *GM Crops and Food*, 11(4), 215–241.

Caraan, J. A. (2022, October 25). Philippine govt approves *Bt* Eggplant for commercial cultivation. https://cafs.uplb.edu.ph/bt_news/philippine-govt-approves-bt-eggplant-for-commercial-cultivation/.

Carpenter, J. E. (2011). Impact of GM crops on biodiversity. *GM Crops*, 2(1), 7–23.

Cohen, S. N., Chang, A. C., & Hsu, L. (1972). Nonchromosomal antibiotic resistance in bacteria: Genetic transformation of *Escherichia coli* by R-factor DNA. *Proceedings of the National Academy of Sciences of the United States of America*, 69(8), 2110–2114.

Del Prado-Lu, J. L. (2015). Insecticide residues in soil, water, and eggplant fruits and farmers' health effects due to exposure to pesticides. *Environmental Health and Preventive Medicine*, 20(1), 53–62.

Department of Science and Technology - Food and Nutrition Research Institute. (2022). *Philippine Nutrition Facts and Figures: 2018–2019 Expanded National Nutrition Survey (ENNS): Food Consumption Survey*. Department of Science and Technology - Food and Nutrition Research Institute.

Department of Science and Technology Biosafety Committee. (2011). Annual report 2010. https://dost-bc.dost.gov.ph/download/annual-reports/category/17-2010-annual-report.

Department of Science and Technology Biosafety Committee. (2012). Annual report 2011. https://dost-bc.dost.gov.ph/download/annual-reports/category/18-2011-annual-report.

Department of Science and Technology Biosafety Committee. (2013). Annual report 2012. https://dost-bc.dost.gov.ph/download/annual-reports/category/19-2012-annual-report.

Department of Science and Technology Biosafety Committee. (2014). Annual report 2013. https://dost-bc.dost.gov.ph/download/annual-reports/category/20-2013-annual-report.

Department of Science and Technology Biosafety Committee. (2014). *The Philippines BiosafetyGuidelines for Contained Use of Genetically Modified Organisms*. https://bch.dost.gov.ph/bch-laws-and-regulations?download=11:the-philippines-biosafety-guidelines-for-contained-use-of-genetically-modified-organisms.

Department of Science and Technology Biosafety Committee. (2015). Annual report 2014. https://dost-bc.dost.gov.ph/download/annual-reports/category/26-2014-annual-report.

Department of Science and Technology Biosafety Committee. (2016). Annual report 2015. https://dost-bc.dost.gov.ph/download/annual-reports/category/29-2015-annual-report.

Department of Science and Technology Biosafety Committee. (2017). Annual report 2016. https://dost-bc.dost.gov.ph/download/annual-reports/category/32-2016-annual-report.

Department of Science and Technology Biosafety Committee. (2018). Annual report 2017. https://dost-bc.dost.gov.ph/download/annual-reports/category/34-2017-annual-report.

Department of Science and Technology Biosafety Committee. (2019). Quarterly report, 2018. https://dost-bc.dost.gov.ph/download/annual-reports/category/35-2018-quarterly-report.

Department of Science and Technology Biosafety Committee. (2020). Semi-annual report 2019. https://dost-bc.dost.gov.ph/download/annual-reports/category/36-2019-semi-annual-report.

Department of Science and Technology Biosafety Committee. (2021). Annual report 2020. https://dost-bc.dost.gov.ph/download/annual-reports/category/48-2020-annual-report.

Department of Science and Technology Biosafety Committee. (2022). Annual report 2021. https://dost-bc.dost.gov.ph/download/annual-reports/category/49-2021-annual-report.

Domingo, E. (2016, July 28). SC reverses ruling on *Bt* 'talong' tests. INQUIRER.net. https://newsinfo.inquirer.net/800262/sc-reverses-ruling-on-bt-talong-tests.

Domingo, J. L., & Giné Bordonaba, J. (2011). A literature review on the safety assessment of genetically modified plants. *Environmental International*, 37(4), 734–742.

Doudna, J. A., & Charpentier, E. (2014). Genome editing: The new frontier of genome engineering with CRISPR-Cas9. *Science*, 346(6213), 1258096.

Dy, R. T. (2017, September 18). Benchmarking ASEAN farm productivity and growth. INQUIRER business. https://business.inquirer.net/237009/benchmarking-asean-farm-productivity-growth.

European Commission. (2019a). Genetically Modified Organisms (GMOs). [European Union website]. food.ec.europa.eu/plants/genetically-modified-organisms_en'.

European Commission. (2019b). Genetically modified feed. https://food.ec.europa.eu/food/animal-feed/genetically-modified-feed_en.

Feed the Future Insect-Resistant Eggplant Partnership. (2023, February 1). Philippines approves Bt eggplant for commercial cultivation. https://bteggplant.cornell.edu/2023/02/01/philippines-approves-bt-eggplant-for-commercial-cultivation/#:~:text=After%2015%20years%20of%20testing,and%20shoot%20borer%20(EFSB).

Fernandez-Cornejo, J., Wechsler, S., Livingston, M., & Mitchell, L. (2014). Genetically engineered crops in the United States, ERR-162 (p. 60). U.S. Department of Agriculture, Economic Research.

Fuchs, M., & Gonsalves, D. (2007). Safety of virus-resistant transgenic plants two decades after their introduction: Lessons from realistic field risk assessment studies. *Annual Review of Phytopathology*, 45(1), 173–202.

Funk, C. (2020, August 27). *Public opinion about genetically modified foods and trust in scientists*. Pew Research Center Science & Society. https://www.pewresearch.org/science/2016/12/01/public-opinion-about-genetically-modified-foods-and-trust-in-scientists-connected-with-these-foods/.

Garg, M., Sharma, N., Sharma, S., Kapoor, P., Kumar, A., Chunduri, V., & Arora, P. (2018). Biofortified crops generated by breeding, agronomy, and transgenic approaches are improving lives of millions of people around the world. *Frontiers in Nutrition*, 12,1–33.

Giudice, G., Moffa, L., Varotto, S., Cardone, M. F., Bergamini, C., De Lorenzis, G., Velasco, R., Nerva, L., & Chitarra, W. (2021). Novel and emerging biotechnological crop protection approaches. *Plant Biotechnology Journal*, 19(8), 1495–1510.

Gonsalves, D. (1998). Control of papaya ringspot virus in papaya: A case study. *Annual Review of Phytopathology*, 36(1), 415–437.

Hautea, D. M., Taylo, L. D., Masanga, A. P. L., Sison, M. L. J., Narciso, J. O., Quilloy, R. B., Hautea, R. A., Shotkoski, F. A., & Shelton, A. M. (2016). Field performance of *Bt* eggplants (*Solanum melongena* L.) in the Philippines: Cry1Ac expression and control of the eggplant and shoot borer (*Leucinodes Orbonalis* Guenée). *PLoS One*, 11(6), e0157498.

Hsieh, Y.-T., & Pan, T.-M. (2006). Influence of planting papaya ringspot virus resistant transgenic papaya on soil microbial biodiversity. *Journal of Agricultural and Food Chemistry*, 54(1), 130–137.

International Rice Research Institute (IRRI). (2019, December 18). Philippines approves Golden Rice for direct use as food and feed, or for processing. https://www.irri.org/news-and-events/news/philippines-approves-golden-rice-direct-use-food-and-feed-or-processing.

International Rice Research Institute (IRRI). (2021, July 22). Philippines becomes first country to approve nutrient-enriched "Golden Rice" for planting. https://www.irri.org/news-and-events/news/philippines-becomes-first-country-approve-nutrient-enriched-golden-rice.

Kanchiswamy, C. N., Malnoy, M., & Maffei, M. E. (2015). Chemical diversity of microbial volatiles and their potential for plant growth and productivity. *Frontiers in Plant Science*, 6, 151.

Keese, P. (2008). Risks from GMOs due to horizontal gene transfer. *Environmental Biosafety Research*, 7(3), 123–149.

Klümper, W., & Qaim, M. (2014). A meta-analysis of the impacts of genetically modified crops. *PLoS One*, 9(11), e111629.

Kremer, R. J., Means, N. E., & Kim, S. J. (2005). Glyphosate affects soybean root exudation and rhizosphere microorganisms. *International Journal of Environmental Analytical Chemistry*, 85(15), 1165–1174.

Krimsky, S. (2015). An illusory consensus behind GMO health assessment. *Science, Technology, and Human Values*, 40(6), 883–914.

Kuiper, H. A., Kleter, G. A., Noteborn, H. P., & Kok, E. J. (2001). Assessment of the food safety issues related to genetically modified foods. *The Plant Journal*, 27(6), 503–528.

Kupferschmidt, K. (2013, August 13). Activists destroy "golden rice" field trial. *Science*. https://www.science.org/content/article/activists-destroy-golden-rice-field-trial.

Laursen, L. (2013). Greenpeace campaign prompts Philippine ban on *Bt* eggplant trials. *Nature Biotechnology*, 31(9), 777–778.

Li, X.-G., Liu, B., Heia, S., Liu, D.-D., Han, Z.-M., Zhou, K.-X., Cui, J.-J., Luo, J.-Y., & Zheng, Y.-P. (2009). The effect of root exudates from two transgenic insect-resistant cotton lines on the growth of *Fusarium oxysporum*. *Transgenic Research*, 18(5), 757–767.

Luci-Atienza, C. (2021, November 4). DOST-FNRI: 1 out of 6 Pinoy kids 'most affected' by vitamin A deficiency. *Manila Bulletin*. https://mb.com.ph/2021/11/03/dost-fnri-1-out-of-6-pinoy-kids-most-affected-by-vitamin-a-deficiency/.

Maghari, B. M., & Ardekani, A. M. (2011). Genetically modified foods and social concerns. *Avicenna Journal of Medical Biotechnology*, 3(3), 109.

Mallikarjuna Swamy, B. P., Marundan Jr, S., Samia, M., Ordonio, R. L., Rebong, D. B., Miranda, R., Alibuyog, A., Rebong, A. T., Tabil, M. A., Suralta, R. R., Alfonso, A. A., Biswas, P. S., Kader, M. A., Reinke, R. F., Boncodin, R., & MacKenzie, D. J. (2021). Development and characterization of GR2E Golden rice introgression lines. *Scientific Reports*, 11(1), 2496.

Marvier, M., McCreedy, C., Regetz, J., & Kareiva, P. (2007). A meta-analysis of effects of *Bt* cotton and maize on nontarget invertebrates. *Science*, 316(5830), 1475–1477.

McGrath, M. (2013, August 9). "Golden rice" GM trial vandalized in the Philippines. *BBC News*. https://www.bbc.com/news/science-environment-23632042.

Mendoza, E. M. T., Garcia, R. N., Sahagun, M., & Laurena, A. C. (2009). *Biosafety Regulations in the Philippines: A Review of the First Fifteen Years, Preparing for the Next Fifteen*. National Academy of Science and Technology (NAST) Philippines.

Nap, J. P., Metz, P. L., Escaler, M., & Conner, A. J. (2003). The release of genetically modified crops into the environment: Part I. Overview of current status and regulations. *The Plant Journal: For Cell and Molecular Biology*, 33(1), 1–18.

Naranjo, S. E. (2011). Impacts of *Bt* transgenic cotton on integrated pest management. *Journal of Agricultural and Food Chemistry*, 59(11), 5842–5851.

National Committee on Biosafety of the Philippines. (2021). DOST-DA-DENR-DOH-DILG JDC NO. 1 S2021. https://bch.dost.gov.ph/dost-da-denr-doh-dilg-jdc-no-1-s2021

Navasero, M. V., Candano, R. N., Hautea, D. M., Hautea, R. A., Shotkoski, F. A., & Shelton, A. M. (2016). Assessing potential impact of *Bt* eggplants on non-target arthropods in the Philippines. *PLoS One*, 11(10), e0165190.

Nawaz, M. A., Golokhvast, K. S., Tsatsakis, A. M., Lam, H.-M., & Chung, G. (2020). GMOs, biodiversity and ecosystem processes. In *GMOs. Topics in Biodiversity and Conservation* (Vol. 19, pp. 3–17). Springer.

Nicolia, A., Manzo, A., Veronesi, F., & Rosellini, D. (2013). An overview of the last 10 years of genetically engineered crop safety research. *Critical Reviews in Biotechnology*, 34(1), 77–88.

Owens, B. (2018). Golden Rice is safe to eat, says FDA. *Nature Biotechnology*, 36(7), 559–560.

Panopio, J. A., & Navarro, M. J. (2011). Drama and communication behind Asia's first commercialized *Bt* corn. *Communication Challenges and Convergence in Crop Biotechnology*, 43–80.

Pellegrino, E., Bedini, S., Nuti, M., & Ercoli, L. (2018). Impact of genetically engineered maize on agronomic, environmental and toxicological traits: A meta-analysis of 21 years of field data. *Scientific Reports*, 8(1), 3113.

Peralta-Yahya, P. P., Zhang, F., del Cardayre, S. B., & Keasling, J. D. (2012). Microbial engineering for the production of advanced biofuels. *Nature*, 488(7411), 320–328.

Philippine Fiber Industry Development Authority (PhilFIDA). (2023, March). Bt cotton to catalyze the revival of the Philippine cotton industry. https://philfida.da.gov.ph/index.php/news-articles/196-bt-cotton-to-catalyze-the-revival-of-the-philippine-cotton-industry.

Philippine Statistics Authority. (2023). *2017–2021 Crops Statistics of the Philippines*. Philippine Statistics Authority.

Polinag, R. L. A. (2020, November 24). Philippine biosafety regulatory gaps and initiatives. FFTC Agricultural Policy Platform (FFTC-AP).

Ponce de Leon, I. Z., Custodio, P. A., & David, C. (2019). Depicting science in a public debate: The Philippine legal challenge against GMO eggplant. *Science Communication*, 41(3), 291–313.

Purcell, J. P., & Perlak, F. J. (2004). Global impact of insect-resistant (*Bt*) cotton. *AgriculturistsBioForum*, 7(1–2), 27–30.

Qaim, M. (2010). Benefits of genetically modified crops for the poor: Household income, nutrition, and health. *New Biotechnology*, 27(5), 552–557.

Qaim, M., & de Janvry, A. (2003). Genetically modified crops, corporate pricing strategies, and farmers' adoption: The case of Bt Cotton in Argentina. *American Journal of Agricultural Economics*, 85(4), 814–828.

Qaim, M., & Kouser, S. (2013). Genetically modified crops and food security. *PLoS One*, 8(6), e64879.

Quisumbing, L. A. (2003). Current and emerging issues on genetically modified organisms. *Philippine Law Journal*, 78, 362.

Rafeeq, H., Afsheen, N., Rafique, S., Arshad, A., Intisar, M., Hussain, A., Bilal, M., & Iqbal, H. M. N. (2023). Genetically engineered microorganisms for environmental remediation. *Chemosphere*, 310, 136751.

Raman, R. (2017). The impact of genetically modified (GM) crops in modern agriculture: A review. *GM Crops and Food*, 8(4), 195–208.

Raven, P. H. (2010). Does the use of transgenic plants diminish or promote biodiversity? *New Biotechnology*, 27(5), 528–533.

Rivera, D. (2023, May 4). Commercial *Bt* cotton propagation eyed. *The Philippine Star*. https://www.philstar.com/business/2023/05/04/2263549/commercial-bt-cotton-propagation-eyed.

Rola, A. C., Chupungco, A. R., Elazegui, D. D., Tagarino, R. N., Nguyen, M. R., & Solsoloy, A. D. (2010). Consequences of Bt cotton technology importation. *Philippine Agricultural Scientist*, 93(1), 9–21.

Shelton, A. M. (2021). *Bt* eggplant: A personal account of using biotechnology to improve the lives of resource-poor farmers. *American Entomologist*, 67(3), 52–59.

Shelton, A. M., Hossain, M. J., Paranjape, V., Azad, A. K., Rahman, M. L., Khan, A. S. M. M. R., Prodhan, M. S. H., Rashid, M. A., Majumder, R., Hossain, M. A., Hussain, S. S., Huesing, J. E., & McCandless, L. (2018). *Bt* eggplant project in Bangladesh: History, present status, and future direction. *Frontiers in Bioengineering and Biotechnology*, 6, 106.

Sneller, C. H. (2003). Impact of transgenic genotypes and subdivision on diversity within elite North American soybean germplasm. *Crop Science*, 43(1), 409–414.

Srinivasan, R. (2008). Integrated pest management for eggplant fruit and shoot borer (*Leucinodes Orbonalis*) in South and Southeast Asia: Past, present and future. *Journal of Biopesticides*, 1(2), 105–112.

Srinivasan, R. (2009). *Insect and Mite Pests on Eggplant: A Field Guide for Identification and Management.* AVRDC - The World Vegetable Center.

Supreme Court of the Philippines. (2023). SC issues writ of Kalikasan on genetically modified rice and eggplant products. https://sc.judiciary.gov.ph/sc-issues-writ-of-kalikasan-on-genetically-modified-rice-and-eggplant-products/.

Tababa, J. (2023, May 3). Three genetically modified crops approved in the Philippines. *Manila Bulletin.* https://mb.com.ph/2023/5/3/three-genetically-modified-crops-approved-in-the-philippines/.

Tecson Mendoza, E. M., Laurena, A. C., & Botella, J. R. (2008). Recent advances in the development of transgenic papaya technology. *Biotechnology Annual Review*, 14, 423–462.

The International Service for the Acquisition of Agri-biotech Applications (ISAAA). (2017). *Global Status of Commercialized Biotech/GM Crops in 2017: Biotech Crop Adoption Surges as Economic Benefits Accumulate in 22 Years.* The International Service for the Acquisition of Agri-biotech Applications (ISAAA).

Thorp, K. R., Ale, S., Bange, M. P., Barnes, E. M., Hoogenboom, G., Lascano, R. J., McCarthy, A. C., Nair, S., Paz, J. O., Rajan, N., Reddy, K. R., Wall, G. W., & White, J. W. (2014). Development and application of process-based simulation models for cotton production: A review of past, present, and future directions. *Journal of Cotton Science*, 18(1), 10–47.

Tripathi, S., Suzuki, J. Y., Ferreira, S. A., & Gonsalves, D. (2008). Papaya ringspot virus-P: Characteristics, pathogenicity, sequence variability and control. *Molecular Plant Pathology*, 9(3), 269–280.

Turrini, A., Sbrana, C., & Giovannetti, M. (2015). Belowground environmental effects of transgenic crops: A soil microbial perspective. *Research in Microbiology*, 166(3), 121–131.

U.S. Food and Drug Administration. (2021). How GMOs are regulated in the United States. https://www.fda.gov/food/agricultural-biotechnology/how-gmos-are-regulated-united-states.

Verma, V., Negi, S., Kumar, P., & Srivastava, D. K. (2021). Global status of genetically modified crops. In D. K. Srivastava, A. K. Thakur, & P. Kumar (Eds.), *Agricultural Biotechnology: Latest Research and Trends* (pp. 305–322). Springer.

Villegas, V. N. (2001). Approaches in developing Ringspot virus-resistant papaya. *Transactions of the National Academy of Science and Technology*, 23, 63–70.

Vistan, E. C. L. (2018). GMO scared: Postscripts on the Philippines' first major legal battle on GMOs. *Philippine Law Journal*, 91, 291.

Vrbnicanin, S., Bozic, D., & Pavlovic, D. (2017). Gene flow from herbicide-resistant crops to wild relatives. In G. W. Cussans, J. C. Caseley, & R. K. Atkin (Eds.), *Herbicide Resistance in Weeds and Crops*. InTech Open.

Webb, T. L., & Hong, E. (2021). GMO Medicines and hospital pharmacy practice: A review. *Journal of Pharmacy Practice and Research*, 51(3), 203–210.

World Health Organization. (2020). Genetically modified organisms: Key facts. https://www.who.int/news-room/q-a-detail/genetically-modified-organisms-(gmos).

Xu, Y., Shan, Y., Lin, X., Miao, Q., Lou, L., Wang, Y., & Ye, J. (2021). Global patterns in vision loss burden due to vitamin A deficiency from 1990 to 2017. *Public Health Nutrition*, 24(17), 5786–5794.

# 13 Reinvigorating Vigilance

## *Updates on Emerging and Re-emerging Infectious Diseases in the Philippines*

*Anelyn Reyes*
Department of Pediatrics
Faculty of Medicine and Surgery
University of Santo Tomas
Manila, Philippines

*Don Eliseo Lucero-Prisno III*
Department of Global Health and Development
London School of Hygiene and Tropical Medicine
London, UK

*Jesus Alfonso Catahay*
Department of Medicine
Saint Peter's University Hospital
New Brunswick, New Jersey, USA

*Connie Gibas*
Department of Pathology and Laboratory Medicine
University of Texas Health
San Antonio, Texas, USA

*Mayan Uy-Lumandas*
Department of Virology
Research Institute for Tropical Medicine
Muntinlupa City, Philippines

*Sheriah Laine de Paz-Silava*
Department of Medical Microbiology
College of Public Health
University of the Philippines
Manila, Philippines

DOI: 10.1201/9781003426219-13

*Mary Jane Flores*
Department of Biology
De La Salle University
Manila, Philippines

*Timothy Hudson David Culasino Carandang*
Pamantasan ng Lungsod ng Maynila College of Medicine
Manila, Philippines

*Ourlad Alzeus Tantengco*
Department of Biology, De La Salle University
Department of Physiology
University of the Philippines College of Medicine
Manila, Philippines

*Fresthel Monica Climacosa*
Department of Medical Microbiology
College of Public Health
University of the Philippines
Manila, Philippines

*Llewelyn Moron-Espiritu*
Department of Biology
De La Salle University
Manila, Philippines

*Adriel Pastrana*
Faculty of Medicine and Surgery
University of Santo Tomas
Manila, Philippines

*Jessica Joyce De Guia*
Department of Biology
De La Salle University
Manila, Philippines

*Daisy Ilagan-Tagarda*
Department of Internal Medicine
Faculty of Medicine and Surgery
University of Santo Tomas
Manila, Philippines

*Kin Israel Notarte*
Department of Pathology
Johns Hopkins University School of Medicine
Maryland, USA

## 13.1 INTRODUCTION

The Philippines, an archipelagic country in Southeast Asia, is inhabited by approximately 110 million people, a substantial number of whom travel globally as migrant workers or tourists. Its rising popularity as a tourist destination exposes the country to potential importation of infectious diseases, making it particularly vulnerable to outbreaks (Mendoza, 2021). The country's geographical location within the tropics, combined with its densely populated regions and extensive international mobility, amplifies the risks of emerging and re-emerging infectious diseases that pose significant threats to public health (World Health Organization, 2022d). Consequently, the country faces a concerning infectious disease crisis marked by high morbidity rates, necessitating constant vigilance from the government and healthcare authorities in combating these threats.

Despite being highly preventable, infectious diseases continue to burden the Philippines, influenced by various interconnected factors contributing to their emergence and potential recurrence. Addressing this requires a comprehensive understanding of the current infectious disease landscape and the underlying determining factors, including changes in ecological patterns, urbanization, globalization, and climate change, all of which drive disease transmission and put the population at risk. The tropical climate in the Philippines provides an ideal environment for the propagation and transmission of various infectious agents, particularly vector-borne diseases (Sumi et al., 2017). Furthermore, the dense urban population and the presence of rural communities contribute to the spread of infectious diseases such as waterborne, foodborne, airborne respiratory infections, and other communicable diseases.

Over the years, the Expanded Program on Immunization (EPI) in the Philippines has made remarkable progress in improving child survival rates through mass immunization initiatives by the government, despite occasional inconsistencies in promptness and coverage among children (Ulep & Uy, 2019). However, the country faces a significant challenge with the resurgence of vaccine-preventable diseases due to declining vaccine confidence following the Dengvaxia controversy in 2017 (The Lancet Infectious Diseases, 2019). This incident has eroded public trust in vaccination programs, jeopardizing efforts to achieve and maintain herd immunity. As a result, the Philippines ranks fifth globally with zero vaccine uptake among approximately a million children (UNICEF, 2023), significantly threatening public health's future with an increased likelihood of fatal disease upsurges if vaccine hesitancy is not addressed.

As the country experiences demographic shifts and increased global connectivity, the risk of infectious diseases becomes more prominent. Effectively addressing the complexities of emerging and re-emerging infectious diseases in the country demands a comprehensive understanding of their dynamics, evaluation of the impact of past vaccination efforts, and proactive measures to restore public trust in immunization, all of which are crucial in safeguarding public health and preventing potential outbreaks. In this chapter, we discussed emerging and re-emerging infectious diseases relevant to the Philippines and the government's initiatives to address them, providing comprehensive updates on the current infectious disease landscape and highlighting the critical importance of enhancing targeted interventions.

## 13.2 INFLUENZA

Throughout the 21st century, the family Orthomyxoviridae, encompassing influenza viruses, has demonstrated a dynamic and constantly evolving nature, resulting in worldwide pandemics and outbreaks, including the 1918 Spanish Flu, 1957 Asian Flu, 1968 Hongkong Flu, 2009 Swine Flu, and various other flu outbreaks. The mutability and frequent genetic reassortment of influenza viruses, leading to antigenic changes in the viral surface glycoproteins, present significant challenges in efforts to control these viruses (Riedel et al., 2019). . Two crucial surface glycoproteins, Hemagglutinin (HA) and Neuraminidase (NA), play a key role in determining the antigenic variation of influenza viruses and influencing host immunity. Antigenic drift occurs when point mutations accumulate in the gene, resulting in amino acid changes in the protein. These sequence alterations can modify antigenic sites on the molecule, enabling the virion to evade recognition by the host's immune system. This process allows new antigenic variants to proliferate. However, it takes two or more mutations for a new, epidemiologically significant strain to emerge (Riedel et al., 2019). On the other hand, antigenic shift involves major changes in the HA or NA sequence due to genetic reassortment between human, swine, and avian influenza viruses. Such shifts lead to drastic alterations in viral surface proteins (Riedel et al., 2019).

Influenza is transmitted through airborne droplets or by direct contact with contaminated surfaces or hands. The incubation period typically ranges from one to four days. Viral shedding begins a day prior to symptom onset, peaks within 24 hours, remains high for one to two days, and then gradually decreases over the following five days. However, in young children and individuals with weakened immune systems, viral shedding may persist for several weeks. It is extremely rare to detect infectious virus in the bloodstream. Common symptoms include cough, nasal congestion, sore throat, and headache, accompanied by fever, muscle aches, fatigue, and loss of appetite. Fever generally lasts for three to five days, while cough and weakness may persist for two to four weeks after the main symptoms subside.

Most cases can be diagnosed clinically. Diagnosis relies on identification of viral antigens or viral nucleic acid in specimens, isolation of the virus, or through reverse transcriptase-polymerase chain (RT-PCR) reaction. Nasopharyngeal swabs and nasal aspirate or lavage fluid are the best specimens for diagnostic testing and should be obtained within three days after the onset of symptoms.

Globally, there is a decrease in the incidence of influenza during the COVID-19 pandemic decreasing steeply in January after a peak in the late 2022. The World Health Organization (WHO) noted that detections in 2022 were predominantly influenza A (H3N2). By early 2023, there were higher proportions of influenza A (H1N1) pdm09 and B virus. In Europe, influenza B predominated after an initial influenza A wave. The United States noted influenza A (H1N1) in higher proportion just like in East Asia (China) and Central Asia (Kazakhstan). In Southeast Asia, influenza activity remained elevated mainly due to influenza B detections in Malaysia. Influenza A (H3N2) are preponderant in Singapore, Korea, India, and Nepal (World Health Organization, 2023c). In the Philippines, predominantly A (H1N1)pdm09 and fewer influenza B detections were reported.

With the ever-changing COVID-19 crisis as the backdrop, waning population immunity against influenza, and reduced genetic diversity of circulating influenza viruses add to the unpredictability of future outbreaks. These create challenges in the development of control strategies (Dhanasekaran et al., 2022). According to Lee et al. (2022), understanding the rebound of influenza in the post-COVID-19 pandemic periods holds important clues for epidemiology and control. However, COVID-19 mitigation has been highly heterogeneous between countries, ranging from containment to achieve elimination on one end to mitigation with strategic relaxation on the other.

The successful implementation of the Pandemic Influenza Preparedness (PIP) framework along with the strengthening of health systems, operational procedures, and continued technical collaboration with global centers of excellence helped strengthen preparedness to response to epidemics of other high-threat pathogens based on the influenza model (Wijesinghe et al., 2020). The political commitment reflected in the Delhi Declaration on emergency preparedness signed by ministers of health in September 2019, supported by the five-year regional strategic plan to strengthen public health preparedness and response—2019–2023, is a catalyst for guidance and support in developing a broad, long-term strategic plan for preparedness and response to high-threat pathogens in the region. As pandemic viruses emerge, countries and regions face different risks at different times. For that reason, countries need to develop their own national risk assessments based on local circumstances, taking into consideration the information provided by WHO (Wijesinghe et al., 2020).

The Global Influenza Program monitors influenza activity worldwide and publishes an update every two weeks. The updates are based on available epidemiological and virological data sources, including FluNet (reported by the WHO Global Influenza Surveillance and Response System), FluID (epidemiological data reported by national focal points), and influenza and other respiratory virus reports from WHO Regional Offices and Member States. During the COVID-19 pandemic, FluNet has also been receiving updates on testing of samples obtained from routine influenza surveillance systems for SARS-CoV-2. Completeness can vary among updates due to availability and quality of data available at the time when the update is developed (World Health Organization, 2023c). The experiences and lessons learned from the implementation of the PIP framework have provided insights that can be used to strengthen preparedness for epidemics (Wijesinghe et al., 2020) like influenza. Legislations, policies, and plans should be widely disseminated.

In a developing country like the Philippines, there is a 'National Policy on Infection Prevention and Control (IPC) in Healthcare Facilities' known as Administrative Order (AO) 2016-0002. This is in line with AO 2022-0038 Health Sector Strategy for 2023–2028. This policy provides comprehensive guidance to both public and private health facilities for the establishment, implementation, monitoring, and evaluation of effective infection prevention and control programs. The government mandates the presence of Hospital Epidemiology and Surveillance Units (HESU) in all hospitals, which are required to submit weekly reports to the City Epidemiology Surveillance Units (CESU). Key functions of public health surveillance encompass case detection, registration, reporting, confirmation, and data management, among others. The weekly surveillance results are made available online through the Department of Health (DOH) platform. Notably, the reporting system has been further enhanced in response to the COVID-19 pandemic, with the appointment of Disease Surveillance Officers in all hospitals and centers.

Within Southeast Asia, ten national influenza laboratories have been recognized by the Global Influenza Surveillance and Response System (GISRS). Since 2005, the Philippines has been an active member of GISRS and has made notable advancements in molecular testing capacity, biosafety measures, biosecurity protocols, and safe sample transportation. The Research Institute for Tropical Medicine (RITM) serves as the designated National Influenza Center in the country. The Influenza-like Illness/Severe Acute Respiratory Infections (ILI/SARI) surveillance efforts aim to monitor local virus circulation, identify patterns of transmission, assess clinical manifestations, identify high-risk groups, and establish a robust framework for a comprehensive pandemic early warning and monitoring system. Following preliminary analysis, samples are subsequently sent to WHO Collaborating Centers for advanced antigenic and genetic analysis. The recommendations made by the WHO for the composition of influenza vaccines each year, as well as relevant risk assessment activities, are based on the findings derived from these surveillance efforts. The Philippine National Influenza Center (PNIC) framework enables a well-coordinated response to the ongoing pandemic, utilizing data provided by the RITM, the sentinel sites, and surveillance units.

In many developing countries, including the Philippines, the lack of local manufacturing plants for flu medicines and vaccines poses a significant challenge in ensuring their availability. To address this issue, countries like India have taken proactive measures by manufacturing Oseltamivir as part of their pandemic preparedness efforts and distributing it free of cost through the public health system (Wijesinghe et al., 2020). This commendable practice can serve as a model for other countries to follow.

To curb the spread of communicable diseases across borders, proper monitoring of Point of Entries (PoEs) is essential. In the Philippines, an online registration system has been implemented for travelers and returning residents, enabling the screening of symptoms. This step aims to enhance border control measures and prevent the entry of infectious diseases into the country. To establish preparedness and ensure an effective government response, the Interagency Task Force for the Management of Emerging Infectious Diseases (IATF-EID) was created in the Philippines. Its mandate is to assess, monitor, contain, control, and prevent the spread of potential pandemics within the country.

Global risk assessments serve as valuable guidance for countries during different phases of a pandemic, allowing them to revise and adapt their national plans based on lessons learned (Wijesinghe et al., 2020). This approach empowers countries to tailor their response actions according to their own experiences and specific circumstances. It is crucial to regularly revisit guidelines and be prepared for the potential impact of influenza outbreaks. Comprehensive control studies should be developed and regularly updated to delve deeper into aspects such as epidemiology, seasonality, transmission, immunity, control measures, and vaccination strategies. Furthermore, all countries must prioritize the monitoring of biosafety and biosecurity compliance to prevent unintended dissemination of infectious agents.

## 13.3 DENGUE

Dengue infection is caused by the dengue virus (DENV), which belongs to the Flaviviridae family. Dengue viruses are transmitted from person to person by bites from *Aedes* mosquitoes (Murugesan & Manoharan, 2020). The virus has four known serotypes, namely DENV 1–4 (Henchal & Putnak, 1990). Infection with each serotype only confers lifelong immunity for the causative serotype; any person can get infected by dengue four times. However, reinfection with a different serotype causes severe disease due to antibody-dependent enhancement (Halstead, 2014). The symptomatology of dengue infection varies and can range from asymptomatic infection to dengue hemorrhagic fever and dengue shock syndrome, which, if not treated properly, can cause severe complications and death (Wong et al., 2022). There is no specific treatment for dengue. Most mild cases are managed at home with supportive treatment to prevent pain, treat fever, and maintain proper hydration. However, more severe cases would require hospitalization and admission to the intensive care unit (Tayal et al., 2023).

Dengue infection poses a significant public health challenge in the Philippines, particularly affecting the pediatric population. The disease is endemic across the country, with a higher prevalence observed in children aged 5 to 9 years (Undurraga et al., 2017). However, there is limited available data on the distribution and seroprevalence of different dengue virus (DENV) serotypes and their associations with disease severity and outcomes. A systematic review indicated that all four DENV serotypes exist in the Philippines. Earlier studies suggested that DENV-1 and DENV-2 were the more prevalent serotypes among pediatric patients. However, more recent investigations have shown a predominance of DENV-3, while different studies have reported DENV-4 prevalence ranging from 0% to 20% (Bravo et al., 2014; Velasco et al., 2014).

From 2016 to 2020, the Philippine Department of Health (DOH) reported 1,147,425 dengue cases and 16,844 deaths (Humanitarian Data Exchange, 2023). Over the past three decades, there has been a 30% increase in dengue incidence in the country. Similarly, the mortality rates associated with dengue have also risen by 29.2% (Wartel et al., 2017). In recent years, the Philippines has faced multiple dengue outbreaks in various regions (Cabresos, 2020). The situation in 2019 was particularly severe, leading to the declaration of a national epidemic due to the alarming number of reported dengue cases. Specifically, between January and July 20, 2019, 146,062 dengue cases were recorded in the Philippines.

Efforts to combat dengue infection and prevent outbreaks in the Philippines have been implemented through various initiatives (Ong et al., 2022). In 1993, DOH launched the National Dengue Prevention and Control Program, which was nationally implemented in 1998. This program adopts an integrated vector control approach, incorporating health education, environmental sanitation, community mobilization, immediate reporting of dengue cases to local health authorities, and vector surveillance (Nerissa & Dominguez, 1997). Additionally, DOH introduced the ABKD (Aksyon Barangay Kontra Dengue) program to eradicate dengue. This program promoted ovicidal/larvicidal trap systems to trap *Aedes* mosquitoes, serving as an early warning signal to prevent dengue outbreaks (Solidum & Solidum, 2013).

Vaccination is another preventive measure employed. The first licensed dengue vaccine, Dengvaxia, was introduced in the Philippines in 2016. However, two years after its introduction and following the vaccination of over 800,000 children, the vaccine developer, Sanofi, announced that it may cause more severe disease in individuals without prior dengue infection. This revelation resulted in public outrage and led to lawsuits against government officials and executives from Sanofi and the distributor, Zuellig Pharma (Ong et al., 2022). The controversy surrounding the Dengvaxia vaccine in the Philippines significantly impacted public confidence in vaccines, particularly among Filipino parents (Fatima & Syed, 2018). The concerns and issues raised regarding the safety and efficacy of the vaccine led to a decrease in trust and confidence in vaccination programs (Dayrit et al., 2020).

Controlling and managing dengue in the Philippines face challenges related to the unavailability of diagnostic tests, particularly in rural areas (Guzman & Martin, 2018). This limited access to diagnostic tools poses difficulties in accurately differentiating between primary and post-primary dengue infections. Currently, the diagnostic tools used in the country for probable dengue cases include assessing leukopenia with or without thrombocytopenia, conducting dengue NS1 antigen tests, and dengue IgM antibody tests. Confirmatory tests such as viral culture isolation, which require a BSL-2 laboratory, and RT-PCR tests are available only in tertiary hospitals (Agrupis et al., 2019; Artika & Ma'roef, 2017; Biggs et al., 2022). In rural areas where these confirmatory tests are not accessible, physicians often rely on symptomatology and laboratory tests such as thrombocytopenia, leukopenia, and hemoconcentration to diagnose. However, these tests have limitations in terms of diagnostic accuracy, leading to a higher risk of misdiagnosis, particularly in rural areas of the Philippines (Guzman & Martin, 2018; DOH, 2018a). These challenges highlight the need for improved access to diagnostic tests and more accurate tools for diagnosing dengue in resource-limited settings. Enhanced availability and accessibility of reliable diagnostic methods are crucial for timely and accurate diagnosis, leading to appropriate patient management and effective disease control.

Dengue has significant health implications and imposes a substantial financial and economic burden on patients and the Philippine government. According to a study conducted in 2015, the average direct medical cost of a hospitalized dengue case was $772.46 in private hospitals and $387.84 in public hospitals. For cases treated solely in an ambulatory setting, the average cost was $79.43 in the public sector and $168.31 in the private sector (Edillo et al., 2015). A more recent study found that dengue infection resulted in an estimated total cost of $139 million for hospitalized cases and $19 million (PhP 827 million) for ambulatory cases, borne by the public payer. Additionally, the average annual productivity losses were estimated at $19 million, and 50,622 disability-adjusted life years were projected to be lost (Cheng et al., 2018). The figures underscore the need for effective preventive measures, early detection, and appropriate management strategies to mitigate the financial consequences of dengue and alleviate the burden on individuals and the healthcare system.

## 13.4 ZIKA DISEASE

Zika virus primarily affects tropical and subtropical regions, including the Philippines. It is mainly transmitted through the bite of *Aedes aegypti* mosquitoes, which also transmit dengue, chikungunya, and yellow fever. These mosquitoes are active during the day, particularly in the early morning and late afternoon or evening. Apart from mosquito bites, Zika virus can also be transmitted during pregnancy, through blood transfusion and organ transplantation, and *via* sexual contact.

Common symptoms of Zika virus infection include fever, rash, conjunctivitis, muscle and joint pain, malaise, and headache. Pregnant individuals infected with Zika virus are at risk of complications such as microcephaly and other congenital abnormalities in the developing fetus. Fetal loss, stillbirth, and preterm birth are also possible outcomes. In adults and older children, Zika virus infection can trigger neuropathy, myelitis, and Guillain-Barre syndrome (World Health Organization, n.d.).

In February 2016, the WHO declared Zika-related microcephaly a Public Health Emergency of International Concern (PHEIC). However, in November of the same year, the emergency declaration was lifted. Zika virus disease is classified as a weekly reportable disease under the Philippine Integrated Disease Surveillance and Response System (PIDSR). Suspected cases are reported on a weekly basis, while confirmed cases are reported within 24 hours (Technical Guidelines, Standards and Other Instructions for Reference in the Implementation of Zika Virus (ZIKV) Disease Surveillance, 2016). Between 2016 and 2023, a total of 78 cases have been reported in the Philippines, resulting in a solitary fatality. Notably, the highest number of Zika cases was observed from 2016 to 2017, with 56 reported cases in 2016 including one fatality and 17 reported cases in 2017 (Department of Health - Epidemiology Bureau, 2017).

Confirmatory testing for Zika disease involves the use of nucleic acid amplification test (NAAT) to detect the presence of the Zika virus. The WHO recommends utilizing urine samples as the preferred specimen for NAAT testing, although other samples such as blood, saliva, CSF, tissues, amniotic fluid, semen, and rectal swabs can also be used (Centers for Disease Control and Prevention, 2019b; World Health Organization, 2022b). As there are no specific antiviral medications available for treating Zika disease, the primary approach to management revolves around providing supportive care (Centers for Disease Control and Prevention, 2019c).

Clinical samples from suspected Zika cases are handled in a BSL-2 facility. In the Philippines, the country has implemented an Aedes-borne Viral Diseases Prevention and Control Program that includes an 'Enhanced 4S' strategy. This strategy involves searching and destroying breeding sites, seeking early consultation, practicing self-protection, and supporting fogging/spraying in hotspot areas where there is an increase in cases registered for two consecutive weeks to prevent an impending outbreak (DOH, 2018a; DOH, 2018b). Local Government Units (LGU) and other stakeholders conduct massive campaigns to remove non-essential containers, handle essential containers and stagnant water, and implement the daily four o'clock habit to search for and destroy water-holding containers and mosquito breeding sites. The absence of mosquitoes and their reduction are monitored using Ovi-Larval Traps (Guidelines for the Nationwide Implementation of the Enhanced 4S-Strategy against Dengue, Chikungunya and Zika, 2018). The DOH advises individuals to take measures for self-protection against mosquito bites, such as wearing light-colored clothing, applying insect repellent, using screen doors and windows or insecticide-treated screens/curtains, and employing spatial repellents (Guidelines for the Nationwide Implementation of the Enhanced 4S-Strategy against Dengue, Chikungunya and Zika, 2018). Additionally, the WHO recommends that individuals with Zika virus infection and their sexual partners in regions with active transmission of Zika virus receive counseling and information on the risks of sexual transmission, available contraceptive methods, and prevention of adverse pregnancy and fetal outcomes (World Health Organization, 2022e).

## 13.5 HUMAN IMMUNODEFICIENCY VIRUS

Around 33 new cases of human immunodeficiency virus (HIV) are diagnosed in the Philippines daily (Department of Health, 2022a). In the Western Pacific Region, the country has the fastest growing HIV epidemic with disease incidence more than tripling since 2010 (Gangcuangco, 2019; HIV and AIDS Data Hub for Asia Pacific, 2021). HIV is primarily transmitted through sexual contact by an infected person who can be asymptomatic, complicating early detection. Although efficacious antiretroviral therapies (ART) are freely available in 160 treatment facilities (Department of Health, 2020d), trivialized barriers exist across different societal issues, disproportionately affecting vulnerable populations. Equally as important as understanding the public health aspect of HIV control is to study the biomedical intricacies of the disease.

In recent years, genotyping studies have revealed a changing molecular epidemiology of circulating viral genotypes in the country. Genotype CRF01_AE constitutes 77% of infections while genotypes B and C only compose 22% and 1%, respectively (Salvaña et al., 2017). This is in contrast with other countries like the United States and Europe, where the B genotype dominates the epidemic. CRF01_AE, unfortunately, has been associated with shorter time to death and increased transmission efficiency, possibly contributing to the exponential increase in new cases in the area (Taylor et al., 2008). Moreover, being infected with the CRF01_AE genotype was associated with higher viral loads (>100,000 copies/mL), lower CD4 counts (<50 cells/mm3), and transmitted drug resistance (Salvaña et al., 2017).

In 2018, the Philippines Department of Health (DOH) proposed a change in confirmatory testing from Western blot to the Rapid HIV Diagnostic Algorithm (rHIVDA). This was done following the advice of the World Health Organization (WHO) and Centers for Disease Control and Prevention (CDC) in lieu of false positive and indeterminate results using Western blot. The algorithm includes two immunoassays and three rapid diagnostic tests for further validation of sensitivity and specificity in the local population (Health Technology Assessment Study Group - Health Policy Development and Planning Bureau, 2018).

The DOH also made a shift in the offered ART in August 2021. Efavirenz-based treatments were replaced with tenofovir/lamivudine/dolutegravir regimen following evidence of better viral suppression and lower odds of treatment-related adverse events. The DOH offers fully subsidized treatment to people living with HIV (PLHIV) (Health Technology Assessment Study Group - Health Policy Development and Planning Bureau, 2021).

To address emerging public health challenges of HIV, the DOH launched the 6th HIV & AIDS Medium Term Plan (6th AMTP) for the years 2017–2022. The plan emphasized four main targets, namely, increasing knowledge of HIV transmission among key populations, preventing new infections, increasing testing and treatment coverage among PLHIV, and eliminating mother-to-child transmission (Philippine National AIDS Council, 2016). In the country's HIV care cascade, the UNAIDS' 90-90-90 target and strategies were also adapted. This includes testing of vulnerable groups, enrollment of PLHIV to ART, treatment of infected individuals, monitoring of viral load, and achieving viral suppression (Department of Health, 2020d). Clinical care for PLHIV is given through registered treatment hubs in major cities and municipalities in different parts of the country (Gangcuangco, 2019).

Unfortunately, the country fell short in most of the UNAIDS targets as of 2020. Only 68% (target: 90%) of the estimated infected individuals know their status. Of those who were diagnosed, only 61% (target: 90%) were on ART. Among those under ART, only 17% were tested for viral load with 94% achieving viral suppression (target: 90%) (Department of Health - Epidemiology Bureau, 2020). The number of new infections did not decrease either, but continued to rise exponentially (Department of Health - Epidemiology Bureau, 2020). While the shift of focus to the COVID-19 pandemic could have affected the implementation of HIV programs, multiple public health barriers also remain to be addressed.

Fear of stigmatization is a main barrier to achieve HIV control. Hence, HIV self-testing (HIVST) is a potential solution to address anticipated HIV testing stigma among individuals who may not undergo traditional facility-based testing services (Sison et al., 2022). In Thailand, the majority of men who have sex with men (MSM) were willing to use HIVST with online supervision (Samoh et al., 2021), notably, having the potential to engage high proportions of first-time testers (Phanuphak et al., 2018). This is important in countries such as the Philippines wherein only 68% of PLHIV are aware of their status. Although HIVST is widely acceptable in both high (Kularadhan et al., 2022) and low- to mid-income countries (Rivera et al., 2021), linkage to HIV care, defined as visiting an HIV care provider within 30 days after learning that they were positive (Centers for Disease Control and Prevention, 2019a), remains to be a major challenge. An obvious barrier to a successful linkage is if a PLHIV chooses not to report the result of the test. Even within the same country, reporting rates are different (Eustaquio et al., 2022; Rosadiño et al., 2023) for which the reasons remain to be investigated. Nevertheless, linkage to care is vital to HIV control for it is a precursor to timely antiretroviral initiation and viral suppression (Dombrowski et al., 2015) as well as social and psychological services (Sanga et al., 2019). This underscores the importance for countries to ensure health service delivery systems are in place if HIVST scale-up plans are on the way.

Knowing that successes direct the way to what works and failures highlight obstacles to overcome is essential in designing plans for HIV control in the country. As the DOH launched the 7th AMTP, all stakeholders and partners must act holistically in addressing health and its social determinants. The multisectoral strategic plan aims to achieve this by cultivating a sustainable and supportive environment responsive to the needs of PLHIV (Department of Health, 2022c).

## 13.6 HUMAN PAPILLOMAVIRUS

Human papillomavirus (HPV) infection is the most important causative factor of cervical cancer (Bosch et al., 2002). The virus, through its E6 and E7 proteins, interact with tumor suppressor genes, promote genomic instability, and lead to malignant transformation in the cervical epithelium (Howie et al., 2009; McLaughlin-Drubin & Müenger, 2009).

Cervical cancer ranks fourth among the most common cancers among women worldwide. Around 604,000 new cases and 342,000 deaths have been reported globally in 2020. This is also the fourth leading cause of cancer death among women (Sung et al., 2021). Unfortunately, 90% of cases are found in low-to-middle-income countries, like the Philippines, where the overall 5-year survival rate is usually below 50% (Bruni et al., 2023).

In the Philippines, 39.6 million women are at risk for cervical cancer. The country has an annual burden of 7,897 cases and consequent 4,052 deaths (Bruni et al., 2023; Nang et al., 2023). A hospital-based case-control study in 1998 reported that HPV DNA could be detected in 93.8% of patients with squamous cell carcinoma, 90.9% among those with adenocarcinoma, and 9.2% among controls. The most common genotypes were HPV 16, 18, and 45 (Ngelange et al., 1998). However, the molecular epidemiology of HPV genotypes in recent years, especially at the community level, is yet to be determined (de Paz-Silava et al., 2023).

All sexually active people are at risk for HPV infection. While in most cases HPV infection clears spontaneously, persistent infection of the high-risk types is associated with cervical cancer. Other factors associated with cervical cancer are also observed in contracting high-risk HPV. These include HIV-positive status and multiple sexual partners (Nang et al., 2023).

In 2005, visual inspection with acetic acid (VIA) was introduced as the initial screening approach in the country. This is usually followed by biopsy or colposcopy with pap smear as confirmatory tests when positive for VIA (Guerrero, 1982). In 2012, the Clinical Practice Guidelines of the Society of Gynecologic Oncologists of the Philippines recommended conventional cytology or liquid-based cytology for cervical cancer screening, which remains to be the preferred screening test to this day (Society of Gynecologic Oncologists of the Philippines, 2019). This, however, is in contrast with the WHO recommendations to utilize HPV DNA-based tests as the primary screening method for cervical cancer.

Unfortunately, the screening coverage in the country was low at 7.7% in 2003. More recent data reveals a modest increase to 36.8% and 13.9% in urban and rural areas, respectively (Imoto et al., 2020; Lintao et al., 2022). These figures show, however, that the screening coverage in the country still remains much lower than the WHO target of 70% (World Health Organization, 2023a).

The HPV vaccination program was only introduced in 2015 through the DOH National Immunization Program. The quadrivalent vaccine containing HPV 6, 11, 16, and 18 was used to immunize girls aged 9–10 years old from ten out of 81 provinces in the country (Department of Health, 2015). In 2017, the School-based Human Papillomavirus (HPV) Immunization was launched in partnership with the Department of Education to reach more schoolgirls (Manila Bulletin, 2017). Despite these initiatives, the vaccination coverage in 2020 remained very low at 23% and 5% for the first and final dose, respectively. The most recent data in 2021 even shows much lower statistics at 4% for the first dose and 0% for the second dose (World Health Organization, 2022c). The poor outcomes of the HPV vaccination program can be explained by the government's shift of focus to the COVID-19 pandemic, on top of many other challenges in HPV vaccination rollout. However, a new initiative to increase global access to HPV vaccine was launched last January 2023 by the HPV Vaccine Acceleration Program Partners Initiative (HAPPI) Consortium (John Snow Inc., 2023). Hence, scale-up in HPV vaccine implementation in the Philippines may be achievable.

Successful scaling up or implementing HPV vaccination strongly depends on its integration with the national cancer control plan of the country. It should also be introduced as part of a coordinated comprehensive strategy to prevent cervical cancer and other HPV-related diseases (World Health Organization, 2022a). This includes education on reducing behaviors that increase the risk of HPV infection and a functioning service delivery component. Likewise, strengthening health promotion capacity, building healthy settings, and developing health literacy are essential in achieving HPV vaccination coverage (World Health Organization, 2023b).

## 13.7 TUBERCULOSIS

*Mycobacterium tuberculosis*, the causative agent of tuberculosis, is an acid-fast, nonmotile, slow-growing, and strictly aerobic bacilli affecting immunocompromised patients primarily in low-income countries. Its potential for respiratory transmission through aerosolization necessitates biosafety level 3 (BSL-3) facilities for safe handling. Four National TB Prevalence Surveys (NTPS) have been done to quantify the burden of tuberculosis in the Philippines, though a pioneer survey done in Minglanilla, Cebu, showed that the prevalence of smear positive cases is 400 per 100,000 (Department of Health & Philippine Coalition Against Tuberculosis, 2004). The prevalence of active pulmonary TB did not differ from the time it was measured in 1983 compared to 1997, though there was a decline of 37% for smear-positive cases and 25% for culture-positive cases (Tupasi et al., 1999). The directly observed treatment, short course (DOTS) only covered 2% of the population in 1996 (Tupasi et al., 1999). Inadequate DOTS coverage, delayed treatment, and unstable supply of drugs contributed to the lack of decline in cases (Tupasi et al., 1999). The greatest reduction of bacteriologically confirmed TB was recorded in the period between 1997 and 2007, with the prevalence dropping from 960 per 100,000 to 660 per 100,000 (World Health Organization, 2021b). This success is attributed to increased partnership with the private sector, training of healthcare workers, and retooling of TB diagnostics (World Health Organization, 2021b). However, the most recent NTPS in 2016 showed that there was no evidence demonstrating a decrease in bacteriologically confirmed TB. In fact, this national survey recorded 1,159 cases per 100,000, the highest prevalence among all national TB surveys (World Health Organization, 2021b). More sensitive screening and diagnostic methods, diagnostic delays, case detection gaps, and lack of improvement in poverty alleviation and insurance coverage account for the rise of cases (World Health Organization, 2021b).

Similarly, three national Drug Resistance Surveys (DRS) have been conducted. The baseline prevalence of drug resistance obtained in the 2004 DRS is 5.7%. Resistance was observed more frequently in re-treatment cases (21%) than in new cases (4%) (Department of Health, 2023). A decrease in the prevalence of drug resistance among new cases was observed in 2012 (2%), though its prevalence in re-treatment cases remained at 21% (Department of Health, 2023). The most recent DRS in 2018 showed a reduction of drug-resistant cases in new (1.2%) and previously treated cases (12%) (Department of Health, 2022b). The incorporation of MDR-TB into the National Tuberculosis Program (NTP) by the Department of Health (DOH) in 2008 has been instrumental in the generally decreasing number of drug-resistant cases since 2004 (Quelapio et al., 2010).

The Philippine Tuberculosis Society, Inc. (PTSI), a private organization established in 1910, is pioneering in the Philippine effort against tuberculosis (Department of Health & Philippine Coalition Against Tuberculosis, 2004). Its primary hospital, the Quezon Institute Tuberculosis, has been serving TB patients since 1918. Its impact nationally was expanded through mandated funding from the Philippine Charity Sweepstakes Office (PCSO) (Department of Health & Philippine Coalition Against Tuberculosis, 2004). PTSI is also involved in education, information dissemination, case finding, and treatment in far-flung areas through its Community Outreach Program and Education (COPE) Initiative in eight field branches across the country (Philippine Tuberculosis Society Inc, 2022).

The Philippine Coalition Against Tuberculosis (PhilCAT) is a coalition of government and non-government organizations championing public-private collaboration in tuberculosis control. It was established in 1993 under the auspices of PCCP, DOH, Philippine Society for Microbiology and Infectious Disease (PSMID), PTSI, Cure TB, and the American College of Chest Physicians (ACCP)-Philippine Chapter (Department of Health & Philippine Coalition Against Tuberculosis, 2004). In 2003, PhilCAT, in coordination with DOH, started developing the Comprehensive and Unified Policy (CUP) for Tuberculosis Control in the Philippines to fill in the need for a national policy uniting both sectors (Philippine Coalition Against Tuberculosis, 2021). This policy formalized and operationalized the collaboration between DOH and other departments in implementing the NTP, provided guidance for private physicians in TB management aligned with the NTP, aligned the TB benefits policies of the Social Security System (SSS), Government Service Insurance System (GSIS), and Employees' Compensation Commission (ECC), and pioneered a TB outpatient benefits package in the Philippine Health Insurance Corporation (Department of Health & Philippine Coalition Against Tuberculosis, 2004).

TB prevention consists mainly of BCG vaccination and the frequently neglected tuberculosis preventive treatment (TPT). The BCG vaccination program, which was introduced in the country under the funding of between 1951 and 1952, became compulsory for children eight years and below after the issuance of Presidential Decree 996 in 1976 (Department of Health & Philippine Coalition Against Tuberculosis, 2004). TPT, on the other hand, involves the identification of inactive or latent TB infection (LTBI) and treatment with a three-month isoniazid and rifapentine regimen (3HP) (Department of Health, 2020b). Recent studies show that TPT can prevent progression of LTBI to active disease (Moonan et al., 2018). Its inclusion in the Philippine Strategic TB Elimination Plan 2020–2023 seeks to improve its low coverage (Department of Health, 2020b; Moonan et al., 2018).

The National TB Control Program was implemented by the DOH nationwide in 1978 and was further strengthened in 1987 by the establishment of sputum microscopy centers in the majority of rural health units (RHUs) and the production and distribution of short course chemotherapy drugs (SCC) (Department of Health & Philippine Coalition Against Tuberculosis, 2003). SCC, which is the administration of rifampicin and isoniazid for nine months and initial supplementation with ethambutol, proved to be comparable and more acceptable than the standard chemotherapy for 18 months (British Thoracic and Tuberculosis Association, 1975). In 1993, the WHO global tuberculosis program began promoting the Directly Observed Short Course (DOTS) strategy in response to rising cases (Otu, 2013). It consists of political commitment, case detection by sputum smear microscopy among symptomatic people, six to eight month regimen of short-course chemotherapy (SCC) with first-line anti-TB drugs administered under direct observation, stable supply of all essential anti-TB drugs, and standardized recording and reporting system (Department of Health & Philippine Coalition Against Tuberculosis, 2003). This strategy proved to be successful in China, Vietnam, and Tanzania since it increased compliance with medications, improved the accuracy of diagnosis, and provided free medications to patients of lower socioeconomic status (Department of Health & Philippine

Coalition Against Tuberculosis, 2003; Otu, 2013). The Collaboration in Rural and Urban Sites to Halt Tuberculosis (CRUSH TB) Project launched by the DOH in 1996 served as the pilot test which provided results crucial for the implementation of the DOTS strategy nationwide (Department of Health & Philippine Coalition Against Tuberculosis, 2003).

However, success brought by the DOTS strategy was hampered by increasing rates of drug resistance in the mid-1990s. Multidrug-resistant tuberculosis emerged from developing countries implementing respective tuberculosis programs, the United States because of increasing TB-HIV coinfection, and eastern Europe due to socioeconomic crisis and unstable health systems (Acosta et al., 2014). In 1999, WHO responded by establishing the Working Group on DOTS-Plus for MDR-TB to study the programmatic treatment of MDR-TB (Blöndal, 2007). The DOTS-Plus strategy builds on the original DOTS strategy and extends by covering rational use of second-line drugs. The Green Light Committee (GLC), a subgroup of the Working Group, negotiated with pharmaceutical companies to reduce the cost of second-line drugs (Blöndal, 2007). They also provide guidance on the proper use of these drugs (Blöndal, 2007). The first DOTS-Plus program in the Philippines was started in 1999 by Makati Medical Center, with the support of the Tropical Disease Foundation and approval of GLC in 2000 (Mora et al., 2022). Initial results were promising, showing that management of MDR-TB in resource-limited setting is feasible and cost-effective (Tupasi et al., 2006). As a result, MDR-TB services were later integrated into the NTP in 2008, despite initial doubts (Portero & Rubio, 2006; Quelapio et al., 2010).

The passage of Republic Act 10767, the Comprehensive TB Elimination Plan of 2016 or the TB Law, paved the way for the implementation of mandatory notification for TB and the drafting of a national strategic plan for tuberculosis elimination, the Philippine Strategic TB Elimination Plan (PhilSTEP) (Department of Health, 2020b). Other provisions of RA 10767 include insurance benefits for multidrug-resistant TB, adequate funding for the NTP, and integration of TB control in education, among others (Tan & others, 2019). Despite the extensive reach of this legislation, ACHIEVE (2019) perceives a lack of community engagement in the national response and inadequate protection against TB-related stigma.

PhilSTEP 2017–2022 has been updated to PhilSTEP 2020–2023 to identify priority activities that will contribute to achieving case finding and treatment targets and add one year of implementation to serve as basis for the development of the next round of support from the Global Fund to Fight AIDS, Tuberculosis, and Malaria (Department of Health, 2020b). The update divides 14 strategies to four major areas of the TB care—screening, testing and diagnosis, treatment, and prevention (Department of Health, 2020b). Some of the strategies include rapid diagnostic test (RDT) expansion and utilization, line probe assay/drug susceptibility test (LPA/DST) optimization, TB-HIV collaboration, and adoption of short TB preventive treatment (TPT) (Department of Health, 2020b). As PhilSTEP reaches its end, the upgraded Philippine Acceleration Action Plan for Tuberculosis (PAAP-TB) aims to continue and hasten the battle by assigning the TB-NCC as the steering committee that will track progress of multi-sectoral and intersectoral efforts.

## 13.8 MENINGOCOCCAL DISEASE

*Neisseria meningitidis* is known to cause primarily two clinical syndromes: meningococcal meningitis and meningococcal septicemia, which is also known as meningococcemia (Jameson et al., 2020). The gram-negative encapsulated aerobic diplococci can be handled safely at BSL-2 (Buchan et al., 2019). Non-blanching petechial and/or purpuric rash is pathognomonic of meningococcal disease (Jameson et al., 2020; Nadel & Kroll, 2007). Meningococcal meningitis alone, which accounts for 30–50% of cases, presents as fever, sometimes associated with rash and presenting with neurological signs like headache, neck stiffness, changes in sensorium, and convulsions (Fabay, 2010; Jameson et al., 2020). Meningococcemia, on the other hand, is seen in 20% of cases and manifests as a sudden onset of fever with rash, frequently progressing to shock (Fabay, 2010; Jameson et al., 2020).

Seven epidemics of meningococcal disease in the Philippines were reported from 1988 to 2020, the largest of which was recorded in 2004–2006 in the Cordillera region with 418 cases (Raguindin et al., 2020). Specifically, Baguio General Hospital and Medical Center logged over 150 cases and 40 deaths in a span of five months (end of September 2004 to February 2005), far exceeding the average of 100 cases annually (Fabay, 2010). The cause was identified to be a penicillin-sensitive strain of *N. meningitidis* serogroup A subtype A1.9 (Fabay, 2010). The case guidelines for its management and control were developed in 2005, with emphasis on contact, early detection, and early treatment with benzyl penicillin (Department of Health, 2005). Likewise, the Pediatric Infectious Disease Society of the Philippines (PIDSP) and Child Neurology Society of the Philippines (CNSP) recommend penicillin for meningococcal meningitis (Pediatric Infectious Disease Society of the Philippines & Child Neurology Society of the Philippines, 2015). The Research Institute for Tropical Medicine (RITM) was assigned in 2012 to be the National Reference Laboratory for Emerging and Re-Emerging Diseases (EREID), which included leptospirosis, meningococcemia, anthrax, pertussis, and diphtheria as diseases of interest in diagnostic surveillance (Department of Health, 2012; Lupisan, 2015).

## 13.9 LEPTOSPIROSIS

First reported in the Philippines in 1932 (Basaca-Sevilla et al., 1986), leptospirosis is an endemic zoonotic disease in the country with an annual average of 680 cases and 40 deaths (Philippine Society for Microbiology and Infectious Diseases, 2010). Outbreaks follow heavy rainfall and flooding typically occurring from July to October. The disease is caused by the Gram-negative, pathogenic species of the genus *Leptospira* that thrive in warm and humid environments. Leptospirosis is believed to be the most widespread zoonosis affecting both developed and developing countries with median annual global incidence ranging from 14 cases per 100,000 population to 975 cases per 100,000 population in the tropics (Leptospirosis Burden Epidemiology Reference Group, 2011).

Pathogenic leptospires, commonly categorized to serovars, propagate in the kidneys of a wide range of mammalian hosts, with rodents (e.g., *Rattus norvegicus*) as the most important source (Haake & Levett, 2015). These maintenance hosts shed the bacteria in the urine, contaminating soil and water environments. Humans become infected through direct contact with infected animals or indirect exposure to contaminated environments. As such, leptospirosis was first associated with specific occupations like mining and farming (Bharti et al., 2003). More recently, global warming, increasing intensity of storms due to climate change, rise in global travel, and engagement in adventure sports are recognized contributory factors to the global emergence of leptospirosis (Hartskeerl et al., 2011). Nevertheless, the disease is strongly linked to poverty wherever poor housing and inadequate sewage disposal and water treatment may lead to rodent exposure.

Clinically, the infection may present as non-specific symptoms of fever, headache, and myalgia, leading to a significant number of cases to be misdiagnosed. Studies have shown that leptospirosis may account for as high as 20% of fevers of unknown origin (Leptospirosis Burden Epidemiology Reference Group, 2011). The World Health Organization (WHO) developed the Modified Faine's scoring to make a presumptive diagnosis; however, a cross-sectional study in a local hospital in the Philippines demonstrated its poor sensitivity and specificity (Brato et al., 1998). Although the definitive diagnosis of leptospirosis is done by culture isolation and direct visualization of the leptospires, the reference standard serologic test of microscopic agglutination test (MAT) is still the widely used diagnostic approach. In this method, sera are reacted with live, serovar-specific leptospiral suspensions and then examined for the presence of agglutination under darkfield microscopy. The procedure thus requires a well-equipped BSL-2 laboratory with highly trained laboratory workers who regularly maintain and prepare live leptospiral cultures. This entails repeated exposures to the bacteria which posed significant biosafety risks among laboratory workers (Mendoza et al., 1979; Levett, 2001). Hence, MAT is limited to a few reference or expert centers.

Treatment with doxycycline and penicillin G remains to be the drug of choice for mild and moderate to severe leptospirosis, respectively. However, Jarisch-Herxheimer reactions, resulting from cytokine activation of degenerating spirochetes, have been reported in patients receiving penicillin G (Bharti et al., 2003). Specific recommendations for selective pre- and post-exposure prophylaxis using antibiotics are in place (Philippine Society for Microbiology and Infectious Diseases, 2010) and its careful implementation is crucial to reduce the risk of emergence of antimicrobial resistance. A local study has shown that *Leptospira* isolates from the country remain to be sensitive to these first-line antibiotics (Chakraborty et al., 2010).

In the Philippines, several small seroepidemiological surveys among humans and animals were conducted in the 1960s to the 1980s mainly in Luzon. Rats from Metro Manila were exhibited to harbor in their kidneys the serovars of Pyrogenes, Bataviae, and Javanica as well as Manilae, a newly identified serovar then, which was named after where the rats were caught (Aragon & Famatiga, 1965). Water buffaloes were shown to have antibodies against Tarassovi, Sejroe, and Poi (Carlos et al., 1970). The first human survey was done in an outbreak in Sablayan, Mindoro, demonstrating that 9% of the residents in the agricultural penal colony had antibodies to leptospires (Basaca-Sevilla et al., 1981). High-risk workers including abattoir employees, dog pound personnel, and fish inspectors in Manila were found to have significant antibody titers against serovars Pyrogenes, Bataviae, and Pomona (Arambulo et al., 1972), while agricultural workers from Central Luzon had the greatest antibody response against serovars Shermani, Pyrogenes, Hardjo, and Fort-Bragg (Padre et al., 1988). In clinical settings, patients diagnosed with leptospirosis were positive for serovars Manilae, Bataviae, Pyrogenes, Grippotyphosa, Pomona, Icterohaemorrhagiae, and Javanica (Alora et al., 1973).

More recent surveys done between 2000 and 2018 have shown similar serovars to be prevalent among animals such as rats in Metro Manila and Laguna (Villanueva, Ezoe, Baterna, Yanagihara, Muto, Koizumi, Fukui, Okamoto, Masuzawa, Cavinta, Gloriani, et al., 2010), water buffaloes, pigs, and dogs in Luzon (Villanueva et al., 2018). The same is true for human sera testing among abattoir workers (Tabo et al., 2018) and patients from major hospitals in Manila (Masuzawa et al., 2001). This suggests that these serovars may have persisted for decades and have been causing leptospirosis in the country. The study of Villanueva et al. (2010) using a Geographic Information System (GIS) revealed that the same serovars, particularly the three predominant groups (i.e., Pyrogenes, Bataviae, and Grippotyphosa), were isolated from human and rat samples collected from the same area. The bacteria may have been maintained among wild rats and may have been transmitted to humans indicating a strong Leptospira-rat-human relationship in Luzon. Pathogenic species were likewise isolated in waters collected from Nueva Ecija (72%) and Metro Manila (21%) (Saito et al., 2013b). Soil samples from rice fields and garbage areas in Central Luzon and Metro Manila yielded up to 44% positivity rates of pathogenic leptospires (Tantengco & Gloriani, 2017). Saito et al. (2013a) demonstrated the existence of pathogenic strains in the soil even months after a storm surge, suggesting that leptospirosis outbreaks in affected areas may still occur.

Despite several outbreaks, national guidelines and statistics for leptospirosis are lacking until after typhoon Ketsana (locally known as Typhoon Ondoy) in 2009 where suspected cases of leptospirosis peaked at 2,292 and 178 deaths were recorded in Metro Manila. Interim clinical practice guidelines on disease prevention were drafted by the Philippine Leptospirosis Task Force. These guidelines were later finalized in 2010 as the Philippine Clinical Practice Guidelines (CPG) on the Diagnosis, Management, and Prevention of Leptospirosis in Adults. Nationwide statistics on leptospirosis started being reported in 2010. The Research Institute for Tropical Medicine (RITM) was assigned in 2017 to be the National Reference Laboratory (NRL) for Emerging and Re-Emerging Diseases (EREID), which included leptospirosis as one of the diseases of interest (Department of Health, 2017b). In the 2020 mandate by the DOH, RITM was assigned distinctly as the NRL for leptospirosis, aside from being the NRL for EREID, highlighting an increased understanding of the impact of leptospirosis on the health of Filipinos (Department of Health, 2000).

Ultimately, reducing exposure by controlling infection sources remains to be the primary strategy to control and prevent leptospirosis. Rodent control efforts may offer short-term benefits. Appropriate housing construction together with comprehensive flood mitigation and environmental sanitation may be difficult to implement but should be recognized as important public health strategies. At the individual level, protective measures to prevent high-risk exposures such as avoiding wading in the floods and wearing personal protective gear (e.g., wearing boots, goggles, gloves, and overalls) depend on occupational or daily life activities (Haake & Levett, 2015). Vaccination is another extremely important means to reduce the incidence and transmission of leptospirosis. Vaccines for cattle, pigs, and dogs are commercially available, while vaccines for humans were locally developed in Cuba, France, China, and Russia. However, the protection afforded by these vaccines is short-lived and serovar-specific, hence revaccination to retain immunity is needed (Hartskeerl et al., 2011). A cornerstone in all these strategies is an effective surveillance system for both humans and animals. Finally, addressing leptospirosis requires holistic and concerted efforts from sectors dealing with human health, animal health, and environmental health. Such a strategy is embedded in the collaborative, multidisciplinary approach of One Health (Pham & Tran, 2022). A deep sense of solidarity among different stakeholders as promoted by the One Health approach is strongly desired to provide comprehensive, effective, and long-lasting solutions for the control and prevention of leptospirosis at the local, regional, and global levels.

## 13.10 ASPERGILLOSIS

*Aspergillus* is a common fungal genus that can cause various infections in humans, including invasive aspergillosis (IA) and allergic bronchopulmonary aspergillosis (ABPA) (Segal, 2009). This fungal pathogen exhibits a high level of pathogenicity, allowing it to invade blood vessels, cartilages, and even bones, resulting in a range of infections such as allergies, asthma, pulmonary infections, and thrombosis (Dagenais & Keller, 2009; Gu et al., 2021). Among the various species of *Aspergillus*, *A. fumigatus* is the most common etiologic agent of the disease. It is a widespread fungus that produces small conidia measuring approximately 2 to 3 µm in diameter, enabling their entry into the lung alveoli (Latgé, 1999). The small size of the conidia enables them to remain highly buoyant, facilitating their easy dispersion through strong air currents without the need for specialized mechanisms (Latgé, 1999). In immunocompetent individuals, inhalation of *Aspergillus* conidia does not result in adverse effects. However, repeated exposure in immunosuppressed individuals can lead to the proliferation of the fungus within lung cavities (Chabi et al., 2015).

The occurrence of *Aspergillus* infections can vary among different populations and regions. IA is frequently observed in individuals with weakened immune systems, such as those undergoing hematopoietic stem cell transplantation or solid organ transplantation (Cadena et al., 2021). Non-traditional patient groups, such as those with chronic obstructive pulmonary disease (COPD), influenza-related pneumonia, and individuals infected with SARS-COV-2, have reported cases of IA (Cadena et al., 2021; Hammond et al., 2020; Koehler et al., 2019). In patients who have undergone solid organ transplantation, lung transplant recipients have the highest risk of invasive pulmonary aspergillosis, with 10–15% of patients developing this condition, especially at the surgical anastomotic site (N. Singh & Paterson, 2005). A recent study examined the occurrence of IA among influenza pneumonia patients admitted to ICUs in Belgium and the Netherlands over seven influenza seasons and found that

IA incidence was high (>10%, regardless of the season) among ICU patients with influenza pneumonia, irrespective of the influenza virus subtype (Cadena et al., 2021). It was also noted that the incidence of COVID-19-associated pulmonary aspergillosis (CAPA) varies; however, it is linked to increased mortality rates among critically ill COVID-19 patients, with some reports indicating a 16–25% increase (Cadena et al., 2021). It was estimated that 50% of the global cases of IA are in Asian countries, with Vietnam having 14,523 cases, Korea with 2,150 cases, the Philippines with 3,085 cases, and Thailand with 941 cases annually (Bongomin et al., 2017). These findings highlight the significant burden of *Aspergillus* infections worldwide, necessitating the implementation of appropriate preventive and management strategies.

Invasive aspergillosis, allergic bronchopulmonary aspergillosis, and chronic pulmonary aspergillosis are the types of infections that have been reported in the Philippines (Batac & Denning, 2017). However, data on the prevalence and incidence is still limited and scarce in the country. Surveillance programs are scarcely implemented in the Philippines for different types of mycoses. Hence, it is imperative to strategize a nationwide approach in the active and sustained surveillance of fungal infections to increase awareness and to provide baseline information on the epidemiology of mycoses, considering that it is difficult to delineate the epidemiology of such infections and diagnosis is solely based on clinical manifestations and microbiological criteria. Proper laboratory investigation and protocols in private and government hospitals are needed to confirm the diagnosis of *Aspergillosis* and other mycoses (Handog & Dayrit, 2005). This also implies the role of the stakeholders providing hospital and laboratory personnel who are equipped with technical skills to improve laboratory diagnosis and good clinical practice.

The clinical data on *Aspergillus* infections have shown a broad spectrum of ailments ranging from mild allergic reactions to severe invasive infections. Although the most severe manifestation is IA (Patterson et al., 2016), *Aspergillus* is also associated with another clinical condition called allergic bronchopulmonary aspergillosis (ABPA), which is characterized by hypersensitivity reactions in individuals with underlying respiratory conditions like asthma or cystic fibrosis (Agarwal et al., 2013). Chronic pulmonary aspergillosis (CPA) is a different form of *Aspergillus* infection that primarily affects individuals with pre-existing lung damage, such as tuberculosis or sarcoidosis (Patterson et al., 2016). ABPA is an allergic response in patients with underlying respiratory conditions, such as asthma, while CPA is characterized by chronic lung cavities and gradual progression. *Aspergillus* spp. can also cause localized infections, such as sinusitis or otomycosis. Antifungal therapy is the typical treatment, and the prognosis depends on several factors, including the patient's immune status and the extent of infection (Patterson et al., 2016).

*Aspergillus* spp. are commonly found in the environment and can pose a biosafety hazard, particularly in laboratory and healthcare settings. These species are capable of generating airborne spores that can be inhaled and cause respiratory infections (Patterson et al., 2016). Therefore, it is crucial to implement suitable biosafety measures to minimize the risk of exposure and subsequent infection. The recommended BSL-2 is typically used to handle *Aspergillus* spp. in laboratory settings. This level involves the use of personal protective equipment such as gloves, masks, and laboratory coats to prevent direct contact and inhalation of spores. Additionally, containment measures like laminar flow hoods, biological safety cabinets, and HEPA filters are employed to control the dissemination of spores. Proper training of laboratory personnel in handling and disposal procedures, as well as regular cleaning and decontamination of workspaces, are crucial in maintaining a safe working environment. Adhering to these biosafety practices helps mitigate the risk of exposure to *Aspergillus* and ensures the protection of laboratory workers and healthcare personnel (Meechan & Potts, 2020).

Preventing *Aspergillus* infections requires a combination of general infection control measures and specific interventions that target high-risk populations. In healthcare settings, strict adherence to infection control practices, such as hand hygiene, proper cleaning and disinfection of surfaces, and implementation of airborne precautions, can help reduce the transmission of *Aspergillus* spp. (Centers for Disease Control and Prevention & Infectious Diseases Society of America, 2000). Individuals with compromised immune systems, such as transplant recipients and those undergoing chemotherapy, may benefit from prophylactic antifungal therapy to prevent *Aspergillus* infections (Patterson et al., 2016). In certain agricultural or occupational settings where exposure to *Aspergillus* spores is common, the use of respiratory protective devices, such as masks or respirators, can be effective in reducing the risk of inhalation (Occupational Safety and Health Administration, 2007). Education and awareness programs targeted at high-risk populations can also promote preventive measures, such as avoiding mold-contaminated environments and practicing good respiratory hygiene (Centers for Disease Control and Prevention & Infectious Diseases Society of America, 2000). By implementing a comprehensive approach that includes both general infection control measures and targeted interventions, the incidence of *Aspergillus* infections can be significantly reduced.

## 13.11 CANDIDIASIS

Candidiasis is a prevalent fungal infection across the globe, caused by different species of *Candida*. The incidence and risk factors of this infection vary among different populations and geographical regions, resulting in diverse epidemiology. As an overview, oral candidiasis is a prevalent condition that affects individuals of various age groups. Studies indicate that the condition has varying prevalence rates, ranging from 20 to 50% in healthy infants, 7–60% in denture wearers, and up to 95% in individuals with HIV/AIDS (Williams & Lewis, 2011). Vulvovaginal candidiasis is a genital yeast infection affecting 70–75% of women at least once during their lifetime (Gonçalves et al., 2016). Invasive candidiasis, which includes bloodstream infections, is a significant manifestation of the condition, with an estimated incidence of 1.4– 10.05 cases per 100,000 people in Europe, and affecting around 25,000 people in the United States per year (Parslow & Thornton, 2022; Pfaller et al., 2011). Although the incidence of candidiasis in Western countries has either stabilized or decreased, reports of this disease continue to increase in many South Asian countries. Its incidence was noted to be 20–30 times higher in South Asia than Western countries. In India, the reported incidence of candidiasis is approximately 1–12 cases per 1,000 hospital admissions, in comparison to the 0.5–0.8 per 1,000 admissions in Australia, 0.8 per 1,000 discharges in the United States, and 0.2–0.5 per 1,000 discharges in European countries (Singh & Chakrabarti, 2017).

Limited research has been conducted in the Philippines regarding the tracking and detection of fungal infections, specifically candidiasis, resulting in a lack of comprehensive data on its occurrence and prevalence. With the rising number of HIV/AIDS cases in the country, there is an increased population of immunocompromised individuals who are susceptible to fungal infections. Previous estimates have indicated that *C. albicans* is one of the most prevalent causative agents of fungal infections in the Philippines, commonly isolated from clinical samples (Bulmer et al., 1999). In a comprehensive 4-year investigative study conducted by Handog and Dayrit (2005) with a focus on identifying the prevalence of fungal infections in the Philippines, candidiasis was identified as the most frequent opportunistic fungal infection. Various predisposing factors described in existing literature, such as tuberculosis, surgery, cancer, indwelling catheters, long-term use of antibiotics, AIDS, and impaired immune responses, were associated with the development of candidiasis (Handog & Dayrit, 2005). Recent data from the Philippines showed an annual incidence of 1,968 cases of candidemia, 3,467 cases of oral candidiasis, and 1,481,899 cases of recurrent vulvovaginal candidiasis (Batac & Denning, 2017). These results show how candidiasis is relatively common and emphasize the need for further research and effective control strategies to counteract its global burden.

Candidiasis is a type of fungal infection that can take on different clinical forms depending on where the infection occurs. Invasive candidiasis, specifically when it affects the bloodstream, can result in serious illness and a high risk of death. *Candida albicans* is the most frequent species associated with bloodstream infections, followed by *C. glabrata*, *C. parapsilosis*, and other non-albicans *Candida* spp. (Pfaller & Diekema, 2007). Candidiasis can cause mild mucocutaneous infections, such as oral thrush and vaginal yeast infections, or severe disseminated infections that affect multiple organs, including the liver, spleen, and kidneys. Risk factors for invasive candidiasis include immunosuppression, prolonged hospitalization, use of broad-spectrum antibiotics, indwelling medical devices, and invasive procedures (Pfaller & Diekema, 2007; Singh & Chakrabarti, 2017). Timely diagnosis and appropriate antifungal therapy are essential for managing candidiasis, as delayed treatment can lead to worse outcomes.

The rise in fungal infections over the past decade has contributed to a significant increase in drug-resistant strains of *C. albicans*. Consequently, the implementation of antifungal susceptibility testing has become crucial for monitoring *Candida* that has evolved a multitude of mechanisms as means of surviving against antifungals. The presence of these resistant strains emphasizes the significance of monitoring and surveillance of emerging yeast-resistant strains, especially within the Philippines, where these protocols are not regularly conducted (Moron et al., 2017). It is imperative to establish routine protocols in laboratories, both in private and government hospitals, for the isolation and identification of *Candida* from clinical samples in the Philippines. These protocols are essential in facilitating the diagnosis of the disease and selecting the appropriate treatment for systemic and opportunistic candidiasis.

The biosafety level (BSL) needed for managing *Candida* spp. is determined by the specific activities and potential exposure risk. In a clinical laboratory environment, handling *Candida* spp. requires compliance with at least BSL-2 practices. However, in circumstances involving highly concentrated or aerosol-generating procedures, additional precautions may be necessary, such as BSL-3 practices (Meechan & Potts, 2020). In healthcare facilities across Southeast Asia, the management of *Candida* is typically in accordance with BSL-2 standards, which are consistent with global guidelines. These standards encompass the utilization of standard precautions, suitable PPE, and correct disinfection and sterilization methods (Meechan & Potts, 2020; World Health Organization, 2020b). In order to manage the biothreat posed by *Candida* spp. and reduce the risk of outbreaks in Southeast Asia, it is imperative to implement biosecurity measures. These measures should aim to prevent the intentional or unintentional release of *Candida* strains that may have increased virulence or resistance. To achieve this goal, it is necessary to adhere strictly to laboratory biosecurity protocols, which include controlled access to laboratories, secure storage of microbial cultures, and proper disposal of infectious waste. By implementing effective biosecurity practices, the risk of *Candida*-related outbreaks can be minimized (World Health Organization, 2020b).

Preventing candidiasis requires a combination of general hygiene practices and specific measures to minimize the risk of infection. Maintaining good personal hygiene is particularly important in the genital and oral areas to reduce the likelihood of candidiasis. For vulvovaginal candidiasis, avoiding irritants like scented soaps or douches and practicing proper genital hygiene can help prevent recurrent infections (Centers for Disease Control and Prevention, 2023). In healthcare settings, adherence to infection control practices, including hand hygiene, appropriate use of PPE, and disinfection and sterilization of medical equipment, is crucial to prevent the spread of *Candida* infections (Centers for Disease Control and Prevention, 2023). For individuals at high risk of invasive candidiasis, such as those with immunocompromised conditions, prophylactic antifungal therapy may be considered in certain cases (Pappas et al., 2016). Overall, an approach that combines hygiene practices, infection control measures, and targeted interventions for high-risk individuals is necessary to effectively prevent candidiasis.

## 13.12 MUCORMYCOSIS

Mucormycosis is a type of fungal infection that occurs when fungi belonging to the order Mucorales invade the body. The disease is most commonly caused by fungal species belonging to the following genera: *Mucor* spp., *Rhizopus* spp., *Lichtheimia* spp., *Rhizomucor* spp., *Cunninghumella* spp., *Saksenaea* spp., and *Apophysomyces* spp. (Prakash & Chakrabarti, 2019). Infections caused by Mucorales have high mortality rates and typically affect immunocompromised individuals, including those with conditions such as diabetes mellitus with ketoacidosis, organ transplant recipients, and patients undergoing chemotherapy and immunosuppressive therapy (Ghuman & Voelz, 2017). The fungus invades the body by breaching the physical barriers of the skin or respiratory system and encounters the components of the innate immune system, including macrophages, neutrophils, and dendritic cells. If the innate immune system fails to eliminate the fungal spores, the fungus is able to germinate and proliferate (Radotra & Challa, 2022).

The prevalence of mucormycosis varies depending on the region and patient population. Certain risk factors, such as uncontrolled diabetes mellitus, hematological malignancies, solid organ transplantation, long-term use of corticosteroids, iron overload, and immunosuppressive therapies, increase the likelihood of developing the infection (Skiada et al., 2020). The prevalence of diabetes mellitus is highest among risk factors in the Asian continent, whereas hematological malignancies and transplantation are the primary risk factors in European countries and the United States (Prakash & Chakrabarti, 2019).

In recent years, there has been an observed increase in the incidence of mucormycosis, particularly among individuals with compromised immunity. Notably, a rising trend of mucormycosis cases has been reported in India, with an incidence rate of 0.14 cases per 1,000 population (Skiada et al., 2020). In the United States, a study revealed that the prevalence of hospitalizations related to mucormycosis was estimated to be 0.12 per 10,000 discharges from January 2005 to June 2014. However, when the definition of mucormycosis was broadened to include cases not requiring the use of amphotericin B or posaconazole, the prevalence increased to 0.16 per 10,000 discharges (Kontoyiannis et al., 2016). Another study conducted in Belgium reported an increase in mucormycosis cases from 0.019 cases per 10,000 patient-days in 2000 to 0.48 cases per 10,000 patient-days in 2009 (Saegeman et al., 2010). In the Philippines, it is estimated that mucormycosis affects around 20 individuals annually, although this figure may be underestimated due to limited epidemiologic fungal disease surveillance in the country (Batac & Denning, 2017). It is important to note that the spread and occurrence of mucormycosis can vary based on the location and demographics of patients. To gain a more comprehensive understanding of the epidemiological trends and risk factors associated with mucormycosis, further research and surveillance efforts are crucial in different regions.

Clinical data on mucormycosis indicate that it has the potential for high rates of morbidity and mortality. A retrospective study conducted in India revealed that the overall mortality rate associated with mucormycosis was 46% (Chakrabarti et al., 2006). Another study conducted in a tertiary care center in the United States reported an overall mortality rate of 54% among patients with this disease (Roden et al., 2005). The disease primarily affects the sinuses and lungs but can also present as disseminated infection, skin and soft tissue involvement, or gastrointestinal infection (Spellberg et al., 2005). To improve clinical outcomes in patients with mucormycosis, prompt diagnosis, early initiation of appropriate antifungal therapy, and aggressive surgical debridement are essential.

Filamentous fungi causing mucormycosis present unique biosafety challenges in healthcare settings. Due to their potential for severe infections and the ability to produce abundant spores, it is essential to implement appropriate biosafety measures. The primary transmission route of mucormycosis is through inhalation of fungal spores present in the environment (Kauffman, 2004). Therefore, controlling airborne transmission is critical to ensure biosafety. To minimize the risk of exposure to infectious spores, measures such as proper ventilation, air filtration, and the use of PPE should be implemented. Additionally, effective cleaning and disinfection of surfaces and equipment are essential to prevent contamination and spread of the fungus (Meechan & Potts, 2020). The biosafety level (BSL) that is suitable for handling mucormycosis relies on the laboratory procedures and the possibility of infectious materials being aerosolized. Mucormycosis is commonly considered a BSL-2 organism due to the moderate risk it poses to lab workers and its potential to cause severe infections in individuals with a weakened immune system. However, activities that involve high-risk, such as producing aerosols or working with significant amounts of the organism, may require additional biosafety measures, such as BSL-3 containment (Meechan & Potts, 2020). Strict adherence to standard infection control practices and guidelines, along with appropriate handling and disposal of infected materials, are necessary to ensure the safety of healthcare workers and prevent nosocomial transmission of mucormycosis (Kauffman, 2004).

A comprehensive approach is necessary to prevent mucormycosis by reducing the risk of exposure to fungal spores and minimizing factors that make individuals more susceptible. Addressing underlying conditions that increase the risk of infection such as uncontrolled diabetes, immunosuppression, or iron overload is crucial (Spellberg et al., 2005). Diabetic patients should maintain good glycemic control, and immunocompromised individuals should receive appropriate antifungal prophylaxis when indicated (Cornely et al., 2019). Additionally, it is important to implement measures to reduce exposure to environmental fungal spores, especially in healthcare settings, by maintaining a clean and hygienic environment through strict infection control practices, proper ventilation, and regular cleaning and disinfection of surfaces (Spellberg et al., 2005). Education and awareness programs targeting high-risk populations, healthcare professionals, and those involved in handling organic materials can also play a vital role in preventing mucormycosis by promoting knowledge about the disease, its risk factors, and preventive measures (Spellberg et al., 2005).

## 13.13 CRYPTOCOCCOSIS

Cryptococcosis represents a significant global health burden primarily affecting immunocompromised individuals, such as patients with advanced HIV infection, solid organ transplant recipients, patients on immunosuppressive therapy, and individuals taking immunosuppressive medications (Gushiken et al., 2021). The global incidence of HIV-associated cryptococcal meningitis reached 223,100 in 2014, 162,500 (72%) of which occurred in sub-Saharan Africa. It is estimated that cryptococcal meningitis causes 181,100 deaths globally per year, and is responsible for 15% of AIDS-related mortality (Rajasingham et al., 2017). In Southeast Asia, *Cryptococcus* spp. were found to be the second and third most isolated pathogens from the blood of HIV-infected individuals (Harris et al., 2012). A study by Chariyalertsak et al. (2001) reported that the incidence of cryptococcal meningitis in HIV-infected individuals in Thailand is 14.7 cases per 100 person-years. Out of the numerous species under the genus *Cryptococcus*, only two have been reported to be pathogenic, *C. neoformans* and *C. gattii*. These two are commonly associated with infections in immunocompromised patients and have been recently seen to cause severe disease in healthy individuals. With its major target being the central nervous system, it can be deadly once left untreated and undiagnosed (Notarte et al., 2023).

Cryptococcosis can present with a wide range of clinical presentations, especially by the respiratory tract, as initiated by the inhalation of yeast cells which are readily aerosolized (Notarte et al., 2023). Infected patients may be asymptomatic or may present with flu-like symptoms, including cough, dyspnea, and chest pain. However, in immunocompromised patients, this may progress to the infection of the CNS resulting in cryptococcal meningitis, accounting for the significant proportion of cases. Disseminated cryptococcosis may involve several organs involving the skin, bones, or viscera (Perfect et al., 2010).

Biosafety plays a very important role in handling *Cryptococcus* spp. in the laboratory because of their potential in causing infections. *C. neoformans* and *C. gattii* are found mainly in the environment through bird droppings or certain trees. It is recommended by the CDC to have a BSL-2 facility in handling clinical specimens containing *Cryptococcus* species. Adherence to protocols may be difficult in resource-limited settings, such as in Southeast Asia. However, the WHO still advocates for the use of BSL-2 precautions when handling these species (World Health Organization, 2019). In addition, the National Guidelines for Laboratory Biosafety in Thailand have provided their respective laboratories comprehensive recommendations in handling, managing, storing, and transporting specimens suspected to contain the pathogens (Soisangwan, 2021).

It is difficult to prevent cryptococcosis due to its ubiquitous nature. Hence, an important factor in its prevention is early screening, diagnosis, and management of HIV infections, solid organ transplant recipients, and in patients on immunosuppressive interventions and medications (Gushiken et al., 2021). The prompt initiation of providing antiretroviral therapies and prophylaxis in HIV-infected individuals and appropriate antifungal therapies to immunocompromised patients is essential in reducing the incidence of mortality associated with cryptococcal infections (Rajasingham et al., 2017; World Health Organization, 2018).

The Philippines is one of the Asian countries that faces challenges in accessing advanced mycological diagnostic tools, resulting in limited diagnostic methods such as microscopy, histopathology, and imaging for identifying fungal diseases. As a consequence, treatment approaches are often based on empirical methods. In 2020, the Asia Fungal Working Group (AFWG) conducted a web-based survey to identify the challenges faced by physicians in Asian countries when dealing with invasive fungal diseases. The survey revealed that respondents from the Philippines, Malaysia, China, Indonesia, Taiwan, Singapore, Thailand, and India highlighted significant factors such as technical training, availability of local guidelines, access to antifungal drugs, and research capacity as crucial aspects for the effective management of invasive mycoses, including cryptococcosis. To address these challenges, it was suggested that country-specific training programs be implemented, national guidelines be established, and improvements be made in laboratory diagnostic tests. Furthermore, the establishment of an antifungal stewardship program in the country could enhance the efficiency of diagnosis and treatment. Over the past decade,

AFWG has made efforts to address these issues by organizing courses and providing hands-on training in medical mycology. These initiatives aim to enhance the technical expertise and skills of microbiologists in countries like the Philippines (Tan et al., 2020).

## 13.14 SOIL-TRANSMITTED HELMINTHIASIS

Soil-transmitted helminthiasis (STH) is an infection caused by any of these main species, namely, common roundworm (*Ascaris lumbricoides*), whipworm (*Trichuris trichiura*), and hookworms (*Necator americanus* and *Ancylostoma duodenale*). Infection from roundworms and whipworms is transmitted through ingestion of infective eggs passed from infected people's feces, while hookworms are through the skin penetration of its infective larva to humans. Pre-school children, school-age children, women of reproductive age, and individuals whose occupations have possible exposure to various intestinal parasites are at risk for these worms (World Health Organization, 2023d). Based on the scoping review by Mationg et al. (Mationg et al., 2021), STH remains a significant public health problem in the Philippines due to its stubborn occurrence, recording a range prevalence of 24.9–97.4% (2004–2018) and remained high among preschool-age children (2–4 years old) and school-age children (5–12 years old), even reaching a prevalence of 67.4% (2003). The identified infection rates complement the national parasite survey conducted by the Research Institute of Tropical Medicine (RITM) in 2016–2018 among school-age children, which showed an overall cumulative prevalence of 28.4% (Luchavez, 2018).

Moreover, in 2020, RITM reported that one in five non-school-age children and adults are infected with STH, entailing that more infections with this parasite are found among non-school-age children (Research Institute for Tropical Medicine, 2020). In an unpublished study by De Guia (2019), high intestinal helminth prevalence was observed in ages 15 and above (37.2%) than those combined from preschool and school-age children (34.1%). The initial strategy of the WHO in controlling and preventing STH infections is the regular administration of anthelminthic medicines to preschool-age and school-age children (Montresor et al., 1998). Therefore, this has been the primary basis and approach of the Department of Health (DOH) since it launched its control program in 1999. The program advocated the mass treatment of children aged 2–14 using albendazole (400 mg) and mebendazole (500 mg) drugs for three years and a frequency of at least twice a year.

This health service is integrated into the Garantisadong Pambata program which started in 1999 to guarantee the deworming service for pre-school-age children at the community level as delivered by trained barangay health workers while a teacher supervised by the school nurse or midwife approach for school-age children (Department of Health, 1999, 2020c). In 2006, the Philippines DOH modified its initial control program into Integrated Helminth Control Program (IHCP) due to a slow and low improvement in reducing STH among school-age children and due to a constant high infection rate among children aged 1–5. IHCP aimed to increase deworming coverage by having mass deworming and selective deworming. Children 1–12 years categorized as mass-targeted deworming, in which this mass chemotherapy will be done consecutively for three years and provided twice a year, while selective deworming targets groups of adolescent females, pregnant women, soldiers, farmers, food handlers/operators, and indigenous peoples in which the deworming occurred yearly (Department of Health, 2006). The mass-targeted deworming strategy of DOH has expanded by collaborating with the Department of Education (DepEd) to increase the number of treated individuals. It materialized by establishing the National School Deworming Day (NSDD), which first rolled out on July 29, 2015, and resulted in 81% treatment coverage from kindergarten to Grade 6 (Department of Education, 2015; Villaverde et al., 2016).

One of the critical interventions laid out by IHCP is water, sanitation, and hygiene (WASH) strategies patterned on UNICEF WASH strategies (2006–2010). Local Government Units (LGUs) conducted and spearheaded these strategies by constructing sanitary toilets, providing clean and safe water, and working with other concerned agencies to monitor and ensure environmental sanitation in sewage and waste disposal (Department of Health, 2006). There is also implementation of WASH in school through the WinS (Comprehensive Water, Sanitation, and Hygiene in Schools) program, which highly advocates deworming, proper hand washing, provision of safe water, toilet, hand washing, and drainage facilities, environmental sanitation, and hygiene, and sanitation education (Department of Education, 2016). The collaborative efforts of DOH with DepEd and LGUs have resulted in deworming coverage of 84.5% for public school children aged 5–18 years old and 74.6% for preschool children in January 2017 (Department of Health, 2017a). However, the COVID-19 pandemic brought about the deferment of mass deworming both in communities and schools. Thus, during this time, a community-based individual approach was implemented where deworming services are provided house-to-house among individuals ages 1–19 by the Barangay Health Workers (Department of Health, 2020a).

World Health Organization (1996) recommends conducting a baseline parasitological survey at the beginning of any STH control programs. Significantly, data from it provides an estimated number of individuals infected with STH, which is the primary basis for having appropriate intervention and control strategies against the said intestinal parasitism (Montresor et al., 2002). Constant and efficient efforts to eliminate and interrupt STH transmission also require effective surveillance, particularly at the community level (Murphy et al., 2023). Non-participation, available equipment, and the need for trained staff are some challenges restricting effective surveillance (Murphy et al., 2023; World Health Organization, 2021a). Moreover, the WHO suggests surveying the prevalence of parasitic infections every 3–5 years to evaluate the impacts and outcomes of the mass deworming strategy (World Health Organization, 2011, 2017a). In 2010, DOH incorporated strategies for monitoring and surveillance of STH infections in its Disease Prevention and Control Program (DPCP).

Such strategies include orientation and training to Local Health Units in monitoring the nutritional and morbidity indicators of the community and training the medical technologists on diagnostic techniques endorsed by WHO. It also stated the involvement of DepEd in the baseline and follow-up parasitological surveys (Department of Health, 2010). Also, the said department tasked RITM to conduct a national parasite survey in 2013 and 2016–2018 (Luchavez, 2018; Tangcalagan et al., 2022). However, no recent data follow the said parasitological surveys, mainly STH infections, that the DOH directly initiated. Thus, there is a need for more updated data on the impact and effectiveness of IHCP at the national level. Instead, issues and challenges regarding consistency and sustainability in efforts of STH infections emerge (Labana et al., 2021).

Countries like Japan and China recognize the significance of regularly obtaining data on parasitic prevalence. With this, Japan has controlled and eliminated STH, while China is nearly approaching elimination status (Hasegawa et al., 2020; Zhu et al., 2020). Japan has controlled these helminth infections through periodical mass treatment with mass stool examination, night-soil control, and direct STH prevention. The Parasitosis Prevention Law in 1931 implemented the following activities nationwide to treat and control STH infections: routine mass stool screening, mass deworming, night-soil treatment, and health education (Hasegawa et al., 2020; Kobayashi et al., 2006). Administered epidemiological studies guide their approach to mass treatment to determine the frequency of deworming, testing, and target groups (Kobayashi et al., 2006). On the other hand, establishing monitoring spots nationwide in China overcomes their limitation in decreasing the helminth infection rates (Chen & Zang, 2015). Based on national surveillance, from 47.0% ascariasis, 18.8% trichuriasis, and 17.2% hookworm infection in 1999 to 1.4%, 0.8%, and 0.5% in 2020 (Zhu et al., 2020).

The Philippines needs to establish an effective surveillance system for STH infections. Regular mass stool collection and examination must become part of controlling and eliminating the parasites. This kind of parasite survey is easy and possible if proper health education on preventing helminth infections is well addressed and delivered at school and community levels. Several studies have strongly recommended that close surveillance of parasitic infections must be regularly implemented and conducted (Delos Trinos et al., 2019; Ladia et al., 2020; Mationg et al., 2017). One of the potential tools for monitoring is creating National GIS risk maps utilizing the data obtained at the municipal/district level. This tool will only be feasible in the country if there is centralized and well-structured STH surveillance (Ross et al., 2017).

## 13.15 SCHISTOSOMIASIS

Schistosomiasis (SCH) is one of the neglected tropical diseases (NTDs) identified by the World Health Organization (WHO). It is considered as a disease of poverty commonly found in tropical and subtropical regions (World Health Organization, 2020a). NTDs are included as targets for global action in the United Nations Sustainable Development Goal 3.3, which aims ‘to end epidemics caused by NTDs’ by 2030 to ‘ensure healthy lives and promote well-being for all at all ages’ (World Health Organization, 2017a, 2020a).

SCH is caused by oriental blood flukes under the genus *Schistosoma*. Among the seven known species, *S. japonicum* was found to be the most pathogenic to humans and animals. Transmission occurs through skin penetration of infective cercariae on exposure to snail-infested waters (Colley et al., 1998). In 2017, approximately 143 million people were infected with schistosomiasis, and in the Philippines, SCH affects approximately 12 million people, with 2.5 million directly exposed to the infection during domestic, agricultural, occupational and recreational activities, as well as their interaction with large mammals, such as carabaos and cattle (Department of Health, 2018c).

The WHO aims to eliminate SCH as a public health problem by lowering the prevalence of heavy intensity infection to <1% by 2025 (World Health Organization, 2013). The WHO recommends an integrated approach to overcome the global impact of SCH which includes: (1) preventive chemotherapy (PC); (2) innovative and intensified disease management; (3) veterinary public health services; (4) vector ecology and management; and (5) provision of safe water, sanitation, and hygiene (WASH) (World Health Organization, 2017b).

In the Philippines, the implementation of these strategies is spearheaded by the Department of Health (DOH) through the SCH Control and Elimination Program (SCEP) (Department of Health, 2007). The SCEP aims to interrupt the transmission of SCH by reducing the incidence of infection in humans, animals, and snails to zero by 2025 (Department of Health, 2018c). However, achieving the target remains a challenge despite decades of implementation of the SCEP. Despite the efforts on schistosomiasis control for decades in the Philippines, some challenges remain in various dimensions. It is therefore imperative to have adequate and updated information to enhance policy and service delivery toward control and elimination. Considering the zoonotic nature of schistosomiasis, adopting the One Health framework which necessitates a multifaceted approach is necessary to influence policies and interventions that may accelerate control and elimination in the country.

A review of the status of schistosomiasis japonica control in the Philippines in the context of human health, animal public health, vector ecology and management, environmental health, and sociocultural dimensions was reported by Belizario et al. (2022). For human health, the infection is associated with hepatomegaly, splenomegaly, portal hypertension, and variceal bleeding (Olds et al., 1996). These symptoms may lead to undernutrition, poor physical and cognitive development among children, and decreased adult productivity, contributing to persisting poverty (King & Dangerfield-Cha, 2008). Central nervous system involvement may also occur when schistosome eggs reach the brain, presenting with seizures, focal neurologic deficits, and/or acute encephalopathy. This form of SCH accounts for up to 3–5% of all cases (Berkowitz et al., 2015). Mortality from schistosomiasis japonica, however, is usually a result of infectious complications of untreated patients (Ferrari & Moreira, 2011).

Microscopic examination of stool processed using the Kato-Katz (KK) technique remains the standard diagnostic procedure for routine surveillance of SCH in countries in the Western Pacific Region (World Health Organization, 2017c). However, the KK technique is known to have limited sensitivity, particularly in detecting low-intensity infections (Zhang et al., 2009). More sensitive techniques such as the FLOTAC and McMaster flotation are becoming more popular (Knopp et al., 2009). Immunological methods such as assays for the detection of circulating anodic antigen and circulating cathodic antigen as well as enzyme-linked immunosorbent assay (ELISA) antigen tests can be utilized for more sensitive and accurate diagnosis of SCH. However, drawbacks in the use of sero-diagnostics for SCH surveillance are the high cost and its invasiveness. Quantitative polymerase chain reaction (qPCR) has a high level of sensitivity in the detection of SCH in low transmission settings, making it ideal for use in confirming transmission interruption of the infection (He et al., 2018). Loop-mediated isothermal amplification (LAMP) assay may also be considered as it is a simple, rapid, and sensitive detection technique and it can amplify a large amount of *S. japonicum* DNA more rapidly than PCR without the need for sophisticated equipment (Xu et al., 2010).

SCH remains endemic in 28 provinces across 12 regions in the Philippines, with endemicity centered mostly in Central and Southern Philippines (Department of Health, 2019). Mass drug administration (MDA) with anthelmintic praziquantel (PZQ) distributed to a target population regardless of infection status alongside other supplementary measures had been implemented by the government since 1991 for morbidity control of SCH.

As regards the veterinary public health aspect, the zoonotic potential of schistosomiasis japonica must be considered as bovines, particularly cattle and carabaos (water buffaloes), contribute to the transmission of the disease to humans. These animals used in agricultural activities serve as important reservoirs of SCH which may excrete a significant number of viable eggs per day into the environment. Manure-based fertilizers may also increase human exposure to bovine feces (Gordon et al., 2015; Wu et al., 2010). Detection of SCH in animal reservoirs employs laboratory techniques such as miracidium hatching technique (MHT), Danish Bilharziasis Laboratory (DBL) technique, KK technique, ELISA, formalin-ethyl acetate sedimentation (FEA-SD) technique, PCR, and qPCR. The MHT, DBL, and KK have low sensitivities in detecting schistosomiasis japonica in carabaos, while ELISA, FEA-SD, and qPCR have higher sensitivity at 82.6%, 90.9%, and 100%, respectively. Most of these techniques have satisfactory specificity, reaching as high as 100% (Gordon et al., 2015; Wu et al., 2010).

Low prevalence of SCH in carabaos was reported in past coprological surveys in the Philippines, however, more sensitive techniques record a prevalence close to 90%, therefore considering carabaos as major reservoirs of SCH (Fernandez Jr et al., 2007). In the provinces of Western Samar and Northern Samar, the prevalence in carabaos by qPCR was 90.9% and 80.0%, respectively (Gordon et al., 2015). The prevalence of SCH in carabaos using qPCR was 51.5% and 46.0% in the provinces of Leyte (Wu et al., 2010) and Cagayan (Angeles et al., 2012), respectively. A prevalence of 48.6% for cattle and 60.5% for water buffaloes were recorded in South Cotabato using FEA-SD technique (Tenorio & Molina, 2020). A recent study reported an 8.6% prevalence of SCH among the sampled water buffaloes in Maragusan, Davao de Oro (Navarro et al., 2021).

The national policy on animal SCH has been included in the SCH Control and Elimination Program (SCEP) Strategic Plan with the implementation of policy and training at the local level scheduled to begin in 2023. The DOH provides financial assistance for most animal public health activities that cover the capacity building of veterinary health teams in priority areas and joint planning and implementation of priority operational research (Department of Health, 2019; World Health Organization, 2017c). The current DOH SCEP strategic plan also included the following activities for the control of SCH in animal reservoirs: (1) conduct of annual animal surveillance, (2) biannual deworming of animals, and (3) mandatory testing of outgoing animals from endemic areas (Belizario et al., 2022; Department of Health, 2019).

Vector ecology and management is one of the strategies recommended by the WHO to control NTDs. The presence of the snail intermediate host, *Oncomelania hupensis quadrasi*, in freshwater habitats is an important factor in the persistence of *S. japonicum* in the Philippines. These snails are amphibious in nature, mostly aquatic and are commonly found in wet soil surfaces, wet swamps, wet rice fields, ponds, and banks of streams (World Health Organization, 2017a). Fishermen and farmers who are always in direct contact with snail-infested water bodies and children who frequently play in the surrounding areas are at higher risk for SCH (World Health Organization, 2021a).

The conventional crushing method, where snails are crushed and mounted between two glass slides and examined under a light microscope for cercariae or sporocysts, is recommended in determining SCH in *O. h. quadrasi* (Madsen et al., 2008). However, this technique has limited sensitivity in areas where there is low parasite burden, aborted development of sporocysts, and prepatent infection. More sensitive techniques such as LAMP and PCR are useful for monitoring and confirming SCH in snails in low infection areas. Despite the diagnostic accuracy of these novel techniques, microscopy remains the most widely used technique because of its technical simplicity, applicability in the field, and ease of application in resource-poor settings (Weerakoon et al., 2015). The molecular analysis by qPCR of environmental DNA involves assessing soil samples (Calata et al., 2019) for the presence of *O. h. quadrasi* snails or freshwater samples for the presence of *S. japonicum* and *O. h. quadrasi* snails may also be utilized (Fornillos, Sato, et al., 2019).

Malacological surveys provide data on snail density, distribution, and infection rate (Fornillos, Fontanilla, et al., 2019). A survey conducted in 147 sites in Samar Province revealed that 55 sites harbored infected snails (Madsen et al., 2008) while five of the 11 sites surveyed in Leyte Province harbored infected *O. h. quadrasi* snails (Fornillos, Sato, et al., 2019).

Snail control and surveillance are included in the SCEP activities such as annual snail mapping, environmental modifications (e.g., monthly or quarterly clearing of snail colonies), cementing of pathways or water walling, earth filling, and construction of canals and footbridges (Belizario et al., 2022; Department of Health, 2019).

Equally important in crafting strategies for control of SCH is Environmental Health. Contact with freshwater during fishing, farming, swimming, washing, bathing, and recreational activities promotes the transmission of SCH. Transmission is also facilitated by environmental pathways through dams and irrigation systems (Garchitorena et al., 2017). Access to safe water and adequate sanitation and hygiene facilities are also associated with SCH transmission. The defecation behavior and unsanitary disposal of waste were also found to be strongly associated with infection status (Exum et al., 2019). Hygiene practices, particularly soap use, also play a role in SCH prevention as soap is toxic to cercariae, miracidia, and specific freshwater snails (Okwuosa & Osuala, 1993).

The Department of Education launched the WASH in Schools Program in 2019 to promote good hygiene and sanitation practices among schoolchildren in all elementary and secondary schools nationwide. Key strategies include the provision and maintenance of water and sanitary facilities and daily supervision on handwashing (Policy Guidelines for the Comprehensive Water, Sanitation and Hygiene in Schools (WINS), 2016). The following approaches to the provision of safe WASH are in place: (1) access to sanitation facilities and management of fecal waste to reduce human excreta in the environment; (2) safe water supply to prevent consumption of contaminated water and enable personal hygiene practices; (3) water resource, wastewater, and solid waste management for vector control and contact prevention; and (4) hygiene measures such as handwashing with soap, laundry, food hygiene, and face washing for overall personal hygiene (World Health Organization, 2013).

Sociocultural factors may affect the initiatives to control SCH. Adequate knowledge on the transmission, clinical manifestation, and risk factors of SCH is vital in controlling and preventing infection. Misconceptions and gaps in knowledge may cast doubt on health programs, which may serve as barriers or limit participation in intervention activities. Examples of these are the notions that participation is not needed if one is asymptomatic or cannot take anthelmintics in the absence of a prior stool examination (Amarillo et al., 2008). Selective treatment was also sometimes preferred over mass treatment by some individuals (Leonardo et al., 2002). Fear of possible side effects of praziquantel such as nausea and vomiting may also impede to MDA compliance. In a knowledge, attitude, and practice (KAP) study conducted in Guimaras on the soil-transmitted helminthiasis control program, some parents doubted that the teachers administering the drugs had the capacity to manage its side effects (Parikh et al., 2013).

The One Health approach is a key to effective control and elimination of schistosomiasis given its zoonotic nature. The multidimensional nature of SCH necessitates a multidisciplinary approach requiring cooperation and collaboration among the various disciplines involved such as human, animal, and environmental health. Each dimension may already have programs in place which only needs a strong and effective implementation or develop an integrated and innovative approaches to address the concerns of the different disciplines.

## 13.16 CONCLUSION

Various emerging and re-emerging infectious diseases continue to be a significant public health concern within the Philippine archipelago. The Philippine government is well aware of this ongoing issue and is actively engaged in combatting these infections through diverse management programs and preventive measures, including vaccinations, vector control, and community health education empowerment. This chapter provides a comprehensive discussion on the different emerging and re-emerging infectious diseases in the country, aiming to understand their epidemiology, symptomatology, diagnosis, and treatment. The chapter emphasizes the government's recognition of the problem and its proactive approach in mitigating the impact of infectious diseases, implementing programs guided by international guidelines, particularly from the World Health Organization (WHO), which have proven instrumental in their successful implementation. However, despite these efforts, the country faces challenges in controlling outbreaks due to various contributing factors, including antimicrobial resistance, a rapidly growing population, and the expansion of urbanization leading to unsanitary living conditions. Therefore, it is crucial for everyone to remain vigilant and actively participate in supporting the government's national programs as the country continues to grapple with the issues of emerging and re-emerging infectious diseases.

## REFERENCES

Acosta, C. D., Dadu, A., Ramsay, A., & Dara, M. (2014). Drug-resistant tuberculosis in Eastern Europe: Challenges and ways forward. *Public Health Action*, *4*(2), S3–S12.

Action for Health Initiatives Inc. (2019). Situational analysis of tuberculosis elimination program in greater Manila using Communities, Rights, and Gender (CRG) tools. *ACHIEVE, Inc.* https://stoptb.org/assets/documents/communities/CRG/TB CRG Assessment Philippines.pdf.

Agarwal, R., Chakrabarti, A., Shah, A., Gupta, D., Meis, J. F., Guleria, R., Moss, R., Denning, D. W., & Group, A. C. A. I. W. (2013). Allergic bronchopulmonary aspergillosis: Review of literature and proposal of new diagnostic and classification criteria. *Clin Exp Allergy*, *43*(8), 850–873.

Agrupis, K. A., Ylade, M., Aldaba, J., Lopez, A. L., & Deen, J. (2019). Trends in dengue research in the Philippines: A systematic review. *PLOS Negl Trop Dis*, *13*(4), e0007280. https://doi.org/10.1371/journal.pntd.0007280.

Alora, B., Nambayan, A., Perez, J., Famatiga, E., & Alora, A. (1973). Leptospirosis in Santo Tomas University Hospital: Analysis of 17 cases (1967–1971). *Phil J Microbiol Infect Dis*, *11*(1), 11–22.

Amarillo, M. L. E., Belizario, V. Y., Sadiang-Abay, J. T., Sison, S. A. M., & Dayag, A. M. S. (2008). Factors associated with the acceptance of mass drug administration for the elimination of lymphatic filariasis in Agusan del Sur, Philippines. *Parasit Vectors*, *1*(1), 1–12.

Angeles, J. M. M., Goto, Y., Kirinoki, M., Asada, M., Leonardo, L. R., Rivera, P. T., Villacorte, E. A., Inoue, N., Chigusa, Y., & Kawazu, S. (2012). Utilization of ELISA using thioredoxin peroxidase-1 and tandem repeat proteins for diagnosis of Schistosoma japonicum infection among water buffaloes. *PLOS Negl Trop Dis*,6(8);1–6.

Aragon, P. R., & Famatiga, E. G. (1965). Studies on leptospirosis. II. Laboratory evidence of human infection. *Acta Med Philipp*, *1*, 196–198. https://pubmed.ncbi.nlm.nih.gov/14319109/.

Arambulo, P. V. 3rd, Topacio, T. M. Jr., Famatiga, E. G., Sarmiento, R. V., & Lopez, S. (1972). Leptospirosis among abattoir employees, dog pound workers, and fish inspectors in the city of Manila. *Southeast Asian J Trop Med Public Health*, *3*(2), 212–220. https://pubmed.ncbi.nlm.nih.gov/5082849/.

Artika, I. M., & Ma'roef, C. N. (2017). Laboratory biosafety for handling emerging viruses. *Asian Pac J Trop Biomed*, *7*(5), 483–491. https://doi.org/10.1016/j.apjtb.2017.01.020.

Basaca-Sevilla, V., Cross, J. H., Alcantara, A., Barrito, B., Pastrana, E., Sevilla, J., Balagot, R., & Dungca, S. (1981). An outbreak of leptospirosis in an agricultural penal colony in the Philippines. *Phil J Microbiol Infect Dis*, *10*, 73–82.

Basaca-Sevilla, V., Cross, J. H., & Pastrana, E. (1986). Leptospirosis in the Philippines. *Southeast Asian J Trop Med*, *17*(1), 71–74.

Batac, M. C. R., & Denning, D. (2017). Serious fungal infections in the Philippines. *Eur J Clin Microbiol Infect Dis*, *36*(6), 937–941. https://doi.org/10.1007/s10096-017-2918-7.

Belizario, V. Y., de Cadiz, A. E., Navarro, R. C., Flores, M. J. C., Molina, V. B., Dalisay, S. N. M., Medina, J. R. C., & Lumangaya, C. R. (2022). The status of schistosomiasis japonica control in the Philippines: The need for an integrated approach to address a multidimensional problem. *Int J One Health*, *8*(1), 8–19.

Berkowitz, A. L., Raibagkar, P., Pritt, B. S., & Mateen, F. J. (2015). Neurologic manifestations of the neglected tropical diseases. *J Neurol Sci*, *349*(1–2), 20–32.

Bharti, A., Nally, J. E., Ricaldi, J. N., Matthias, M. A., Diaz, M. M., Lovett, M. A., Levett, P. N., Gilman, R. H., Willig, M. R., Gotuzzo, E., Vinetz, J. M., & Peru-United States Leptospirosis Consortium. (2003). Leptospirosis: A zoonotic disease of global importance. *Lancet Infect Dis*, *3*(12), 757–771. http://infection.thelancet.com.

Biggs, J. R., Sy, A. K., Ashall, J., Santoso, M. S., Brady, O. J., Reyes, M. A. J., Quinones, M. A., Jones-Warner, W., Tandoc, A. O., Sucaldito, N. L., Mai, H. K., Lien, L. T., Thai, H. Do, Nguyen, H. A. T., Anh, D. D., Iwasaki, C., Kitamura, N., Van Loock, M., Herrera-Taracena, G., … Hibberd, M. L. (2022). Combining rapid diagnostic tests to estimate primary and post-primary dengue immune status at the point of care. *PLOS Negl Trop Dis*, *16*(5), e0010365.

Blöndal, K. (2007). Barriers to reaching the targets for tuberculosis control: Multidrug-resistant tuberculosis. *Bull World Health Organ*, *85*(5), 387–390.

Bosch, F. X., Lorincz, A., Muñoz, N., Meijer, C., & Shah, K. V. (2002). The causal relation between human papillomavirus and cervical cancer. *J Clin Pathol*, *55*(4), 244–265.

Brato, D. G., Mendoza, M. T., & Cordero, C. P. (1998). Validation of the World Health Organization (WHO) criteria using the Microscopic Agglutination Test (MAT) as the gold standard in the diagnosis of leptospirosis. *Phil J Microbiol Infect Dis*, *27*(125), 28.

Bravo, L., Roque, V. G., Brett, J., Dizon, R., & L'Azou, M. (2014). Epidemiology of dengue disease in the Philippines (2000–2011): A systematic literature review. *PLOS Negl Trop Dis*, *8*(11), e3027.

British Thoracic and Tuberculosis Association. (1975). Short-course chemotherapy in pulmonary tuberculosis. *Lancet*, 2; 119–124.

Bruni, L., Albero, G., Serrano, B., Mena, M., Collado, J., Gómez, D., Muñoz, J., Bosch, F., & de Sanjosé, S. (2023). *Human papillomavirus and related diseases in Philippines.* https://hpvcentre.net/statistics/reports/PHL.pdf?t=1559474177321.

Buchan, B., Relich, R. F., & Mahlen, S. D. (2019). *Interim clinical laboratory guideline for biological safety.* https://asm.org/ASM/media/Policy-and-Advocacy/Biosafety-white-paper-2019.pdf?ext=.pdf.

Bulmer, G. S., Marquez, M. L., Co-Barcelona, L., & Fromtling, R. A. (1999). Yeasts and fluconazole susceptibility in the Philippines. *Mycopathologia*, *146*(3), 117–120.

Cabresos, J. A. (2020). Spatial time series analysis of ongoing Dengue outbreaks in the Philippines. *Int J Infect Dis*, *101*, 254. https://doi.org/10.1016/j.ijid.2020.11.101.

Cadena, J., Thompson, G. R., & Patterson, T. F. (2021). Aspergillosis: Epidemiology, diagnosis, and treatment. *Clin Infect Dis*, *35*(2), 415–434.

Calata, F. I. C., Caranguian, C. Z., Mendoza, J. E. M., Fornillos, R. J. C., Tabios, I. K. B., Fontanilla, I. K. C., Leonardo, L. R., Sunico, L. S., Kawai, S., Chigusa, Y., & others (2019). Analysis of environmental DNA and edaphic factors for the detection of the snail intermediate host Oncomelania hupensis quadrasi. *Pathogens*, *8*(4), 160.

Carlos, E. R., Kundin, W. D., Watten, R. H., Tsai, C. C., Irving, G. S., Carlos, E. T., & Gaitmaitan, O. (1970). Leptospirosis in the Philippines VI. Serologic and isolation studies on carabaos. *Southeast Asian J Trop Med Public Health, 1*(4), 481–482.

Centers for Disease Control and Prevention. (2019a). *Understanding the HIV care continuum.*

Centers for Disease Control and Prevention. (2019b). *Zika virus diagnostic test.* CDC. https://www.cdc.gov/zika/laboratories/types-of-tests.html.

Centers for Disease Control and Prevention (2019c). *Zika virus treatment.* CDC. https://www.cdc.gov/zika/symptoms/treatment.html.

Centers for Disease Control and Prevention. (2023). *Infection prevention and control for Candida auris.* CDC. https://www.cdc.gov/fungal/candida-auris/c-auris-infection-control.html.

Centers for Disease Control and Prevention, & Infectious Diseases Society of America. (2000). Guidelines for preventing opportunistic infections among hematopoietic stem cell transplant recipients. *Biol Blood Marrow Transplant, 6*(6a), 7–83.

Chabi, M. L., Goracci, A., Roche, N., Paugam, A., Lupo, A., & Revel, M. P. (2015). Pulmonary aspergillosis. *Diagn Interv Imaging, 96*(5), 435–442.

Chakrabarti, A., Das, A., Mandal, J., Shivaprakash, M. R., George, V. K., Tarai, B., Rao, P., Panda, N., Verma, S. C., & Sakhuja, V. (2006). The rising trend of invasive Zygomycosis in patients with uncontrolled diabetes mellitus. *Med Mycol, 44*(4), 335–342. https://doi.org/10.1080/13693780500464930.

Chakraborty, A., Miyahara, S., Villanueva, S. Y. A. M., Gloriani, N. G., & Yoshida, S. I. (2010). In vitro sensitivity and resistance of 46 Leptospira strains isolated from rats in the Philippines to 14 antimicrobial agents. *Antimicrob Agents Chemother, 54*(12), 5403–5405. https://doi.org/10.1128/AAC.00973-10.

Chariyalertsak, S., Sirisanthana, T., Saengwonloey, O., & Nelson, K. E. (2001). Clinical presentation and risk behaviors of patients with acquired immunodeficiency syndrome in Thailand, 1994–1998: Regional variation and temporal trends. *Clin Infect Dis, 32*(6), 955–962.

Chen, Y. D., & Zang, W. (2015). Current situation of soil-transmitted nematodiasis monitoring in China and working keys in future. *Zhongguo Xue Xi Chong Bing Fang Zhi Za Zhi, 27*(2), 111–114.

Cheng, K. J. G., Lam, H. Y., Rivera, A. S., Tumanan-Mendoza, B. A., Alejandria, M. M., & Wu, D. B. C. (2018). Estimating the burden of dengue in the Philippines using a dynamic transmission model. *Acta Med Philipp, 52*(2), 153–159. https://doi.org/10.47895/amp.v52i2.427.

Colley, D. G., Addiss, D., & Chitsulo, L. (1998). Schistosomiasis. *Bull World Health Organ, 76*(Suppl), 150–151.

Cornely, O. A., Alastruey-Izquierdo, A., Arenz, D., Chen, S. C. A., Dannaoui, E., Hochhegger, B., Hoenigl, M., Jensen, H. E., Lagrou, K., Lewis, R. E., & others (2019). Global guideline for the diagnosis and management of mucormycosis: An initiative of the European Confederation of Medical Mycology in cooperation with the Mycoses Study Group Education and Research Consortium. *Lancet Infect Dis, 19*(12), e405–e421.

Dagenais, T. R. T., & Keller, N. P. (2009). Pathogenesis of Aspergillus fumigatus in invasive aspergillosis. *Clin Microbiol Rev, 22*(3), 447–465.

Dayrit, M. M., Mendoza, R. U., & Valenzuela, S. A. (2020). The importance of effective risk communication and transparency: Lessons from the dengue vaccine controversy in the Philippines. *J Public Health Policy, 41*(3), 252–267. https://doi.org/10.1057/s41271-020-00232-3.

De Guia, J. J. R. (2019). *Detection of intestinal helminths in Aetas and swine in Brgy. Villa Maria, Porac, Pampanga* [Unpublished Master's Thesis].

de Paz-Silava, S. L. M., Tabios, I. K. B., Tantengco, O. A. G., Climacosa, F. M. M., Velayo, C. L., Lintao, R. C. V., Cando, L. F. T., Perias, G. A. S., Idolor, M. I. C., Francisco, A. G., & others (2023). Determinants of acquisition, persistence, and clearance of oncogenic cervical human papillomavirus infection in the Philippines using a multi-omics approach: DEFEAT HPV study protocol. *Healthcare*, *11*(5), 658.

Delos Trinos, J. P. C. R., Belizario Jr, V. Y., Sison, O. T., Erasmo, J. N., Te, M. J., & Modequillo, M. C. (2019). Child development center-based sentinel surveillance of soil-transmitted helminthiases in preschool-age children in selected local government units in the Philippines. *Acta Trop*, *194*, 100–105.

Department of Education. (2015). *Wash in schools three star approach*. DOH. https://www.deped.gov.ph/wp-content/uploads/2019/07/TSA-WinS-Booklet-Deworming-PRINT-20181031.pdf.

Department of Education. (2016). *Policy guidelines for the comprehensive water, sanitation and hygiene in schools (WINS) program: DO 10, s.2016*. DOH. https://www.deped.gov.ph/wp-content/uploads/2016/02/DO_s2016_10-1.pdf.

Department of Health. (1999). *The soil transmitted helminthiases control program: Guidelines on the implementation of the soil transmitted helminthiases control program*. DOH. https://doh.gov.ph/sites/default/files/health_programs/Implementation of STH Control Program AO 30-F s 99_I.pdf.PDF.

Department of Health. (2000). Designation of national reference laboratories and transfer of corresponding equipment instruments supplies specimens records from the Bureau of Research and Laboratories to the designated national reference laboratories. DOH.

Department of Health. (2005). *Case guidelines on the management and control of meningococcal disease*. DOH.

Department of Health. (2006). *Strategic and operational framework for establishing integrated helminth control program (IHCP)*. DOH.

Department of Health. (2007). *Revised guidelines in the management and prevention of schistosomiasis*. DOH.

Department of Health. (2010). *Guidebook for a disease prevention and control program for soil-transmitted helminth infections and diarrheal diseases*. DOH.

Department of Health. (2012). *National framework of the national health laboratory network*. DOH.

Department of Health. (2015). *Guidelines in the Implementation of the Human Papillomavirus (HPV) Vaccination*. Department of Health.

Department of Health. (2016). Technical guidelines: Standards and other instructions for reference in the implementation of Zika virus (ZIKV) disease surveillance. DOH.

Department of Health. (2017a). *DOH conducts national deworming month to reinforce prevention and control of soil-transmitted helminths*. DOH. https://doh.gov.ph/node/10545.

Department of Health. (2017b). *Emerging and re-emerging infectious diseases program - Policies*. DOH.

Department of Health. (2018a). Amendment to the guidelines for the nationwide implementation of the enhanced 4S-Strategy against Dengue, Chikungunya and Zika. DOH.

Department of Health. (2018b). *Guidelines for the nationwide implementation of the enhanced 4S-strategy against Dengue, Chikungunya and Zika*. DOH.

Department of Health. (2018c). *Schistosomiasis control and elimination program*. DOH. https://ro9.doh.gov.ph/index.php/68-health-programs/68-schistosomiasis-control-and-elimination-program.

Department of Health. (2019). *2019–2025 Strategic plan towards interruption of SCH infection in the Philippines*. DOH.

Department of Health. (2020a). *Delivery of routine deworming services under the integrated helminth control program (HCP) during the COVID-19 pandemic*. DOH. https://doh.gov.ph/sites/default/files/health_programs/dc2020-0302.pdf.

Department of Health. (2020b). *Updated Philippine strategic TB elimination plan phase, 1: 2020–2023*.DOH.

Department of Health. (2020c). *Garantisadong Pambata*. DOH. https://ro1.doh.gov.ph/programs-services/local-health-and-technical-assistance/family-health-cluster/32-programs-projects-services/local-health-and-technical-assistance/family-health/263-garantisadong-pambata#:~:text=A twice a year program,five years old.

Department of Health. (2020d). *HIV drugs supply is stable*. DOH. https://doh.gov.ph/press-release/HIV-DRUGS-SUPPLY-IS-STABLE-DOH.

Department of Health. (2022a). *HIV/AIDS & ART registry of the Philippines*. aidsdatahub.org/resource/hiv-aids-and-art-registry-philippines-february-2022.

Department of Health. (2022b). *Strengthening surveillance of drug-resistant tuberculosis (A health policy brief based on the 2018 Philippine TB drug resistance survey)*.DOH.

Department of Health. (2022c, December 12). *DOH launches the 7th AIDS medium term plan, emphasizes equality to address HIV*. Philippine Information Agency.

Department of Health. (2023). *Magnitude of TB*. DOH. https://ntp.doh.gov.ph/about-tb/magnitutde-of-tb/.

Department of Health - Epidemiology Bureau. (2017). *PH now has 57 Zika cases, 3 more confirmed pregnant cases included in latest report*. Department of Health. https://quarantine.doh.gov.ph/ph-now-has-57-zika-cases-3-more-confirmed-pregnant-cases-included-in-latest-report/

Department of Health - Epidemiology Bureau. (2020). *A briefer on the Philippine HIV estimates 2020*. https://www.aidsdatahub.org/sites/default/files/resource/philippines-briefer-ph-estimates-2020-2021.pdf

Department of Health, & Philippine Coalition Against Tuberculosis. (2003). *Comprehensive and unified policy for TB control in the Philippines*. https://www.philhealth.gov.ph/partners/providers/pdf/ComprehensiveUnifiedPolicy_TB.pdf

Department of Health, & Philippine Coalition Against Tuberculosis. (2004). *Comprehensive and unified policy (C.U.P. 2004) for TB control in the Philippines*. https://oshc.dole.gov.ph/wp-content/uploads/2020/09/Comprehensive-and-Unified-Policy.pdf

Dhanasekaran, V., Sullivan, S., Edwards, K. M., Xie, R., Khvorov, A., Valkenburg, S. A., Cowling, B. J., & Barr, I. G. (2022). Human seasonal influenza under COVID-19 and the potential consequences of influenza lineage elimination. *Nat Commun*, *13*(1), 1721.

Dombrowski, J. C., Simoni, J. M., Katz, D. A., & Golden, M. R. (2015). Barriers to HIV care and treatment among participants in a public health HIV care relinkage program. *AIDS Patient Care STDs*, *29*(5), 279–287.

Edillo, F. E., Halasa, Y. A., Largo, F. M., Erasmo, J. N. V., Amoin, N. B., Alera, M. T. P., Yoon, I.-K., Alcantara, A. C., & Shepard, D. S. (2015). Economic cost and burden of dengue in the Philippines. *Am J Trop Med Hyg*, *92*(2), 360–366. https://doi.org/10.4269/ajtmh.14-0139.

Eustaquio, P. C., Figuracion, R., Izumi, K., Morin, M. J., Samaco, K., Flores, S. M., Brink, A., & Diones, M. L. (2022). Outcomes of a community-led online-based HIV self-testing demonstration among cisgender men who have sex with men and transgender women in the Philippines during the COVID-19 pandemic: A retrospective cohort study. *BMC Public Health*, *22*(1), 1–11.

Exum, N. G., Kibira, S. P. S., Ssenyonga, R., Nobili, J., Shannon, A. K., Ssempebwa, J. C., Tukahebwa, E. M., Radloff, S., Schwab, K. J., & Makumbi, F. E. (2019). The prevalence of schistosomiasis in Uganda: A nationally representative population estimate to inform control programs and water and sanitation interventions. *PLOS Negl Trop Dis*, *13*(8), e0007617.

Fabay, X. C. T. J. (2010). Terror in the air: Meningococcal disease outbreak in the Philippines. *PIDSP*, *11*(1),17–25.

Fatima, K., & Syed, N. I. (2018). Dengvaxia controversy: Impact on vaccine hesitancy. *J Glob Health*, *8*(2), 10312. https://doi.org/10.7189/jogh.08.020312.

Fernandez Jr, T. J., Tarafder, M. R., Balolong Jr, E., Joseph, L., Willingham III, A. L., Belisle, P., Webster, J. P., Olveda, R. M., McGarvey, S. T., & Carabin, H. (2007). Prevalence of Schistosoma japonicum infection among animals in fifty villages of Samar Province, the Philippines. *Vector Borne Zoonotic Dis*, *7*(2), 147–155.

Ferrari, T. C. A., & Moreira, P. R. R. (2011). Neuroschistosomiasis: Clinical symptoms and pathogenesis. *Lancet Neurol*, *10*(9), 853–864.

Fornillos, R. J. C., Fontanilla, I. K. C., Chigusa, Y., Kikuchi, M., Kirinoki, M., Kato-Hayashi, N., Kawazu, S., Angeles, J. M., Tabios, I. K., Moendeg, K., & others (2019). Infection rate of Schistosoma japonicum in the snail Oncomelania hupensis quadrasi in endemic villages in the Philippines: Need for snail surveillance technique. *Trop Biomed*, *36*(2), 402–411.

Fornillos, R. J. C., Sato, M. O., Tabios, I. K. B., Sato, M., Leonardo, L. R., Chigusa, Y., Minamoto, T., Kikuchi, M., Legaspi, E. R., & Fontanilla, I. K. C. (2019). Detection of Schistosoma japonicum and Oncomelania hupensis quadrasi environmental DNA and its potential utility to schistosomiasis japonica surveillance in the Philippines. *PLoS One*, *14*(11), e0224617.

Gangcuangco, L. M. A. (2019). HIV crisis in the Philippines: Urgent actions needed. *The Lancet Public Health*, *4*(2), e84.

Garchitorena, A., Sokolow, S. H., Roche, B., Ngonghala, C. N., Jocque, M., Lund, A., Barry, M., Mordecai, E. A., Daily, G. C., Jones, J. H., & others (2017). Disease ecology, health and the environment: A framework to account for ecological and socio-economic drivers in the control of neglected tropical diseases. *Philos Trans R Soc Lond, B Biol Sci*, *372*(1722), 20160128.

Ghuman, H., & Voelz, K. (2017). Innate and adaptive immunity to mucorales. *J Fungi*, *3*(3), 48.

Gonçalves, B., Ferreira, C., Alves, C. T., Henriques, M., Azeredo, J., & Silva, S. (2016). Vulvovaginal candidiasis: Epidemiology, microbiology and risk factors. *Crit Rev Microbiol*, *42*(6), 905–927.

Gordon, C. A., Acosta, L. P., Gobert, G. N., Jiz, M., Olveda, R. M., Ross, A. G., Gray, D. J., Williams, G. M., Harn, D., Li, Y., & others. (2015). High prevalence of Schistosoma japonicum and Fasciola gigantica in bovines from Northern Samar, the Philippines. *PLOS Negl Trop Dis*, *9*(2), e0003108.

Gu, X., Hua, Y.-H., Zhang, Y.-D., Bao, D., Lv, J., & Hu, H.-F. (2021). The pathogenesis of, Host Defense Mechanisms, and the Development of AFMP4 Antigen as a Vaccine. *Pol J Microbiol*, *70*(1), 3–11.

Guerrero, A. M. (1982). Philippine medicinal plants found effective. *Fil Fam Phys*, *20*(1), 39–41.

Gushiken, A. C., Saharia, K. K., & Baddley, J. W. (2021). Cryptococcosis. *Infect Dis Clin North Am*, *35*(2), 493–514.

Guzman, J. C., & Martin, J. A. (2018). Analyzing the effectiveness and limitations of clinical diagnostic techniques for dengue fever in rural Philippines. *Int J Infect Dis*, *73*, 273. https://doi.org/10.1016/j.ijid.2018.04.4038.

Haake, D. A., & Levett, P. N. (2015). Leptospirosis in humans. *Curr Top Microbiol Immunol*, *387*, 65–97. https://doi.org/10.1007/978-3-662-45059-8_5.

Halstead, S. B. (2014). Dengue antibody-dependent enhancement: Knowns and unknowns. *Microbiol Spectr*, *2*(6). https://doi.org/10.1128/microbiolspec.aid-0022-2014.

Hammond, E. E., McDonald, C. S., Vestbo, J., & Denning, D. W. (2020). The global impact of Aspergillus infection on COPD. *BMC Pulm Med*, *20*(1), 1–10.

Handog, E. B., & Dayrit, J. F. (2005). Mycology in the Philippines, revisited. *Nippon Ishinkin Gakkai Zasshi*, *46*(2), 71–76.

Harris, J. R., Lindsley, M. D., Henchaichon, S., Poonwan, N., Naorat, S., Prapasiri, P., Chantra, S., Ruamcharoen, F., Chang, L. S., Chittaganpitch, M., & others. (2012). High prevalence of cryptococcal infection among HIV-infected patients hospitalized with pneumonia in Thailand. *Clin Infect Dis*, *54*(5), e43–e50.

Hartskeerl, R. A., Collares-Pereira, M., & Ellis, W. A. (2011). Emergence, control and re-emerging leptospirosis: Dynamics of infection in the changing world. *Clin Microbiol Infect*, *17*(4), 494–501. https://doi.org/10.1111/j.1469-0691.2011.03474.x.

Hasegawa, M., Pilotte, N., Kikuchi, M., Means, A. R., Papaiakovou, M., Gonzalez, A. M., Maasch, J. R. M. A., Ikuno, H., Sunahara, T., Ásbjörnsdóttir, K. H., & others (2020). What does soil-transmitted helminth elimination look like? Results from a targeted molecular detection survey in Japan. *Parasit Vectors*, *13*(1), 1–11.

He, P., Gordon, C. A., Williams, G. M., Li, Y., Wang, Y., Hu, J., Gray, D. J., Ross, A. G., Harn, D., & McManus, D. P. (2018). Real-time PCR diagnosis of Schistosoma japonicum in low transmission areas of China. *Infect Dis Pover*, *7*(01), 25–35.

Health Technology Assessment Study Group - Health Policy Development and Planning Bureau. (2018). *Rapid HIV diagnostic algorithm (rHIVda) for the philippines.* AHEAD-HPSR.

Health Technology Assessment Study Group - Health Policy Development and Planning Bureau. (2021). *Tenofovir/lamivudine/Dolutegravir for treatment-naive and treatment-experienced adolescents and adults living with HIV.* https://hta.doh.gov.ph/2021/09/01/tenofovir-lamivudine-dolutegravir-tld-for-treatment-naive-and-treatment-experienced-adolescents-and-adults-living-with-hiv/.

Henchal, E. A., & Putnak, J. R. (1990). The dengue viruses. *Clin Microbiol Rev*, *3*(4), 376–396. https://doi.org/10.1128/cmr.3.4.376.

HIV and AIDS Data Hub for Asia Pacific. (2021). *Key facts on HIV.* HIV and Aids Data Hub for Asia Pacific. https://www.aidsdatahub.org/country-profiles/philippines.

Howie, H. L., Katzenellenbogen, R. A., & Galloway, D. A. (2009). Papillomavirus E6 proteins. *Virology*, *384*(2), 324–334.

Humanitarian Data Exchange. (2023). *Philippine dengue cases and deaths.* https://data.humdata.org/dataset/philippine-dengue-cases-and-deaths?

Imoto, A., Honda, S., & Llamas-Clark, E. F. (2020). Human papillomavirus and cervical cancer knowledge, perceptions, and screening behavior: A cross-sectional community-based survey in rural Philippines. *Asian Pac J Cancer Prev*, *21*(11), 3145.

Jameson, J., Fauci, A., Kasper, D., Hauser, S., Longo, D., & Loscalzo, J. (2020). *Harrison's principles of internal medicine, 20e* (20th ed.). McGraw Hill.

John Snow Inc. (2023, January 27). *New initiative to increase global access to HPV vaccine.* JSI.

Kauffman, C. A. (2004). Zygomycosis: Reemergence of an old pathogen. *Clin Infect Dis*, *39*(4), 588–590.

King, C. H., & Dangerfield-Cha, M. (2008). The unacknowledged impact of chronic schistosomiasis. *Chronic Illn*, *4*(1), 65–79.

Knopp, S., Rinaldi, L., Khamis, I. S., Stothard, J. R., Rollinson, D., Maurelli, M. P., Steinmann, P., Marti, H., Cringoli, G., & Utzinger, J. (2009). A single FLOTAC is more sensitive than triplicate Kato--Katz for the diagnosis of low-intensity soil-transmitted helminth infections. *Trans R Soc Trop Med Hyg*, *103*(4), 347–354.

Kobayashi, A., Hara, T., & Kajima, J. (2006). Historical aspects for the control of soil-transmitted helminthiases. *Parasitol Int*, *55*, S289–S291.

Koehler, P., Bassetti, M., Kochanek, M., Shimabukuro-Vornhagen, A., & Cornely, O. A. (2019). Intensive care management of influenza-associated pulmonary aspergillosis. *Clin Microbiol Infect*, *25*(12), 1501–1509.

Kontoyiannis, D. P., Yang, H., Song, J., Kelkar, S. S., Yang, X., Azie, N., Harrington, R., Fan, A., Lee, E., & Spalding, J. R. (2016). Prevalence, clinical and economic burden of mucormycosis-related hospitalizations in the United States: A retrospective study. *BMC Infect Dis*, *16*(1), 1–6.

Kularadhan, V., Gan, J., Chow, E. P. F., Fairley, C. K., & Ong, J. J. (2022). HIV and STI testing preferences for men who have sex with men in high-income countries: A scoping review. *Int J Environ Res Public Health*, *19*(5), 3002.

Labana, R. V., Romero, V. A., Guinto, A. M., Caril, A. N., Untalan, K. D., Reboa, A. J. C., Sandoval, K. L., Cada, K. J. S., Lirio, G. A. C., Bernardo, I. R. A., & others. (2021). Prevalence and intensity of soil-transmitted helminth infections among school-age children in the Cagayan Valley, the Philippines. *Asian Pac J Trop Med*, *14*(3), 113–121.

Ladia, M. A. J., Belizario Jr, V. Y., Lacuna, J. M., Durano, L. P., & Alonte, A. I. (2020). Schistosomiasis and soil-transmitted helminthiasis morbidity control in selected communities in eastern Visayas, Philippines: Post Haiyan. *Acta Med Philipp*,57(7),24-30.

Latgé, J.-P. (1999). Aspergillus fumigatus and aspergillosis. *Clin Microbiol Rev*, *12*(2), 310–350.

Lee, S. S., Viboud, C., & Petersen, E. (2022). Understanding the rebound of influenza in the post COVID-19 pandemic period holds important clues for epidemiology and control. *Int J Infect Dis*, *122*, 1002–1004.

Leonardo, L. R., Acosta, L. P., Olveda, R. M., & Aligui, G. D. L. (2002). Difficulties and strategies in the control of schistosomiasis in the Philippines. *Acta Trop*, *82*(2), 295–299.

Leptospirosis Burden Epidemiology Reference Group. (2011). *Report of the second meeting of the leptospirosis burden epidemiology reference group*. www.who.int.

Levett, P. N. (2001). Leptospirosis. *Clin Microbiol Rev*, *14*(2), 296–326. https://doi.org/10.1128/CMR.14.2.296-326.2001.

Lintao, R. C. V., Cando, L. F. T., Perias, G. A. S., Tantengco, O. A. G., Tabios, I. K. B., Velayo, C. L., & de Paz-Silava, S. L. M. (2022). Current status of human papillomavirus infection and cervical cancer in the Philippines. *Front Med*, *9*, 929062.

Luchavez, J. (2018). *Looking to the future: The NTD laboratory network proceedings of the 6th Neglected Tropical Diseases (NTDs) Stakeholders' Forum*. RITM.

Lupisan, S. (2015). *Preparedness and response to a pandemic or emerging infectious disease outbreak*. https://aseanregionalforum.asean.org/wp-content/uploads/2019/10/Annex-D.1-Preparedness-and-Response-to-a-Pandemic-or-EID-Outbreak-in-the-Philippines.pdf.

Madsen, H., Carabin, H., Balolong, D., Tallo, V. L., Olveda, R., Yuan, M., & McGarvey, S. T. (2008). Prevalence of Schistosoma japonicum infection of Oncomelania quadrasi snail colonies in 50 irrigated and rain-fed villages of Samar Province, the Philippines. *Acta Trop*, *105*(3), 235–241.

Manila Bulletin. (2017, August 19). *School-Based Human Papillomavirus (HPV) Immunization launched in Mandaluyong city*. Manila Bulletin.

Masuzawa, T., Dancel, L. A., Miyake, M., & Yanagihara, Y. (2001). Serological analysis of human leptospirosis in the Philippines. *Microbiol Immunol*, *45*(1), 93–95. https://doi.org/10.1111/j.1348-0421.2001.tb01264.x.

Mationg, M. L. S., Gordon, C. A., Tallo, V. L., Olveda, R. M., Alday, P. P., Reñosa, M. D. C., Bieri, F. A., Williams, G. M., Clements, A. C. A., Steinmann, P., & others. (2017). Status of soil-transmitted helminth infections in schoolchildren in Laguna Province, the Philippines: Determined by parasitological and molecular diagnostic techniques. *PLOS Negl Trop Dis*, *11*(11), e0006022.

Mationg, M. L. S., Tallo, V. L., Williams, G. M., Gordon, C. A., Clements, A. C. A., McManus, D. P., & Gray, D. J. (2021). The control of soil-transmitted helminthiases in the Philippines: The story continues. *Infect Dis Pover, 10*(1), 85.

McLaughlin-Drubin, M. E., & Müenger, K. (2009). The human papillomavirus E7 oncoprotein. *Virology, 384*(2),335–344.

Meechan, P. J., & Potts, J. (2020). *Biosafety in microbiological and biomedical laboratories* (6th ed.). CDC.

Mendoza, M. T., Tan, S., Torres, D., & Tupasi, T. E. (1979). Human leptospirosis: Clinical and laboratory diagnosis. *J Phil ASSN, 55*, 219–224.

Mendoza, R. U. (2021, August 2). *The Philippine economy under the pandemic: From Asian tiger to sick man again?* Brookings Institution.

Montresor, A., Crompton, D. W. T., Gyorkos, T. W., Savioli, L., & World Health Organization. (2002). *Helminth control in school-age children ilA guide for managers of control programmes* (2nd ed.). World Health Organization.

Montresor, A., Crompton, D. W. T., Hall, A., Bundy, D. A. P., Savioli, L., & World Health Organization. Division of Control of Tropical Diseases. Schistosomiasis and Intestinal Parasites Unit. (1998). *Guidelines for the evaluation of soil-transmitted helminthiasis and schistosomiasis at community level : a guide for managers of control programmes.* World Health Organization.

Moonan, P. K., Nair, S. A., Agarwal, R., Chadha, V. K., Dewan, P. K., Gupta, U. D., Ho, C. S., Holtz, T. H., Kumar, A. M., Kumar, N., & others (2018). Tuberculosis preventive treatment: The next chapter of tuberculosis elimination in India. *BMJ Glob Health, 3*(5), e001135.

Mora, C., McKenzie, T., Gaw, I. M., Dean, J. M., von Hammerstein, H., Knudson, T. A., Setter, R. O., Smith, C. Z., Webster, K. M., Patz, J. A., & others. (2022). Over half of known human pathogenic diseases can be aggravated by climate change. *Nat Clim Change, 12*(9), 869–875.

Moron, L. S., Oyong, G. G., Chua, J. C. C., Cabrera, E. C., & others. (2017). Polyphasic identification of clinical Candida albicans in the Philippines. *Curr Res Environ Appl Mycol, 7*(4), 346–355.

Murphy, E., Togbevi, I. C., Ibikounlé, M., Avokpaho, E. F. G. A., Walson, J. L., & Means, A. R. (2023). Soil-transmitted helminth surveillance in Benin: A mixed-methods analysis of factors influencing non-participation in longitudinal surveillance activities. *PLOS Negl Trop Dis, 17*(1), e0010984.

Murugesan, A., & Manoharan, M. (2020). Dengue virus. In *Emerg reemerging viral Pathog* (pp. 281–359). https://doi.org/10.1016/B978-0-12-819400-3.00016-8.

Nadel, S., & Kroll, J. S. (2007). Diagnosis and management of meningococcal disease: The need for centralized care. *FEMS Microbiol Rev, 31*(1), 71–83.

Nang, D. W., Tukirinawe, H., Okello, M., Tayebwa, B., Theophilus, P., Sikakulya, F. K., Fajardo, Y., Afodun, A. M., & Kajabwangu, R. (2023). Prevalence of high-risk human papillomavirus infection and associated factors among women of reproductive age attending a rural teaching hospital in western Uganda. *BMC Womens Health, 23*(1), 1–8.

Navarro, R. C., de Cadiz, A. E., Fronda, J. M., Ong, L. A. D., & Belizario Jr, V. Y. (2021). Prevalence of Schistosoma japonicum infection in water buffaloes in selected areas in Davao del Norte and Davao de Oro, the Philippines. *Int J One Health, 7*(1), 12–18.

Nerissa, M. A., & Dominguez, N. (1997). Current DF / DHF prevention and control programme in the Philippines. *Dengue Bull, 21*, 41.

Ngelange, C., Munoz, N., Bosch, F. X., Festin, M. R., Deacon, J., Jacobs, M. V., Santamaria, M., Meijer, C. J. L. M., & Walboomers, J. M. M. (1998). Causes of cervical cancer in the Philippines: A case-control study. *J Natl Cancer Inst, 90*(1), 43–49.

Notarte, K. I. R., Pastrana, A. M., Ver, A. T. M., Velasco, J. V. L., Gellaco, M. M. L. D., & Pecundo, M. H. (2023). Mycosis in the Philippines: Epidemiology, clinical presentation, diagnostics and interventions. In J. J. Guerrero, T. Dalisay, M De Leon, M. A. Balendres, K. I. Notarte, T. E. Dela Cruz (Eds.), *Mycology in the Tropics* (pp. 213–233). Academic Press.

Occupational Safety and Health Administration. (2007). *Guidance on preparing workplaces for an influenza pandemic*. OSHA.

Okwuosa, V. N., & Osuala, F. O. (1993). Toxicity of washing soaps to Schistosoma mansoni cercariae and effects of sublethal concentrations on infectivity in mice. *Appl Parasitol*, *34*(1), 69–75.

Olds, G. R., Olveda, R., Wu, G., Wiest, P., McGarvey, S., Aligui, G., Zhang, S., Ramirez, B., Daniel, B., Peters, P., & others (1996). Immunity and morbidity in schistosomiasis japonicum infection. *Am J Trop Med Hyg*, *55*(5 Suppl), 121–126.

Ong, E. P., Obeles, A. J. T., Ong, B. A. G., & Tantengco, O. A. G. (2022). Perspectives and lessons from the Philippines' decades-long battle with dengue. *The Lancet Reg Health West Pac*, *24*, 100505. https://doi.org/10.1016/j.lanwpc.2022.100505.

Otu, A. A. (2013). Is the directly observed therapy short course (DOTS) an effective strategy for tuberculosis control in a developing country? *Asian Pac J Trop Dis*, *3*(3), 227–231.

Padre, L. P., Watt, G., Tuazon, M. L., Gray, M. R., & Laughlin, L. W. (1988). A serologic survey of rice-field leptospirosis in Central Luzon, Philippines. *Southeast Asian J Trop Med Public Health*, *19*(2), 197–199.

Pappas, P. G., Kauffman, C. A., Andes, D. R., Clancy, C. J., Marr, K. A., Ostrosky-Zeichner, L., Reboli, A. C., Schuster, M. G., Vazquez, J. A., Walsh, T. J., Zaoutis, T. E., & Sobel, J. D. (2016). Clinical practice guideline for the management of candidiasis: 2016 update by the Infectious Diseases Society of America. *Clin Infect Dis*, *62*(4), e1–e50. https://doi.org/10.1093/cid/civ933.

Parikh, D. S., Totañes, F. I. G., Tuliao, A. H., Ciro, R. N. T., Macatangay, B. J., & Belizario, V. Y. (2013). Knowledge, attitudes and practices among parents and teachers about soil-transmitted helminthiasis control programs for school children in Guimaras, Philippines. *Southeast Asian J Trop Med Public Health*, *44*(5), 744–752.

Parslow, B. Y., & Thornton, C. R. (2022). Continuing shifts in epidemiology and antifungal susceptibility highlight the need for improved disease management of invasive candidiasis. *Microorganisms*, *10*(6), 1208.

Patterson, T. F., Thompson III, G. R., Denning, D. W., Fishman, J. A., Hadley, S., Herbrecht, R., Kontoyiannis, D. P., Marr, K. A., Morrison, V. A., Nguyen, M. H., & others (2016). Practice guidelines for the diagnosis and management of aspergillosis: 2016 update by the Infectious Diseases Society of America. *Clin Infect Dis*, *63*(4), e1–e60.

Pediatric Infectious Disease Society of the Philippines, & Child Neurology Society of the Philippines. (2015). Philippine clinical practice guidelines on the diagnosis and management of acute bacterial meningitis. *PIDSP*, *16*(2), 2–42. http://www.pidsphil.org/home/wp-content/uploads/2017/04/jo49_ja01-1.pdf

Perfect, J. R., Dismukes, W. E., Dromer, F., Goldman, D. L., Graybill, J. R., Hamill, R. J., Harrison, T. S., Larsen, R. A., Lortholary, O., Nguyen, M.-H., & others (2010). Clinical practice guidelines for the management of cryptococcal disease: 2010 update by the Infectious Diseases Society of America. *Clin Infect Dis*, *50*(3), 291–322.

Pfaller, M. A., & Diekema, D. J. (2007). Epidemiology of invasive candidiasis: A persistent public health problem. *Clin Microbiol Rev*, *20*(1), 133–163. https://doi.org/10.1128/cmr.00029-06.

Pfaller, M. A., Messer, S. A., Moet, G. J., Jones, R. N., & Castanheira, M. (2011). Candida bloodstream infections: Comparison of species distribution and resistance to echinocandin and azole antifungal agents in Intensive Care Unit (ICU) and non-ICU settings in the SENTRY Antimicrobial Surveillance Program (2008–2009). *Int J Antimicrob Agents*, *38*(1), 65–69.

Pham, H. T., & Tran, M. H. (2022). One health: An effective and ethical approach to leptospirosis control in Australia. *Trop Med Infect Dis*, *7*(11). https://doi.org/10.3390/TROPICALMED7110389.

Phanuphak, N., Anand, T., Jantarapakde, J., Nitpolprasert, C., Himmad, K., Sungsing, T., Trachunthong, D., Phomthong, S., Phoseeta, P., Tongmuang, S., & others. (2018). What would you choose: Online or offline or Mixed services? Feasibility of online HIV counselling and testing among Thai men who have sex with men and transgender women and factors associated with service uptake. *J Int AIDS Soc*, *21*, e25118.

Philippine Coalition Against Tuberculosis. (2021). *History and trailblazing accomplishments*. PhilCAT.

Philippine National AIDS Council. (2016). *Sixth HIV & AIDS medium term plan 2017–2022: Synergizing the HIV and AIDS Response*. PNAC.

Philippine Society for Microbiology and Infectious Diseases. (2010). *Clinical practice guidelines for leptospirosis*.PSMID.

Philippine Tuberculosis Society Inc. (2022). *Who we are*. PTSI.

Portero, J. L., & Rubio, M. (2006). Multidrug-resistant TB in the Philippines: Totem and taboo. *PLOS Med*, *3*(12), e539.

Prakash, H., & Chakrabarti, A. (2019). Global epidemiology of mucormycosis. *J Fungi*, *5*(1), 26.

Quelapio, M. I. D., Mira, N. R. C., Orillaza-Chi, R. B., Belen, V., Munez, N., Belchez, R., Egos, G. E., Evangelista, M., Vianzon, R., & Tupasi, T. E. (2010). Responding to the multidrug-resistant tuberculosis crisis: Mainstreaming programmatic management to the Philippine National Tuberculosis Programme. *Int J Tuberc Lung Dis*, *14*(6), 751–757.

Radotra, B., & Challa, S. (2022). Pathogenesis and pathology of COVID-associated mucormycosis: What is new and why. *Curr Fungal Infect Rep*, *16*(4), 206–220.

Raguindin, P. F., Rojas, V. M., & Lopez, A. L. (2020). Meningococcal disease in the Philippines: A systematic review of the literature. *Int J Infect Dis*, *101*, 149–150.

Rajasingham, R., Smith, R. M., Park, B. J., Jarvis, J. N., Govender, N. P., Chiller, T. M., Denning, D. W., Loyse, A., & Boulware, D. R. (2017). Global burden of disease of HIV-associated cryptococcal meningitis: An updated analysis. *Lancet Infect Dis*, *17*(8), 873–881.

Research Institute for Tropical Medicine. (2020). *NTD awareness month: Soil-transmitted helminthiasis (STH)*. RITM. https://ritm.gov.ph/january-is-world-neglected-tropical-diseases-ntd-awareness-month/.

Riedel, S., Hobden, J. A., Miller, S., Morse, S. A., Mietzner, T. A., Detrick, B., Mitchell, T. G., Sakanari, J. A., Hotez, P., & Mejia, R. (2019). Orthomyxoviruses. In *Jawetz, Melnick & Adelberg's medical microbiology* (28th ed., pp. 581–593). McGraw-Hill Education. accessmedicine.mhmedical.com/content.aspx?aid=1163277716

Rivera, A. S., Hernandez, R., Mag-Usara, R., Sy, K. N., Ulitin, A. R., O'Dwyer, L. C., McHugh, M. C., Jordan, N., & Hirschhorn, L. R. (2021). Implementation outcomes of HIV self-testing in low-and middle-income countries: A scoping review. *PLoS One*, *16*(5), e0250434.

Roden, M. M., Zaoutis, T. E., Buchanan, W. L., Knudsen, T. A., Sarkisova, T. A., Schaufele, R. L., Sein, M., Sein, T., Chiou, C. C., Chu, J. H., Kontoyiannis, D. P., & Walsh, T. J. (2005). Epidemiology and outcome of Zygomycosis: A review of 929 reported cases. *Clin Infect Dis*, *41*(5), 634–653. https://doi.org/10.1086/432579.

Rosadiño, J. D. T., Pagtakhan, R. G., Briñes, M. T., Dinglasan, J. L. G., Cruz, D. P., Corciega, J. O. L., Pagtakhan, A. B., Regencia, Z. J. G., & Baja, E. S. (2023). Implementation of unassisted and community-based HIV Self-Testing (HIVST) during the COVID-19 pandemic among Men-who-have-sex-with-Men (MSM) and Transgender Women (TGW): A demonstration study in Metro Manila, Philippines. *PLoS One*, *18*(3), e0282644.

Ross, A. G. P., Olveda, R. M., McManus, D. P., Harn, D. A., Chy, D., Li, Y., Tallo, V., & Ng, S.-K. (2017). Risk factors for human helminthiases in rural Philippines. *Int J Infect Dis, 54*, 150–155.

Saegeman, V., Maertens, J., Meersseman, W., Spriet, I., Verbeken, E., & Lagrou, K. (2010). Increasing incidence of mucormycosis in University Hospital, Belgium. *Emerg Infect Dis, 16*(9), 1456.

Saito, M., Villanueva, S. Y. A. M., Chakraborty, A., Miyahara, S., Segawa, T., Asoh, T., Ozuru, R., Gloriani, N. G., Yanagihara, Y., & Yoshida, S. (2013a). Comparative analysis of Leptospira strains isolated from environmental soil and water in the Philippines and Japan. *Appl Environ Microbiol, 79*(2), 601–609.

Saito, M., Villanueva, S. Y. A. M., Chakraborty, A., Miyahara, S., Segawa, T., Asoh, T., Ozuru, R., Gloriani, N. G., Yanagihara, Y., & Yoshida, S. I. (2013b). Comparative analysis of Leptospira strains isolated from environmental soil and water in the Philippines and Japan. *Appl Environ Microbiol, 79*(2), 601–609. https://doi.org/10.1128/AEM.02728-12.

Salvaña, E. M. T., Schwem, B. E., Ching, P. R., Frost, S. D. W., Ganchua, S. K. C., & Itable, J. R. (2017). The changing molecular epidemiology of HIV in the Philippines. *Int J Infect Dis, 61*, 44–50.

Samoh, N., Peerawaranun, P., Jonas, K. J., Lim, S. H., Wickersham, J. A., & Guadamuz, T. E. (2021). Willingness to use HIV self-testing with online supervision among app-using young men who have sex with men in Bangkok. *Sex Transm Dis, 48*(3), e41–e44.

Sanga, E. S., Mukumbang, F. C., Mushi, A. K., Lerebo, W., & Zarowsky, C. (2019). Understanding factors influencing linkage to HIV care in a rural setting, Mbeya, Tanzania: Qualitative findings of a mixed methods study. *BMC Public Health, 19*(1), 1–15.

Segal, B. H. (2009). Aspergillosis. *N Engl J Med, 360*(18), 1870–1884.

Singh, N., & Paterson, D. L. (2005). Aspergillus infections in transplant recipients. *Clin Microbiol Rev, 18*(1), 44–69.

Singh, R., & Chakrabarti, A. (2017). Invasive candidiasis in the Southeast-Asian region. In R. Prasad (Ed.), *Candida albicans: Cellular and molecular biology* (pp. 25–40). Springer International Publishing. https://doi.org/10.1007/978-3-319-50409-4_3.

Sison, O. T., Baja, E. S., Bermudez, A. N. C., Quilantang, M. I. N., Dalmacion, G. V., Guevara, E. G., Garces-Bacsal, R. M., Hemingway, C., Taegtmeyer, M., Operario, D., & others. (2022). Association of anticipated HIV testing stigma and provider mistrust on preference for HIV self-testing among cisgender men who have sex with men in the Philippines. *BMC Public Health, 22*(1), 1–11.

Skiada, A., Pavleas, I., & Drogari-Apiranthitou, M. (2020). Epidemiology and diagnosis of mucormycosis: An update. *J Fungi, 6*(4), 265.

Society of Gynecologic Oncologists of the Philippines. (2019). *The 2019 clinical practice guidelines for the obstetrician-gynecologist.* SGOP.

Soisangwan, P. (2021). Biosafety and biosecurity law in Thailand: From legislation to practice. *J Biosaf Biosecur, 3*(2), 91–98.

Solidum, J. N., & Solidum, G. G. (2013). Assessment and Filling of Gaps of the Anti-dengue Program in The Philippines. *The Asian Conference on the Social Sciences 2013 Official Conference Proceedings Osaka, Japan Assessment*, pp. 406–418.

Spellberg, B., Edwards Jr, J., & Ibrahim, A. (2005). Novel perspectives on mucormycosis: Pathophysiology, presentation, and management. *Clin Microbiol Rev, 18*(3), 556–569.

Sumi, A., Telan, E. F. O., Chagan-Yasutan, H., Piolo, M. B., Hattori, T., & Kobayashi, N. (2017). Effect of temperature, relative humidity and rainfall on dengue fever and leptospirosis infections in Manila, the Philippines. *Epidemiol Infect, 145*(1), 78–86.

Sung, H., Ferlay, J., Siegel, R. L., Laversanne, M., Soerjomataram, I., Jemal, A., & Bray, F. (2021). Global cancer statistics 2020: GLOBOCAN estimates of incidence and mortality worldwide for 36 cancers in 185 countries. *CA Cancer J Clin*, *71*(3), 209–249.

Tabo, N. A., Villanueva, S., & Gloriani, N. (2018). Prevalence of Leptospira-agglutinating antibodies in Abattoir Workers and Slaughtered Animals in Selected Slaughterhouses in Cavite, Philippines. *Philipp J Sci*, *147*(1), 27–35. https://www.researchgate.net/publication/326458823.

Tan, A. (2019). Tuberculosis and universal health care legislations: Pathways to upholding the right to health of every people. *J Glob Health Sci*, *1*(1), e10.

Tan, B. H., Chakrabarti, A., Patel, A., Chua, M. M. M., Sun, P.-L., Liu, Z., Rotjanapan, P., Li, R., Wahyuningsih, R., Chayakulkeeree, M., & others. (2020). Clinicians' challenges in managing patients with invasive fungal diseases in seven Asian countries: An Asia Fungal Working Group (AFWG) survey. *Int J Infect Dis*, *95*, 471–480.

Tangcalagan, D. A., Daga, C. M., Tan, A., Reyes, R. A., Macalinao, M. L. M., Mationg, M. L., Alday, P., Galit, S., Luchavez, J., Erce, E., & others. (2022). The 2013–2015 nationwide prevalence survey of soil-transmitted helminths (STH) and schistosomiasis among school-age children in public schools in the Philippines. *Pediatr Infect Dis Soc Philippines J*, *23*(1), 75–96.

Tantengco, O. A. G., & Gloriani, N. G. (2017). Isolation and genetic detection of pathogenic Leptospira spp. from environmental soils and water in Central Luzon, Philippines. *Asian Pac J Trop Dis*, *7*(12), 748–752. https://doi.org/10.12980/apjtd.7.2017D7-220.

Tayal, A., Kabra, S. K., & Lodha, R. (2023). Management of dengue: An updated review. *Indian J Pediatr*, *90*(2), 168–177. https://doi.org/10.1007/s12098-022-04394-8.

Taylor, B. S., Sobieszczyk, M. E., McCutchan, F. E., & Hammer, S. M. (2008). The challenge of HIV-1 subtype diversity. *N Engl J Med*, *358*(15), 1590–1602.

Tenorio, J. C. B., & Molina, E. C. (2020). Schistosoma japonicum infections in cattle and water buffaloes of farming communities of Koronadal City, Philippines. *Int J One Health*, *6*(1), 28–33.

The Lancet Infectious Diseases. (2019). Infectious disease crisis in the Philippines. *Lancet Infect Dis*, *19*(12), 1265.

Tupasi, T. E., Gupta, R., Quelapio, M. I. D., Orillaza, R. B., Mira, N. R., Mangubat, N. V., Belen, V., Arnisto, N., Macalintal, L., Arabit, M., & others. (2006). Feasibility and cost-effectiveness of treating multidrug-resistant tuberculosis: A cohort study in the Philippines. *PLOS Med*, *3*(9), e352.

Tupasi, T. E., Radhakrishna, S., Rivera, A. B., Pascual, M. L. G., Quelapio, M. I. D., Co, V. M., Villa, M. L. A., Beltran, G., Legaspi, J. D., Mangubat, N. V., & others. (1999). The 1997 nationwide tuberculosis prevalence survey in the Philippines. *Int J Tuberc Lung Dis*, *3*(6), 471–477.

Ulep, V. G., & Uy, J. (2019). *Too early, too late: Timeliness of child vaccination in the Philippines. PIDS Discussion Paper Series*, No. 2019-21, Philippine Institute for Developmental Studies (PIDS), Quezon City.

Undurraga, E. A., Edillo, F. E., Erasmo, J. N. V., Alera, M. T. P., Yoon, I.-K., Largo, F. M., & Shepard, D. S. (2017). Disease Burden of Dengue in the Philippines: Adjusting for Underreporting by Comparing Active and Passive Dengue Surveillance in Punta Princesa, Cebu City. *Am J Trop Med Hyg*, *96*(4), 887–898. https://doi.org/10.4269/ajtmh.16-0488.

UNICEF. (2023). *New data indicates declining confidence in childhood vaccines of up to 25 percentage points in the Philippines during the COVID-19 pandemic*. UNICEF.

Velasco, J. M. S., Alera, M. T. P., Ypil-Cardenas, C. A., Dimaano, E. M., Jarman, R. G., Chinnawirotpisan, P., Thaisomboonsuk, B., Yoon, I. K., Cummings, D. A., & Mammen, M. P. (2014). Demographic, clinical and laboratory findings among adult and pediatric patients hospitalized with dengue in the Philippines. *Southeast Asian J Trop Med Public Health*, *45*(2), 337–345.

Villanueva, S. Y. A. M., Baterna, R. A., Cavinta, L. L., Yanagihara, Y., Gloriani, N. G., & Yoshida, S. (2018). Seroprevalence of leptospirosis among water buffaloes, pigs, and dogs in selected areas in the Philippines, 2007 to 2008. *Acta Med Philipp*, *52*(1), 109–117.

Villanueva, S. Y. A. M., Ezoe, H., Baterna, R. A., Yanagihara, Y., Muto, M., Koizumi, N., Fukui, T., Okamoto, Y., Masuzawa, T., Cavinta, L. L., Gloriani, N. G., & Yoshida, S. I. (2010). Serologic and molecular studies of Leptospira and leptospirosis among rats in the Philippines. *Am J Trop Med Hyg*, *82*(5), 889–898. https://doi.org/10.4269/ajtmh.2010.09-0711.

Villanueva, S. Y. A. M., Ezoe, H., Baterna, R. A., Yanagihara, Y., Muto, M., Koizumi, N., Fukui, T., Okamoto, Y., Masuzawa, T., Cavinta, L. L., & others. (2010). Serologic and molecular studies of Leptospira and leptospirosis among rats in the Philippines. *Am J Trop Med Hyg*, *82*(5), 889.

Villaverde, M. C., Gepte, A. T. D., & Baquiran, R. S. (2016). *Performance assessment of the national objectives for health Philippines 2011–2016*. Ateneo De Manila University.

Wartel, T. A., Prayitno, A., Hadinegoro, S. R. S., Capeding, M. R., Thisyakorn, U., Tran, N. H., Moureau, A., Bouckenooghe, A., Nealon, J., & Taurel, A.-F. (2017). Three decades of dengue surveillance in five highly endemic South East Asian countries. *Asia Pac J Public Health*, *29*(1), 7–16. https://doi.org/10.1177/1010539516675701.

Weerakoon, K. G. A. D., Gobert, G. N., Cai, P., & McManus, D. P. (2015). Advances in the diagnosis of human schistosomiasis. *Clin Microbiol Rev*, *28*(4), 939–967.

Wijesinghe, P. R., Ofrin, R. H., Bhola, A. K., Inbanathan, F. Y., & Bezbaruah, S. (2020). Pandemic influenza preparedness in the WHO South-East Asia Region: A model for planning regional preparedness for other priority high-threat pathogens. *WHO S East Asia J Public Health*, *9*(1), 43–49.

Williams, D., & Lewis, M. (2011). Pathogenesis and treatment of oral candidosis. *J Oral Microbiol*, *3*(1), 5771.

Wong, J. M., Adams, L. E., Durbin, A. P., Muñoz-Jordán, J. L., Poehling, K. A., Sánchez-González, L. M., Volkman, H. R., & Paz-Bailey, G. (2022). Dengue: A growing problem with new interventions. *Pediatrics*, *149*(6), e2021055522. https://doi.org/10.1542/peds.2021-055522.

World Health Organization. (n.d.). *Zika virus disease*. WHO. Retrieved March 24, 2023, from https://www.who.int/health-topics/zika-virus-disease#tab=tab_1.

World Health Organization. (1996). Report of the WHO informal consultation on the use of chemotherapy for the control of morbidity due to soil-transmitted nematodes in humans, Geneva, 29 April to 1 May 1996.

World Health Organization. (2011). *Helminth control in school-age children: A guide for managers of control programmes*. World Health Organization.

World Health Organization. (2013). *Schistosomiasis: Progress report 2001–2011, strategic plan 2012–2020*. World Health Organization.

World Health Organization. (2017a). *Fourth WHO report on neglected tropical diseases: Integrating neglected tropical diseases into global health and development*. WHO.

World Health Organization. (2017b). *Guideline: Preventive chemotherapy to control soil-transmitted helminth infections in at-risk population groups*. World Health Organization.

World Health Organization. (2017c). *Expert consultation to accelerate elimination of Asian Schistosomiasis*. Meeting Report WHO.

World Health Organization. (2018). *Guidelines for the diagnosis, prevention, and management of cryptococcal disease in HIV-infected adults, adolescents and children, March 2018: Supplement to the 2016 consolidated guidelines*. WHO. https://apps.who.int/iris/handle/10665/260399.

World Health Organization. (2019). *Joint external evaluation of IHR core capacities republic of the Philippines: Mission report: 10–14 September 2018*. World Health Organization.

World Health Organization. (2020a). *Ending the neglect to attain the sustainable development goals: A road map for neglected tropical diseases 2021–2030*. WHO.

World Health Organization. (2020b). *Laboratory biosafety manual* (4th ed.). WHO. https://www.who.int/publications/i/item/9789240011311.

World Health Organization. (2021a). *Diagnostic target product profile for monitoring and evaluation of soil-transmitted helminth Control programmes*. WHO.

World Health Organization. (2021b). *National tuberculosis prevalence surveys 2007–2016*. World Health Organization. https://www.who.int/publications/i/item/9789240022430.

World Health Organization. (2022a). Human papillomavirus vaccines: WHO position paper, December 2022. *Wkly Epidemiol Rec*, *97*(50). World Health Organization.

World Health Organization. (2022b). *Laboratory testing for Zika Virus and Dengue virus infections: Interim guidance*. WORLD HEALTH ORGANIZATION.

World Health Organization. (2022c). *Human Papillomavirus (HPV) Vaccination coverage*. WHO.

World Health Organization. (2022d). *One health approach in the Philippines: A multi-sectoral collaboration to improving the health of people and the planet*. WHO.

World Health Organization. (2022e). *Zika virus*. WHO.

World Health Organization. (2023a). *Cervical cancer elimination initiative*. WHO.

World Health Organization. (2023b). *Health promotion*. WHO.

World Health Organization. (2023c). *Influenza update N° 443*. WHO. https://www.who.int/publications/m/item/influenza-update-n--443.

World Health Organization. (2023d). *Soil-transmitted helminth infections*. WHO. https://www.who.int/news-room/fact-sheets/detail/soil-transmitted-helminth-infections.

Wu, H.-W., Qin, Y.-F., Chu, K., Meng, R., Liu, Y., McGarvey, S. T., Olveda, R., Acosta, L., Ji, M.-J., Fernandez, T., & others. (2010). High prevalence of Schistosoma japonicum infection in water buffaloes in the Philippines assessed by real-time polymerase chain reaction. *Am J Trop Med Hyg*, *82*(4), 646.

Xu, J., Rong, R., Zhang, H. Q., Shi, C. J., Zhu, X. Q., & Xia, C. M. (2010). Sensitive and rapid detection of Schistosoma japonicum DNA by loop-mediated isothermal amplification (LAMP). *Int J Parasitol*, *40*(3), 327–331.

Zhang, Y.-Y., Luo, J.-P., Liu, Y.-M., Wang, Q.-Z., Chen, J.-H., Xu, M.-X., Xu, J.-M., Wu, J., Tu, X.-M., Wu, G.-L., & others. (2009). Evaluation of Kato--Katz examination method in three areas with low-level endemicity of schistosomiasis japonica in China: A Bayesian modeling approach. *Acta Trop*, *112*(1), 16–22.

Zhu, H.-H., Huang, J.-L., Zhu, T.-J., Zhou, C.-H., Qian, M.-B., Chen, Y.-D., & Zhou, X.-N. (2020). National surveillance on soil-transmitted helminthiasis in the People's Republic of China. *Acta Trop*, *205*, 105351.

# 14 Biosafety and Biosecurity in the Time of the SARS-CoV-2 Pandemic

*Angelo dela Tonga*
National Training Center for Biosafety and Biosecurity
Institute of Molecular Biology and Biotechnology NIH UP Manila

*Janiza Lianne Foronda*
Department of Virology Research
Institute for Tropical Medicine

*Alexander Sadiasa*
Department of Microbiology Research
Institute for Tropical Medicine

## 14.1 INTRODUCTION

The first case of SARS-CoV-2 human infection was reported in Wuhan, Hubei Province, China, in December 2019. Since then, the virus has spread worldwide causing an epidemic with approximately 770,000,000 confirmed cases including 6,974,473 deaths (WHO, 2023). In the Philippines, the first case of COVID-19 was reported on January 30, 2020, by the Department of Health, and the first local transmission was confirmed on March 7, 2020.

Different approaches have been taken by countries to respond to the pandemic. A critical aspect of the response was to increase testing, resulting in rapid expansion of diagnostic capacity and research capability. However, the increase in demand for diagnostic capacity did not necessarily trigger an equivalent increase in biosafety and biosecurity capacity, especially in low-resource settings such as the Philippines. It is essential that diagnostic laboratories are provided with the capability to work safely and securely on pathogens that have high consequences. This section provides a snapshot of the biosafety and biosecurity aspects of Philippine laboratories during the start of the pandemic. A very different time from the present, SARS-CoV-2 was still a new pathogen that would then unexpectedly explode into a pandemic.



DOI: 10.1201/9781003426219-14

### 14.1.1 Risk Assessment

Risk assessment is one of the pillars of the biorisk management system. It is a procedure wherein a pathogen is evaluated using its characteristics and the multiple aspects in which it will be handled in the laboratory. This enables institutions to determine the risk posed by a particular organism on a specific procedure. Risk assessments are common practice in laboratories; on the other hand, they provide public health information on the possible risks associated with an emerging disease against the general human population (Schröder, 2020).

## 14.2 PATHOGEN INFORMATION

During the first months of the COVID-19 pandemic, necessary information regarding SARS-CoV-2 was still unknown. It was not until May 2020 that WHO released the first laboratory guidance in handling samples for testing (WHO, 2020).

The genome of SARS-CoV-2 has been generated and made available in January 2020. SARS-CoV-2 is related to SARS-CoV-1 pathogen (79%) and has similar host receptor binding properties (Lu et al., 2020). Due to the similarity between the two pathogens, early measures when handling samples were based on SARS-CoV-1. SARS-CoV-2 exhibited rapid spread and a high risk to human health and the economy. Effective therapeutic interventions did not exist during the first half of 2020. These pathogen properties would identify SARS-CoV-2 as a Risk Group 4 pathogen (Schröder, 2020).

The possible host for SARS-CoV-2 has a high likelihood to be bats due to the relatedness of its genome to SARS-like bat viruses (Lu et al., 2020). The host range for SARS-CoV-2 expands as it spreads throughout the world creating additional reservoirs and increasing the probability of SARS-CoV-2 variants. The route of transmission can be direct through sneezing, coughing, and speaking with an infected person. Droplets have the potential to be directly deposited to mucous membranes in proximity. Aerosolized SARS-CoV-2 can remain airborne and be transmitted via direct inhalation. The stability of SARS-CoV-2 particles in droplets on surfaces has a half-life of 8.5, 13.1, and 15.9 hours on cardboard, steel, and plastic surfaces, respectively, while particles in aerosols remain viable in the air for 3 hours making it possible for airborne transmission (van Doremalen et al., 2020). The development of new insights and information on the pathogen in the late months of 2020 resulted in improvements in handling procedures (Kaufer et al., 2020; Rutjes et al., 2023).

Management of COVID-19 requires the laboratory detection of SARS-CoV-2 by nucleic acid amplification testing (NAAT), which usually employs a fluorescence-based real-time reverse transcription-polymerase chain reaction (RT-PCR), typically in upper respiratory swabs (i.e., nasopharyngeal and oropharyngeal swabs). Real-time RT-PCR is the gold-standard method of the laboratory diagnosis of COVID-19 (Chu et al., 2020; Corman et al., 2020).

## 14.3 MITIGATION CONTROLS

### 14.3.1 Infrastructure

The shortage of infrastructure that has all the requirements for licensing has been one of the problems in the early months of COVID-19. The Research Institute for Tropical Medicine, which is the designated National Reference Laboratory (NRL) for Emerging Diseases by the DOH, was the only accredited testing laboratory for the detection of SARS-CoV-2 for the whole country at the initial onset of the disease (DOH, 2000). This resulted in a limited testing capacity and rate, which is inadequate considering the dense population of the Philippines and the demand brought by the urgency of the situation. Infrastructure, as part of engineering control, is expensive and heavily relies on supply chains which have been very challenging during the start of the pandemic. Establishment and accreditation of additional laboratories to aid RITM and augment the testing capacity of the country was slow at the start with only those with established BSL3 laboratories such as UP-NIH and Lung Center of the Philippines. Upon getting to know more about SARS-CoV-2 and COVID-19, and based on the risk assessment, the guidelines set by the Philippine Department of Health (DOH) are to have COVID-19 testing laboratories provide adequate, appropriate, and dedicated areas to safely, effectively, and efficiently conduct the following activities: specimen reception, virus inactivation and nucleic acid extraction (Pre-PCR), reagent storage and handling, RT-PCR, and clerical activities. It will have the facility features of a BSL-2 laboratory with additional requirements such as unidirectional workflow and exhaust in specimen handling and sample preparation room having at least 12 air changes per hour and the addition of pass through box for sample transfer (DOH, 2020). The requirements set by WHO for non-propagative diagnostic laboratory work such as nucleic acid amplification can be conducted in a facility using procedures and set-up of a BSL-2 laboratory (Baclig, 2021; WHO, 2020). This opened up opportunities to increase the establishment of more testing laboratories and other alternative setups.

One of these alternatives that has risen to increase the capacity of testing is through mobile laboratories in remote provinces of the Philippines. These turnkey systems contain equipment to perform the RT-PCR based detection of SARS-CoV-2 (Aster, n.d.; Healthcare Asia, 2021). Moreover, the design has been based on risk assessment and has the specifications according to WHO (Healthcare Asia, 2021). The model of mobile laboratory makes the certification faster since each laboratory has been made with a similar design that has been pre-consulted to DOH. However, mobile laboratories are expensive to set up and do not have the necessary space. These types of infrastructure are more applicable to rapid deployment settings; a well-planned facility that could perform diagnostic activities mandated by DOH is ideal for long-term diagnostic services.

### 14.3.2 Containment Equipment

#### 14.3.2.1 Biosafety Cabinets

Biosafety cabinets (BSC) are one of the features described in the BSL-2 laboratory. Identified as the primary barrier, it is a requirement to use BSCs in procedures that have the risk to generate infectious aerosols such as pipetting liquid suspensions and homogenizing infectious materials (WHO, 2020). Each of the facilities is required to have a Class II A2 BSC for procedures that handle specimens not yet inactivated. Biosafety cabinets are required to have annual certification provided by a certified engineer. A problem that was observed even before the pandemic is the shortage of engineers that can perform certification. This has been more evident due to restrictions on local travel during the pandemic.

#### 14.3.2.2 Centrifuge

Centrifuges are also primary barriers used to process samples for SARS-CoV-2 RT-PCR testing. Sealed rotors or aerosol-tight rotors are components of centrifuge that provide additional safety to users who have a risk of exposure to aerosol-producing procedures. These rotors are recommended to be opened inside the BSC to provide another level of aerosol protection. These types of rotors are not very common in the clinical setting in the Philippines but are now increasing due to the pandemic.

### 14.3.3 Administrative Controls

#### 14.3.3.1 Licensing

The COVID-19 laboratory licensing is a multistage application that ensures that necessary control measures in handling SARS-CoV-2 samples are present in each facility. The licensing body is composed of technical experts from the Research Institute of the Philippines (RITM), Health Facilities and Services Regulatory Bureau (HFSRB), Centers for Health Development-Regulation, Licensing and Enforcement Division (CHD-RLED), and the World Health Organization (Baclig, 2021). RITM and WHO provided recommendations for biosafety, biosecurity, engineering/infrastructure, and laboratory quality practices to help applying laboratories meet regulatory standards. Included in RITM's responsibilities is to provide proficiency test (PT) panels, evaluate PT results, and endorse labs with passing marks to HFSRB and CHD-RLED who will then issue a license to operate (LTO) (Department of Health, 2020).

The application is initiated through the submission of the self-assessment tool and application for a license to operate. This in turn would be the basis of the DOH site assessment team to determine possible gaps in technical and safety protocols and facility design (Figure 14.1). The applicant laboratory would then make the necessary adjustments and resolve gaps found by the assessment team. Once recommendations have been addressed, the next step is proficiency testing using a test panel provided by DOH-RITM. The data interpretation shall then be submitted, and this will be the basis of the capability of the laboratory to perform the test. Issuance of license and full-scale implementation will take place as the final step. Facility monitoring is done by the licensing body to ensure safety and quality of results.

### 14.3.4 Standard Operating Procedures

Explicit knowledge in the form of written documents such as Standard Operating Procedures (SOPs) is important during the time of pandemic. Detailed and validated SOPs provide information to workers on the safe handling of samples. Additionally, SOPs set standard practices that influence the quality of diagnostic tests. Having a specific set of SOPs that covers areas but are not limited to swabbing, specimen transport, procedures for RT-PCR testing, and releasing of results has also been a requirement for COVID-19 laboratory licensing (Baclig, 2021).

Once SOPs have been developed, the next challenge is SOP adherence. Adherence does not happen automatically and should require organizational discipline and management focus (Amare, 2012). It should be a regular exercise for organizations to review SOPs and keep them up to date. Equipment, personnel, and tests in the laboratory change over time, so it is essential for the SOPs to be on point to these changes.

## 14.4 RISK COMMUNICATION

The COVID-19 pandemic is considered a health crisis that poses situations wherein the public is neither prepared nor possesses enough knowledge to deal with uncertainty. The government plays a major role in providing support and guidance to mobilize the public in times of crises. This is where risk communication comes in; risk communication provides the public's need for information in order to mitigate

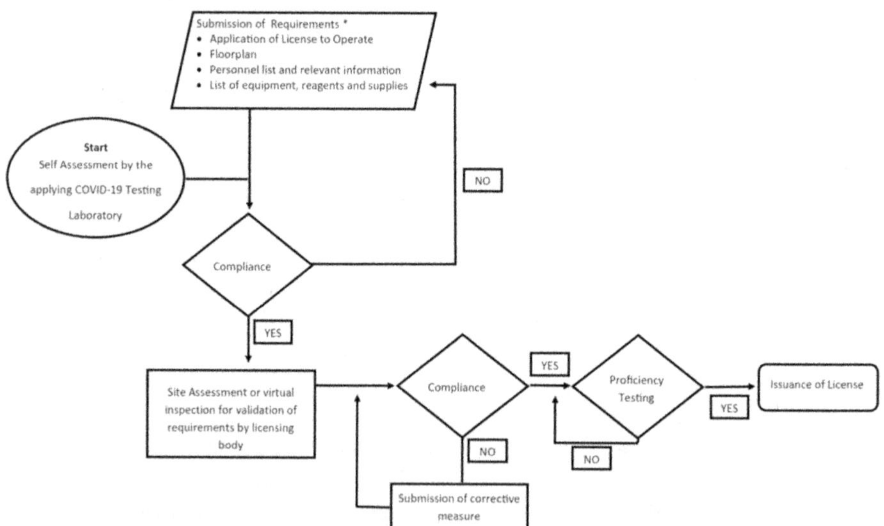


**FIGURE 14.1** Process diagram to obtain DOH accreditation to operate COVID-19 laboratory. * Basic requirements, institutions can submit other requirements that would support the application process.

the effects of an emergency (Flores & Asuncion, 2020). In the COVID-19 pandemic, risk communication information are on the topic of transmission route of the virus and how it can be slowed down or prevented; what to do in cases wherein there are infected people and exposed individuals, number of confirmed cases, and state of healthcare capacity (Flores & Asuncion, 2020). The Philippine government was able to strengthen national risk communication through press briefings, sponsored health-related television, information over the internet, and infographics (Amit, Pepito, & Dayrit, 2021). At the level of Local Government Units (LGU), the use of social media platforms has been an effective outlet for risk communication (Flores & Asuncion, 2020).

### 14.4.1 Biosafety Associations

Biosafety associations are non-profit, apolitical groups of professionals committed to share resources and experiences to further advance Biosafety and Biosecurity. For the Philippines, there is a group of biorisk management professionals which are graduates of Advanced Biosafety Officer's Training all part of the National Training Center for Biosafety and Biosecurity. Most of these professionals were able to lend expertise during the time of the pandemic. Biosafety Associations can provide additional support to country initiatives such as building and assuring competency of biosafety officers and establishment of standards that can be used by laboratories.

### 14.4.2 Laboratory Materials and Reagents

Diagnostic and clinical laboratories require numerous materials and reagents. The Philippines as a developing country rely on importation of these reagents which was severely affected by the pandemic. In the early months of the pandemic there is a shortage of viral RNA extraction kits and RT-PCR reagents. Materials used for biosafety activities such as qualitative fit test kit and biological indicators were not available on the distributors. Government institutions can have a common stash of materials necessary for health emergencies or have a partnership with international distributors to have faster access to these resources. Alternatively, the capability to locally produce reagents would be the ideal move for each country to have a level of sustainability.

## 14.5 PERSONNEL

The focus in the early months of the pandemic is to capacitate the personnel from the subnational laboratories to perform testing, since there is a sudden need to expand the diagnostic capability in the country and the demand for personnel that can perform the procedures for SARS-CoV-2 RT-PCR increased. Specific training in swabbing and RT-PCR testing has been required as part of licensing. As mentioned, the speed of expansion in testing capacity is not in parallel with an increase in biosafety and biosecurity capacity, this was observed during the onset of the increase in laboratory licensing application where one of the causes of failure to secure license is noncompliance to biosafety practices (Baclig, 2021).

Training of manpower via the traditional classroom-based face-to-face training could not be done due to the mandatory lockdown. The National Training Center for Biosafety and Biosecurity developed the Biosafety Education and Awareness Training COVID-19 (BEAT-COVID 19). This program helped in increasing Biosafety and Biosecurity awareness by training 3371 personnel that would be involved in testing and helped more than 100 testing laboratories obtain licenses in 2020 (Cena-Navarro et al., 2022). The RITM Biorisk Management Office, who have initially provided biosafety and biosecurity training through a hands-on approach, has also shifted to online platform. This online program catered to 2560 participants further expanding the knowledge base of laboratories (RITM, 2023).

## 14.6 PERSONAL PROTECTIVE EQUIPMENT (PPE)

Considered as the last possible line of defense against pathogens, PPEs provide a barrier between pathogen and the laboratorian. Guidelines on the PPE appropriate for handling respiratory transmissible pathogens such as SARS-CoV-2 lists respirators or PAPRs (air purifying respirator), disposable gowns, disposable gloves, face shield, and eye protection (Ağalar & Öztürk Engin, 2020; WHO, 2020). The specifications and design of PPE vary on the setting (laboratory or clinical) and rely on local risk assessment. In addition, with PPE design, it is also crucial to provide adequate training to personnel regarding PPE recommendations and possible limitations. Donning and doffing techniques should be standardized since they are crucial for the protection of personnel using the PPE (McCarthy et al., 2020).

One of the gaps that were observed during the pandemic in the Philippines is the lack of knowledge of most institutions regarding fit testing. The need for fit testing was later realized when the fit testing record of personnel handling COVID-19 was included in the list of requirements for licensing.

Another gap is the shortage of the different types of PPE. The Philippines as a developing country does not have the stockpile of PPE (Destura et al., 2021). The shortage is more evident in the provinces where the supply chain is slower. Like other countries, there was not enough supply of respirators and gowns due to delays in the supply chain. This has brought issues and the need for guidelines for the reuse of PPE, especially respirators and gowns (Rimmer, 2020).

## 14.7 DECONTAMINATION

Tons of COVID-19-related biomedical waste were generated daily in developing countries. Manila has been estimated to generate an additional 280 tons of healthcare waste during the pandemic (Tsukiji, Gamaralalage, Pratomo, Onogawa, & Alverson, 2020). The main method of waste decontamination for this type of waste is heat treatment using autoclaves and chemical decontamination. However, even though both methods are effective in deactivating the pathogen, they do not completely destroy the waste and would still need dump sites. The waste treatment is problematic for the

Philippines due to its reliance on dump sites for waste disposal; incineration of waste is prohibited as stated in the 'Philippine Clean Air Act' (Legarda, 1999).

### 14.7.1 Autoclave

The use of autoclave sterilization is the most common method to decontaminate hazardous waste generated from COVID-19 testing in the Philippines. An important aspect of autoclave use for decontamination is to monitor sterilization and perform validation runs. Non-strict compliance with sterilization monitoring can increase the risk of virus leakage during medical waste transfer (Lv et al., 2021).

### 14.7.2 Chemical Disinfectant

Various chemical disinfectants have been proven effective against SARS-CoV-2; the main issue is access to these chemicals during the pandemic. Two types of chemicals can be easily found in homes. First, alcohols, ethanol, and isopropanol have broad-spectrum properties against fungi, bacteria, and viruses. Both these alcohols are capable of destroying coronavirus at a concentration of 70–90% within 30 seconds (Al-Sayah, 2020). Second, chlorine-releasing agents such as bleach can inactivate SARS-CoV-2 at 0.1% sodium hypochlorite within 1 minute (Kampf et al., 2020). Chemical disinfectants vary in efficacy in inactivating SARS-CoV-2; the selection would be important together with obtaining information about the stability and safety for the user. With the use of chemical disinfectants, especially for household bleach, it is recommended to determine the available hypochlorite. In the Philippines, different variants of household bleach contain different concentrations of hypochlorite. Training and documentation of the proper preparation of disinfectant might be trivial but a good practice to avoid false sense of safety when employing these chemicals.

## 14.8 SHIPPING

Shipping of biological samples for SARS-CoV-19 detection happens within laboratories. Reasons for sending samples to other institutions are: biobanking, additional testing for confirmation, alleviate the lag in capacity of one laboratory, and perform downstream whole genome sequencing to determine circulating variants. Recommendations from RITM are like guidelines from the International Air Transport Association (IATA). All biological specimens should be in a triple-packaging system. For air transport samples for diagnostic testing for SARS-CoV-2 infection shall be shipped category B while viral culture and isolates shall be shipped as category A.

## 14.9 PERFORMANCE

A facility will be directly under DOH licensing once it obtains accreditation. With this, the licensing arm of DOH will have access to the information of each laboratory and can perform random audits to determine compliance. It is important to note that biosafety does not end with audits; additional tasks such as internal audits to determine arising gaps in the laboratory can be addressed. Internal policies regarding health monitoring of each personnel working in the laboratory have been practiced to reduce the spread of infection. Retraining of personnel is done using online platforms for biosafety and biosecurity. Proficiency testing is done annually for renewal of license.

## 14.10 BIOSECURITY

The pandemic has revived issues regarding the use of biological materials for malicious intent. One of the greatest security risks is insider threat or individuals who have direct access to samples and have the capability to perform manipulation. There was an increase in the number of personnel who had access to samples during the pandemic and lack of oversight in each laboratory. Enhancing biosecurity pillars such as material control and accountability together with personnel reliability programs would be helpful in preventing issues on insider threat.

Another issue is the rise of misinformation during the time of COVID-19 that created confusion. The use of hydroxychloroquine to treat COVID-19 and later on the use of ivermectin as protection against COVID-19 infection (Alibudbud, 2023; Nelson et al., 2020). The apparent issue on dengue vaccine has also affected the acceptance of the COVID-19 vaccine (Mendoza et., 2021).

Restrictions on data access is one of the issues observed during the pandemic. Patient data are collected and shared with authorities for contact tracing purposes. Aside from the Philippine Data Privacy Act of 2013, there are no other mitigations to protect individual data from being used.

## 14.11 FUTURE PROSPECTS

The COVID-19 pandemic made the whole world realize that no country is fully prepared to prevent and control major infectious disease outbreaks. This was reported by the Global Health Security Index of 2019 wherein it found that countries do not have the necessary preparedness for epidemics or pandemics (Rutjes et al., 2023). The response to such pandemics is more reactive, countries have different rates of developing capacity to address pandemics; countries which have more resources to mobilize, and more established healthcare capacity and technologies are faster than developing countries.

Focusing on the Philippines as one of the developing countries, there are several aspects that can be strengthened to yield improvements in diagnostic response and biorisk management in outbreak settings.

Having a full working framework for the biorisk management system for the whole country would provide the necessary control and oversight on governance of activities related to biorisk management. Previous studies have pointed out the framework can enhance sustainability of biosafety and biosecurity in developing countries (Destura et al., 2021; Heckert et al., 2011). A framework involving the different departments and stakeholders in the Philippines would provide a much faster approach in monitoring outbreaks and cascade mitigations more efficiently.

The shortage of testing laboratories has been one of the gaps in the Philippines that resulted in the slow deployment of testing. An initiative of the DOH, RITM, and WHO post pandemic is to establish subnational laboratories that would expand the country's capacity for diagnostics (WHO, 2022). Currently, there is an establishment of COVID-19 Laboratory Network Quality Assurance Program to ensure compliance of licensed COVID-19 laboratories with regulatory requirements (Baclig, 2021; RITM, 2021). These efforts would ensure that laboratories that can cater to health emergencies are ready and placed in strategic locations in the country.

Infrastructure would provide the facility capacity, but it is still necessary to build human capital. In terms of testing, laboratories should be encouraged to pursue ISO 15189 or 17025. Having certifications drives institutions to continuously improve on processes and invest in human capital. In the case of biosafety and biosecurity, there is an effort by DOH to provide online training to laboratory personnel on biosafety and biosecurity, this effort is to continue expanding knowledge on biorisk management. Biosafety officers will then be recommended for each health laboratory; with enough knowledge and authorized personnel for each institution then improvements can be made moving forward to adopting ISO 35001:2019 which is biorisk management for laboratories and other related organizations.

## REFERENCES

Ağalar, C., & Öztürk Engin, D. (2020). Protective Measures for Covid-19 for Healthcare Providers and Laboratory Personnel. *Turkish Journal of Medical Sciences*, *50*(SI-1), 578–584. https://doi.org/10.3906/sag-2004-132

Al-Sayah, M. H. (2020). Chemical Disinfectants of COVID-19: An Overview. *Journal of Water and Health*, *18*(5), 843–848. https://doi.org/10.2166/wh.2020.108

Alibudbud, R. (2023). A Case of Pharmaceutical Messianism Amidst the COVID-19 Pandemic: An Infodemiological Study of Ivermectin in the Philippines. *Policy, Politics, and Nursing Practice*, *24*(1), 17–25. https://doi.org/10.1177/15271544221139455

Amare, G. (2012). Reviewing the Values of a Standard Operating Procedure. *Ethiopian Journal of Health Sciences*, *22*(3), 205–208. Retrieved from http://www.pubmedcentral.nih.gov/articlerender.fcgi?artid=3511899&tool=pmcentrez&rendertype=abstract%5Cnhttp. www.ncbi.nlm.nih.gov/pubmed/23209355%5Cnhttp://www.pubmedcentral.nih.gov/articlerender.fcgi?artid=PMC3511899

Amit, A. M., Pepito, V. C., & Dayrit, M. (2021). Early Response to COVID-19 in the Philippines. *Western Pacific Surveillance and Response Journal*, *12*(1), 1–5. https://doi.org/10.5365/wpsar

Aster, E. (n.d.). COVID-19 Mobile Diagnostic Labs and Containerized Isolation Medical Centres. Retrieved from http://www.escoglobal.com/product/covid-19-products/covid-19-mobile-diagnostic-labs-and-containerized-isolation-medical-centres/CMDL/

Baclig, M. O. (2021). Biosafety in the Time of Severe Acute Respiratory Syndrome Coronavirus 2 Pandemic: The Philippine Experience. *Applied Biosafety*, *26*(S1), S10–S15. https://doi.org/10.1089/apb.20.0069

Cena-Navarro, R., Vitor, R. J., Canoy, R. J., Dela Tonga, A., Ulanday, G. E., Silva, M. R. C., & Destura, R. V. (2022). Biosafety Capacity Building during the COVID-19 Pandemic: Results, Insights, and Lessons Learned from an Online Approach in the Philippines. *Applied Biosafety*, *27*(1), 42–50. https://doi.org/10.1089/apb.2021.0021

Chu, D. K. W., Pan, Y., Cheng, S. M. S., Hui, K. P. Y., Krishnan, P., Liu, Y., ... Poon, L. L. M. (2020). Molecular Diagnosis of a Novel Coronavirus (2019-nCoV) Causing an Outbreak of Pneumonia. *Clinical Chemistry*, *66*(4), 549–555. https://doi.org/10.1093/clinchem/hvaa029

Corman, V., Landt, O., Kaiser, M., Molenkamp, R., Meijer, A., Chu, D. K., ... Chantal, R. (2020). Detection of 2019 -nCoV by RT-PCR. *Euro Surveill*, *25*(3), 1–8.

Department of Health. (2020). Guidelines in Securing a License to Operate a COVID-19 Testing Laboratory in the Philippines.

Destura, R. V., Lam, H. Y., Navarro, R. C., Lopez, J. C. F., Sales, R. K. P., Gomez, M. I. F. A., ... Ulanday, G. E. (2021). Assessment of the Biosafety and Biosecurity Landscape in the Philippines and the Development of the National Biorisk Management Framework. *Applied Biosafety*, *26*(4), 232–244. https://doi.org/10.1089/apb.20.0070

Destura, R. V., Lam, H. Y., Navarro, R. C., Lopez, J. C. F., Sales, R. K. P., Gomez, M. I. F. A., ... Ulanday, G. E. (2021). Assessment of the Biosafety and Biosecurity Landscape in the Philippines and the Development of the National Biorisk Management Framework. *Applied Biosafety*. https://doi.org/10.1089/apb.20.0070

DOH. (2000). Department order no. 393-E s. 2000 Designation of National Reference Laboratories and Transfer of Corresponding Equipment, Instruments, Supplies, Specimens, and Records from the Bureau f Research and Laboratories to the Designated National Reference Labora.

Flores, R., & Asuncion, X. V. (2020). Toward an Improved Risk/Crisis Communication in This Time of COVID-19 Pandemic; a Baseline Study for Philippine Local Government Units. *Journal of Science Communication*, *2507*(February), 1–9.

Healthcare Asia. (2021). *GMT Manila Defies Odds with Containerized RT-PCR Labs, Wins COVID Management Initiative of the Year from Healthcare*. Asia. Retrieved from https://tinyurl.com/32tbmn7x

Heckert, R., Reed, J. C., Gmuender, F. K., Ellis, M., & Tonui, W. (2011). International Biosafety and Biosecurity Challenges: Suggestions for Developing Sustainable Capacity in Low-Resource Countries. *Applied Biosafety*, *16*(4), 223–230. https://doi.org/10.1177/153567601101600404

Kampf, G., Todt, D., Pfaender, S., & Steinmann, E. (2020). Persistence of Coronaviruses on Inanimate Surfaces and Their Inactivation with Biocidal Agents. *Journal of Hospital Infection*, *104*(3), 246–251. https://doi.org/10.1016/j.jhin.2020.01.022

Kaufer, A. M., Theis, T., Lau, K. A., Gray, J. L., & Rawlinson, W. D. (2020). Laboratory Biosafety Measures Involving SARS-CoV-2 and the Classification as a Risk Group 3 Biological Agent. *Pathology*, *52*(7), 790–795. https://doi.org/10.1016/j.pathol.2020.09.006

Legarda, L. (1999). Republic Act No 8749.

Lu, R., Zhao, X., Li, J., Niu, P., Yang, B., Wu, H., ... Tan, W. (2020). Genomic Characterisation and Epidemiology of 2019 Novel Coronavirus: Implications for Virus Origins and Receptor Binding. *The Lancet*, *395*(10224), 565–574. https://doi.org/10.1016/S0140-6736(20)30251-8

Lv, J., Yang, J., Xue, J., Zhu, P., Liu, L., & Li, S. (2021). Investigation of Potential Safety Hazards during Medical Waste Disposal in SARS-CoV-2 Testing Laboratory. *Environmental Science and Pollution Research*, *28*(27), 35822–35829. https://doi.org/10.1007/s11356-021-13247-4

McCarthy, R., Gino, B., d'Entremont, P., Barari, A., & Renouf, T. S. (2020). The Importance of Personal Protective Equipment Design and Donning and Doffing Technique in Mitigating Infectious Disease Spread: A Technical Report. *Cureus*, *12*(Cdc). https://doi.org/10.7759/cureus.12084

Mendoza, R. U., Dayrit, M. M., Alfonso, C. R., & Ong, M. M. A. (2021). Public Trust and the COVID-19 Vaccination Campaign: Lessons from the Philippines as It Emerges from the Dengvaxia Controversy. *International Journal of Health Planning and Management*, *36*(6), 2048–2055. https://doi.org/10.1002/hpm.3297

Nelson, T., Kagan, N., Critchlow, C., Hillard, A., & Hsu, A. (2020). The Danger of Misinformation in the COVID-19 Crisis. *Missouri Medicine*, *117*(6), 510–512.

Rimmer, A. (2020). Covid-19: Experts Question Guidance to Reuse PPE. *BMJ (Clinical Research Ed.)*, *369*(April), m1577. https://doi.org/10.1136/bmj.m1577

RITM. (2021). COVID-19 Laboratory Nework Quality Assurance Program. Retrieved from https://www.ritmmbleqap.com/

RITM. (2023). *2020–2021 COVID-19 RITM Achievement Report*.https://ritm.gov.ph/wp-content/uploads/2023/02/2020.

Rutjes, S. A., Vennis, I. M., Wagner, E., Maisaia, V., & Peintner, L. (2023). Biosafety and Biosecurity Challenges during the COVID-19 Pandemic and Beyond. *Frontiers in Bioengineering and Biotechnology*, *11*(March), 1–8. https://doi.org/10.3389/fbioe.2023.1117316

Schröder, I. (2020). COVID-19: A Risk Assessment Perspective. *ACS Chemical Health & Safety*, *27*(3), 160–169. https://doi.org/10.1021/acs.chas.0c00035

Tsukiji, M., Gamaralalage, P. J., Pratomo, I., Onogawa, K., & Alverson, K. (2020). *Waste Management During the COVID-19 Pandemic from Response to Recovery*. https://doi.org/10.1016/B978-0-323-95998-8.00008-X

van Doremalen, N., Bushmaker, T., Morris, D., Holbrook, M., Gamble, A., Williamson, B. ., … Gerber, S. (2020). Aerosol and Surface Stability of SARS-CoV-2 as Compared with SARS-CoV-1. *The New England Journal of Medicine*, 382(16),0–3.

WHO. (2020). Laboratory Biosafety Guidance Related to COVID-19. Who. Retrieved from https://www.who.int/docs/default-source/coronaviruse/laboratory-biosafety-novel-coronavirus-version-1-1.pdf?sfvrsn=912a9847_2

WHO. (2022). DOH, RITM, WHO Establihsh Subnational Laboratories to Expand the Country's Capacity in Detecting Vaccine-Preventable Diseases. Retrieved from https://www.who.int/philippines/news/detail/20-01-2022-doh-ritm-who-establish-subnational-laboratories-to-expand-the-country-s-capacity-in-detecting-vaccine-preventable-diseases#:~:text=Lingad Memorial Regional Hospital%2C Western,Cotabato Regional and Me

WHO. (2023). WHO Coronavirus (COVID-19) Dashboard. Retrieved from https://covid19.who.int/?mapFilter=deaths

World Health Organization. (2020). *Laboratory Biosafety Manual 4th ed. Monograph: Biological Safety Cabinets and Other Primary Containment Devices*. Retrieved from https://apps.who.int/iris/bitstream/handle/10665/337957/9789240011335-eng.pdf?sequence=1&isAllowed=y

World Health Organization (WHO). (2020). Technical Specifications of Personal Protective Equipment for COVID-19. *World Health Organization* (V), 1–40. Retrieved from https://www.who.int/publications/i/item/WHO-2019-nCoV-PPE_specifications-2020.1

# 15 Engineering Designs for Biological Laboratories

*Allan Fellizar*
Mariano Marcos Memorial Hospital and
Medical Center Molecular Biology Laboratory
City of Batac, Ilocos Norte

*Bianca Mae Adalem*
Independent Researcher
Metro, Manila, Philippines

*Rodel Jonathan S. Vitor II*
National Training Center for Biosafety and Biosecurity
National Institutes of Health
University of the Philippines
Manila, Philippines
Department of Biology, College of Science
De La Salle University,
Manila, Philippines

*Anna Gibson*
Scilore LLC.
Muntinlupa City
Philippines

*Rohani Cena-Navarro*
National Institutes of Health
University of the Philippines
Manila, Philippines

## 15.1 LABORATORY DESIGN CONSIDERATIONS

Designing laboratories to balance the needs of the end-users and following standards and local laws has been challenging to most, but possible. By adopting a collaborative and informed approach, designers can balance the end-user's needs and preferences while ensuring the laboratory design meets or exceeds the necessary standards and regulations and being mindful of the cost. Regular communication, biorisk, and biosecurity assessment, and expert advice play crucial roles in achieving this balance.

 DOI: 10.1201/9781003426219-15

### 15.1.1 Understanding End-User's Requirements

In any construction project, it is essential to identify and comprehend the end-user's requirements. However, if these needs are poorly communicated, it can result in over-engineered and inefficient laboratories being designed. To avoid this, effective communication is key.

During the design process, end-users must comprehensively understand their laboratory processes, the potential hazards associated with each step, and the assessment of risks and costs for possible mitigation strategies. End-users can begin by listing the laboratory services needed, identifying the flow of biological materials or agents within the laboratory, and reviewing the overall process flow. Subsequently, these services can be categorized into zones, such as the Non-laboratory zone, Laboratory zone, and Laboratory service zone, with further refinement and definition in collaboration with the designers.

It is important to note that zoning these services does not initially require a detailed architectural drawing from the end-users. It can be illustrated using bubble diagrams or a flow chart to represent the various zones and their relationships.

By ensuring effective communication and collaboration between end-users and designers, the design of the laboratory can be tailored to meet specific requirements, optimizing functionality and efficiency while minimizing unnecessary complexities.

### 15.1.2 Zoning

Zoning refers to organizing and partitioning areas within a space according to their intended functions. Traditionally, this involves constructing numerous interior walls to establish rooms with specific purposes. With the increasing demand for adaptable laboratories, it is advisable to minimize the presence of interior walls. Using risk assessment techniques helps determine the necessity of interior walls in laboratories.

Non-laboratory zones primarily serve activities with little to no risk, such as office space or secure agent reception areas. These zones may have less stringent design considerations, such as tiled flooring or minimum HVAC requirements, and are typically accessible.

Depending on the required services and associated hazards, laboratory service zones may adopt similar design considerations but with restricted access. Non-laboratory zones can encompass areas such as the laboratory's electrical and HVAC room, autoclaved waste collection area, and laboratory-side fire exit.

Laboratory zones encompass all the necessary services the laboratory requires, including storage areas, workspaces, autoclaves, and more. Each requirement within this zone may entail distinct design considerations and risk assessments tailored to meet specific needs and mitigate potential hazards.

End-users are advised to communicate their priorities to designers, emphasizing the importance of maintaining a clean-to-dirty flow and imposing appropriate restrictions on each zone. For instance, it is crucial to avoid placing an office or meeting area immediately after a cold room or in close proximity to laboratory benches. Additionally, locating the laboratory service zone in a restricted area is recommended. By correctly identifying and delineating zones for each requirement, end-users can effectively monitor updates in the design process, ensure adherence to their desired laboratory flow, and maintain awareness of hazards associated with each zone. Taking these precautions during the design phase is more cost-effective than implementing mitigations after construction has concluded.

After identifying the requirements and zoning, it is advisable for the end-user to carefully plan the flow of the laboratory operations, starting from receiving agents or samples to processing, storage, and, ultimately, disposal. During this process, conducting a comprehensive risk assessment for each step is crucial while simultaneously updating the equipment list for the designer. Some important questions to consider are:

1. Receiving Biological Sample/Agent: Is it necessary to open the biological sample/agent in a Biosafety Cabinet (BSC), on a benchtop, or within an adequately isolated room when receiving it?
2. Specific Processes: Should certain processes be conducted in an enclosed and isolated area, or is an open bench acceptable?
3. Aerosolization/Airborne Risk: Could the operation cause any materials to become aerosolized or airborne, posing a risk?
4. Equipment Requirements: Does the equipment need to be connected to an Uninterruptible Power Supply (UPS)? Are there any vibration restrictions or specific requirements like filtered water? Should accessibility to the equipment be maintained at all times?
5. Process Exemptions: Are there specific processes that require exemptions from other design considerations due to their nature? For example, allowing floor drains, implementing positive air pressure, or incorporating a dry fire protection system?

By addressing these questions and incorporating the findings into the design process, end-users can ensure that the laboratory layout effectively supports their operational needs while minimizing risks and meeting safety requirements.

Additionally, when choosing a BSC, it's crucial to consider the specific tasks it will be used for. Table 15.1 serves as an illustrative guide for end-users, allowing them to choose the appropriate class of BSCs based on biosafety levels. This resource

**TABLE 15.1**
**BSC Selection through Risk Assessment**

| Risk assessed | Protection Provided | | | |
|---|---|---|---|---|
| **Biosafety Level** | **Personnel** | **Product** | **Environmental** | **BSC Class** |
| 1 to 3 | Yes | No | Yes | I |
| 1 to 3 | Yes | Yes | Yes | II (A1, A2, B1, B2, C1) |
| 4 | Yes | Yes | Yes | III |
| | | | | II (when used in a room with suit or canopy hood) |

proves particularly valuable for laboratories that prioritize adherence to the recommendations outlined by the Centers for Disease Control and Prevention (CDC)—Biosafety in Microbiological and Biomedical Laboratories (BMBL). Connecting with laboratory equipment suppliers is another avenue to tailor the selection to the laboratory's requirements. It's advisable to track reputable companies, such as NSF, that certify laboratory equipment to guarantee quality. On its website, NSF offers an official listing of Biosafety cabinet providers, including products and services.

A sample of laying out the flow of the laboratory processes has been provided below. This can have more detail as end-users continue the design process with their designers.

#### 15.1.2.1 Sample 1: Laboratory Process 1

1. **Room 1**
   Brief description of what processes will be done and sample room requirements
   a. **Step 1**
      i. Equipment needed
         1. Equipment 1
            a. Equipment measurements (L x W x H) or (D x H) and required clearances at the sides, back, front, bottom, and top
            b. Equipment requirements (e.g., electrical requirements from the designer, water requirements, drain requirements, temperature requirements, vibration restrictions, space/clearance restrictions, table top or floor mounted)
            c. Redundancy requirements (e.g., UPS, generator, back-up equipment)
         2. Equipment 2 *[Follow data input in **Equipment 1**]*
         3. *Equipment 3 [Add necessary equipment needed for **Step 1**]*
      ii. Biorisk Assessment
         1. Hazard identification
         2. Potential Mitigation
      iii. Biosecurity requirements
         1. Restriction requirements (e.g., full enclosure, visibility at all times, restricted accessibility)
   b. Step 2 *[Follow data input in **Step 1**]*
      i. Equipment needed *[Follow data input in **Step 1**]*
      ii. Biorisk assessment *[Follow data input in **Step 1**]*
      iii. Biosecurity requirements *[Follow data input in **Step 1**]*
   c. Step 3 *[Add necessary steps to be conducted in **Room 1**]*
   d. Notes
      i. Decontamination strategies
      ii. Maintenance strategies
      iii. Others *[Add necessary notes for **Room 1**]*
2. Room 2 *[Follow data input in **Room 1**]*
3. Room 3 *[Add necessary rooms needed to conduct **Laboratory process 1**]*

Room requirements can be taken from local and international standards and should be provided by designers. However, end-users can also suggest which standards to follow if they require a more stringent design, particularly for specific accreditations. Equipment measurements are vital not only for determining the necessary casework but also for determining the dimensions of the laboratory module. Discussing potential mitigation solutions with the designers during the biorisk assessment is also advised. This enables end-users to evaluate whether the identified risks are acceptable with the recommended engineering controls, considering the associated costs.

When designing the layout of laboratory services, it is crucial to consider the maintenance strategy for the facility. This aspect becomes even more critical for containment labs, where gaseous decontamination methods may be necessary before conducting any maintenance activities. Additionally, decontamination of the laboratories might be required during program changes or as part of the institution's biological safety program or certification processes. Many institutions opt to decontaminate their labs periodically, such as yearly, to ensure compliance with safety standards and certification requirements.

### 15.1.3 Capacity Management

End-users are advised to determine whether their laboratory should operate at 100% of the target capacity from the start of the building's life or at a lower initial capacity with plans for gradual expansion in the future. This decision requires consideration of the equipment that will be introduced over time and recommendations from the designer regarding rooms that could be converted for expansion purposes.

Planning for future expansion can present challenges, as initial design checks may need to be more comprehensive for future needs. Therefore, involving technical end-users who can thoroughly review the expansion plans is crucial. This includes ensuring that the future expansion can accommodate necessary extensions of HVAC systems, electrical lines, plumbing, and other essential infrastructure.

By carefully considering the desired capacity and involving experts in the expansion planning process, end-users can make informed decisions about the initial layout and anticipate future needs, ensuring the laboratory's ability to adapt and grow efficiently while minimizing disruptions and costly modifications.

### 15.1.4 Redundancy Planning

Redundancy planning for laboratories involves implementing backup systems, equipment, and processes to ensure continuity of operations and minimize downtime in the event of failures or disruptions. Here are some key engineering aspects of redundancy planning in laboratories:

1. Redundant Equipment: Laboratories often utilize redundant equipment, which involves having duplicate or backup laboratory and facility instruments, appliances, and critical machinery available. This redundancy strategy is called N+1, where N represents the required number of elements.

For instance, if a space requires two air-conditioning units, providing three units would offer redundancy to account for equipment failures or maintenance shutdowns. In this scenario, one unit may remain on standby while the others operate at reduced capacity, ensuring continuous functionality of the laboratory.

2. Power Redundancy: Laboratories often require a stable power supply. Redundancy planning involves implementing backup power sources, such as uninterruptible power supply (UPS) systems or generators, to provide electricity in case of power outages.
3. Data Redundancy: Laboratories generate and rely on large amounts of data. Redundancy planning involves implementing on-site and off-site data backup and recovery systems to prevent data loss and enable swift recovery in the event of system failures or disasters.

By incorporating redundancy planning into laboratory operations, organizations can enhance reliability, minimize disruptions, and safeguard against potential risks, thereby maintaining the integrity and efficiency of laboratory processes.

### 15.1.5 Construction and Maintenance Costs

Lowering engineering control costs for laboratories in resource-limited countries requires careful planning, efficient use of resources, and strategic decision-making. Here are some strategies that can help reduce construction and maintenance costs:

1. Simplify Design: Opt for a simplified laboratory design that focuses on essential functionality and meets the minimum requirements for safety and functionality. Avoid unnecessary architectural complexities and intricate features that can drive up construction costs.
2. Right-Sizing: Determine the appropriate size of the laboratory based on the intended scope of work and available resources. Avoid oversizing the facility, as larger spaces require more materials and incur higher construction costs.
3. Value Engineering: Engage in value engineering to identify cost-saving opportunities without compromising safety or functionality. This involves assessing alternative materials, equipment, and construction techniques that offer comparable performance at a lower cost. This may mean using different engineering controls for each zone. For example, non-laboratory and laboratory service zones may opt for split-type air conditioning units while the laboratory zone uses centralized air conditioning units.
4. Local Materials and Suppliers: Utilize locally available materials and engage local suppliers to reduce transportation costs and take advantage of competitive pricing.
5. Prioritize Critical Areas: Allocate resources to critical areas directly impacting safety and functionality, such as containment zones or specialized equipment areas. Prioritize these areas during the design and construction

phases to ensure they meet necessary standards while optimizing resource allocation.

6. Phased Approach: Consider adopting a phased construction approach where the laboratory is built in stages or modules. This allows for incremental investment and flexibility to adapt to evolving needs and available resources.
7. Project Management: Employ effective project management practices to ensure efficient use of resources, timely decision-making, and cost control throughout the construction process.
8. Training and Capacity Building: Invest in training and capacity-building programs to enhance local laboratory construction and maintenance expertise. This can help reduce reliance on external consultants or contractors and drive down costs in the long run.

Combining these strategies and adapting them to the specific context of resource-limited countries makes it possible to lower construction costs for laboratories while maintaining necessary standards for safety and functionality.

Laboratory construction expenses can vary significantly, influenced by location, local norms, and the specific services offered. In the Philippines, and potentially in other Low- to Middle-Income Countries (LMIC), challenges arising from the location, such as challenging terrain or limited expandability due to congestion in high-density cities like Manila, may necessitate costly structural expansions or complete site demolition and reconstruction in some instances. The construction costs are also impacted by the range of services provided by the laboratory. In some LMICs, where resources may be concentrated in central cities, other cities requiring clinical, agricultural, or educational laboratories may need help accessing necessary operational services. For instance, waste disposal may be infrequent, necessitating more extensive waste storage, and lower water quality may require on-site water filtration.

Considering local practices, areas in the Philippines prone to floods, storms, and earthquakes may have stricter construction requirements, increasing costs compared to less disaster-affected regions. It is crucial to note that integrating more engineering controls into the design generally leads to higher construction and maintenance expenses. Therefore, adopting a strategic approach focusing on essential engineering controls is key to ensuring safety while managing costs effectively.

Additionally, it is crucial to consider that a larger quantity of equipment will necessitate a more extensive maintenance and replacement plan. By providing the designers with a comprehensive equipment list, similar to ***Sample 1: Laboratory Process 1***, end-users can proactively determine their operation and maintenance schedules and associated costs, leading to better budgeting and planning. Provided in Table 15.2 is a sample of Operations and Maintenance schedule cost table. List is not exhaustive and can be expanded to fit the laboratory needs.

#### 15.1.5.1 Local and International Standards

Tables 15.3, 15.4, and 15.5 are some of the current local and international regulations and Philippine standards that can be valuable for end-users and laboratory designers

**TABLE 15.2**
**Operations and Maintenance (O&M) Schedule and Cost**

| Room O&M | Person-in-charge | Shut Down Time | Frequency | Cost | Notes |
|---|---|---|---|---|---|
| Laboratory process 1 | | | | | |
| Room 1 | | | | | |
| Change portable fire extinguishers | 3rd-party supplier | None | 1 year | ₱1,200.00 | Visually inspect extinguishers monthly |
| Repainting of walls | 3rd party | Days | 5 years | ₱200.00 /sqm | May vary from 5 to 7 years depending on visual inspection |
| *[add necessary O&M for **Room 1**]* | *[Follow data input in **Room 1**]* | | | | |
| Equipment 1 | | | | | |
| Replace lid gaskets | Laboratory personnel 1 | 10 mins | 6 months | ₱6,000.00 | |
| *[add necessary O&M for **Equipment 1**]* | *[Follow data input in **Equipment 1**]* | | | | |
| Equipment 2 | | | | | |
| Annual Calibration | 3rd party | 1 hour | 1 year | ₱3,500.00 | |
| *[add necessary O&M for **Equipment 2**]* | *[Follow data input in **Equipment 1**]* | | | | |
| Biosafety Cabinet | | | | | |
| Daily Maintenance | Laboratory personnel 1 | 5 mins | Every use | Varies *[₱500.00 to ₱1,500.00]* | Visual inspections and general cleaning |
| Decontamination of work area | Laboratory personnel 2 | 10 mins | 1 month | Varies *[₱2,000.00 to ₱5,000.00]* | Decontamination of work area |
| Annual Calibration [Downflow velocity, inflow velocity, airflow smoke pattern, HEPA filter leak, Pressure Decay/Soap Bubble] | Authorized Calibration Engineer - based on country of origin of BSC | 4 hours | 1 year | ₱100,000.00 | Note: Tests for light intensity, vibration, and noise levels vary if required by the laboratory. These tests may not be necessary for some laboratories. |

*(Continued)*

**TABLE 15.2 (CONTINUED)**
**Operations and Maintenance (O&M) Schedule and Cost**

| Room O&M | Person-in-charge | Shut Down Time | Frequency | Cost | Notes |
|---|---|---|---|---|---|
| *[add necessary O&M for* ***Biosafety cabinets****]* | *[Follow data input in* ***Equipment 1****]* | | | | |
| *Room 2* | | | | | |
| *[add necessary rooms for* ***Laboratory process 1****]* | | | | | |
| *[add necessary O&M for* ***Room 1****]* | | | | | |
| *[Follow data input in* ***Room 1****]* | | | | | |
| *Laboratory process 2 [Add necessary* ***Laboratory processes****]* | | | | | |

during the design process. To obtain a thorough compilation, it is recommended to consult with local engineers and legal experts. Local experts are able to provide comprehensive cross-referencing among standards tailored to the specific requirements of the laboratory.

Please note that this list is not exhaustive, and regulations may vary by location and type of laboratory. Laboratories have the option to refer to and integrate guidelines from any applicable standard, taking into account their unique requirements, geographic location, and regulatory context. Consulting with local experts, engineers, and legal professionals is essential to ensure compliance with all relevant regulations and standards.

## 15.2 Risk-Based and Need Assessment Approach to Engineering Controls

The capacity to initiate scientific investigations to respond to an occurrence or imminent threat of an illness or health condition requires suitable facilities. This component of biocontainment sufficiently prepares for as yet unidentified or unrecognized emerging infectious diseases. At the onset of 2020, the COVID-19 pandemic overwhelmed the world and provided the ultimate test to date, of the global biorisk management system. The WHO's strategic preparedness and response plan requires strengthening the diagnostic capacity for SARS-CoV-19 testing in a containment facility, where infection control and prevention and biosafety are embedded in the key performance indicators to monitor implementation of the plan (Rutjes et al., 2023). Building or repurposing a laboratory that is highly reliant on engineering controls and technology presents a major challenge in many resource-limited countries where construction and maintenance costs are restrictive. There is a frequent practice to implement containment solutions which are benchmarked with the 'western standard' in vulnerable areas with high demand for laboratory activity due to disease outbreak and or a continuous high prevalence (Dickmann et al., 2015). This is not always effective since, aside from complex costs, there is a lack of well-trained biocontainment facility engineers, readily available and reliable suppliers, maintenance provision, and other logistic measures. Such standards are depicting a high-resource setting perspective that is relatively distant in different parts of the world and neither achievable nor sustainable in low-resource settings (Hendrickson, 2019; Wheatley, 2018). Rather than depending on a high technology approach, a risk-based and need assessment approach to engineering control is proposed and this strategy shifts from the use of predetermined standards focusing on conditions of biosafety and biosecurity that is cost-effective, practical, locally driven, and sustainable.

### 15.2.1 General Considerations Prior to Conduct of Risk Assessment

The risk-based and need assessment approach builds on a contextual assessment of risks and considers the system and environment in which laboratory capacity is being built (Bouchaut & Asveld, 2020). Cognizant of these risks from conceptualization, design, and construction is important to ensure effective biocontainment. The objective of risk assessment is to identify hazards or risks inherent in

**TABLE 15.3**
**Required Local Approvals**

| Laboratory Type | Agency | Standards or Permits |
|---|---|---|
| Animal Laboratories | Department of Agriculture (DA) | DA AO 40 |
| | Office of the Building Official (OBO) | Presidential Decree (PD) 1096 (i.e. group J Division 1) |
| Chemical Laboratories | Professional Regulation Commission (PRC) | Chemistry Profession Act (Republic Act (RA) 10657) |
| | Philippine National Police (PNP) | Firearm and Explosives Law (RA 9516) |
| | Philippine Drug Enforcement Agency (PDEA) | |
| | Office of the Building Official (OBO) | PD 1096 (i.e., group G) |
| Clinical Laboratories | Department of Environment and Natural Resources–Environmental Management Bureau (DENR-EMB) | Priority Chemicals List<br>Permit to Operate<br>Hazardous Waste Generator Registration<br>Wastewater Discharge Permit |
| | Department of Health (DOH) | DOH Administrative Order (AO) No. 2021-0037<br>DOH AO No. 2008-0007 |
| | Office of the Building Official (OBO) | PD 1096 (i.e., group D Division 2) |
| Educational or School Laboratories | Department of Education (DepED) | DepED Order No. 48 |
| | Office of the Building Official (OBO) | PD 1096 |
| Specialized Laboratory—Soils | DA-Bureau of Soils and Water Management (BSWM) | PD 1435 |

laboratories that may cause harm to laboratory workers, community, and the environment (Burzoni et al., 2020). When risks have been recognized and evaluated, appropriate measures to minimize them can be implemented. This is known as risk mitigation and can include building design features, good microbiological and laboratory practices, personal protective equipment, protocols, and staff training programs, among others. The following may be considered prior to the conduct of risk assessment:

1. The foundation of a good risk assessment is based on the principles of clarity, reliability, transparency, robustness, and rationality, keeping them in mind all throughout the process (Patterson et al., 2014). It should be

**TABLE 15.4**
**Other Local Standards to Be Considered for Any Infrastructure (Not Limited to Laboratories)**

| Agency | Standards |
|---|---|
| Department of Environment and Natural Resources (DENR) | DENR AO No. 2003-0039 |
| Department of Labor and Employment (DOLE) | Occupational Safety and Health Standards<br>• Certain tables in the standard may require cross-referencing with alternative standards to determine the more rigorous set of criteria, such as Table 8 in the Department of Labor and Employment Occupational safety and health standards |
| Local Government Units | Design guidelines |
| National government, its departments, bureaus, offices, and agencies, including state universities and colleges, government-owned and/or controlled corporations, government financial institutions, and local government units | Government Procurement Reform Act (RA 9184) |
| Integrated National Police | PD 1185 |
| Department of Health | PD 856 |
| Department of Public Works and Highways | PD 1096 |
| Department of Transportation and Communications | Batas Pambansa (BP) 344 |

planned and steered to address the central question of what could go wrong and what would be the consequences of such occurrences. Risk analysis must be conducted using established guidelines and approaches that must be able to endure rigors of scientific review.

2. The highlights of the building and its systems influence planning the risk assessment. The process should commence with a comprehensive facility analysis that examines the facility's design and security features, building system, the site and surrounding environment, and community (Patterson et al., 2014). Laboratory operations and activities that are expected to be performed are vital in the scope of risk assessment.
3. Protocols, project plans, and equipment specifications are equally essential and may serve as a basis for defining facility and procedure hazards. Assumptions may be formulated if facility design and operations are not yet established and then improved during the risk assessment process as more information is generated (van Gelder et al., 2021).

## TABLE 15.5
## International Standards to Be Considered

| Agency | Standard | Description |
|---|---|---|
| World Health Organization (WHO) | Laboratory Biosafety Manual (LBM) | The WHO LBM is a globally applicable guide that addresses a wide range of biosafety issues in laboratories.<br>It covers principles, risk assessment, containment measures, and emergency response, with a specific focus on handling pathogens with global health emergency potential.<br>Integrating the WHO LBM during the design phase enables end-users and designers to proactively assess and address biosafety and biosecurity concerns, thereby avoiding costly mitigation measures during or after construction.<br>This approach enhances safety and contributes to cost reduction. |
| Centers for Disease Control and Prevention (CDC) | Biosafety in Microbiological and Biomedical Laboratories (BMBL) | The CDC recommends using the BMBL as a valuable tool for assessing and mitigating risks in biomedical and clinical laboratories.<br>The BMBL emphasizes classifying laboratories into biosafety levels (BSL) and offers detailed guidance on practices, containment, and facility design based on these levels.<br>Like the WHO LBM, the CDC BMBL can be applied during the design phase to improve safety measures and achieve cost reductions. |
| The National Institutes of Health | Design Requirements Manual (DRM) | The DRM serves as a valuable reference for architects, engineers, and contractors, providing essential guidance to ensure that research and laboratory facilities meet the required standards for functionality and safety.<br>It offers detailed guidelines across various areas such as architecture, structure, mechanics, electrical systems, and safety considerations. |
| American National Standards Institute and American Society of Safety Professionals. | Laboratory Ventilation | ANSI/ASSP Z9.5-2022 is a standard that sets forth minimum requirements and best practices for laboratory ventilation systems, with a focus on protecting personnel from physical harm and overexposure to airborne contaminants.<br>Applicable to most laboratories, the standard addresses various aspects, including a laboratory ventilation management plan, fume hoods, exposure control devices, ventilation system design, commissioning and performance testing, training, exhaust system maintenance, and air cleaning.<br>The goal is to ensure the safety of individuals and safeguard property by providing guidelines for the effective management and operation of laboratory ventilation systems. |

(*Continued*)

**TABLE 15.5 (CONTINUED)**
**International Standards to Be Considered**

| Agency | Standard | Description |
|---|---|---|
| American Society of Heating, Refrigerating and Air-Conditioning Engineers (ASHRAE) | Laboratory Design Guide: Planning and Operation of Laboratory HVAC Systems | Focused on heating, ventilation, air conditioning, and refrigeration (HVAC&R) systems, the guide addresses key considerations in laboratory design to ensure optimal functionality, safety, and efficiency.<br>It covers various aspects, including air quality, temperature control, energy efficiency, and sustainability.<br>ASHRAE's Laboratory Design Guide serves as a valuable reference for architects, engineers, and professionals involved in the planning and development of laboratory facilities, offering best practices and industry standards for creating spaces that meet rigorous technical and environmental requirements. |

### 15.2.2 Pathogen-Based Risk Assessment and Biocontainment

The pathogen under investigation is one, if not, the most critical factors for a biocontainment laboratory risk assessment. The range of pathogens should be exhaustive enough to characterize the risk associated with laboratory operations. The list should also include agents of public health concern as well as consider the recommendations of expert scientific advisors to the project. It appears to have a worldwide consensus on the four-risk group classification system (Belgian Biosafety Server, n.d.). Extensive study of the majority of classifications of biological agents performed by national committees of experts shows uniform results. The risk group to which an infectious agent or toxin is classified is the primary, but not only, consideration used in a biological risk assessment to determine the appropriate biosafety level in which a laboratory worker can handle or manipulate the infectious agent or toxin. Depending on the country or an organization, this classification system considered the main factors related to:

1. pathogenicity of the organism
2. mode of transmission and host range
3. availability of effective preventive measures (e.g., vaccines)
4. availability of effective treatment (e.g., antimicrobial agents)
5. other important characteristics of the pathogen to consider are stability, infectious dose, transmissibility, case fatality rate, infectivity period, and endemicity
6. laboratory workforce competency, highly susceptible individuals or hosts, high-risk laboratory procedures (aerosolization, use of sharps, high sample volumes)
7. biosecurity issues related to a weaponizable pathogen

However, some disagreements occur between and even within independent states to assign agents to one risk group. One of the problems arises from the geographical and climatic distribution of the microorganisms, their reservoir, and their vectors, particularly when animal or plant pathogens are of concern. Political and economic situations are also to be factored in. Due to variability of pathogen characteristics, the work conditions (diagnostics, research, or production), host models and systems, and other aspects, it is suggested that these risk group tables should be used only as a guide for comparison of the relative risk levels of the agent.

When risk group classification is not existing in a country, the World Health Organization (WHO) recommends to formulate its own classification by risk group of the agents encountered in that country, in view of the factors mentioned above and based on its local risk assessment (Belgian Biosafety Server, n.d.). Periodic updates are vital when risk groups are developed so that new evidence that supports changes to the risk profile of a pathogen is readily incorporated into a risk group reassignment. Oversight mechanisms should consider the studies being conducted in facilities and how they contribute to or reduce risk and be addressed accordingly. Along with pathogen risk grouping, the Pathogen Data Safety Sheets (PSDS) are comprehensive resources in performing risk assessment. These outline the hazardous properties of pathogens and provide references for work involving these agents in a laboratory setting. These are educational and information resources produced by the Public Health Agency of Canada for laboratory personnel working with human pathogens. When experiments involve reconstructed, engineered, or modified pathogens, risk assessment is done by comparing the novel pathogen to the wild type or previously assessed variant, relating the several modifications to predicted effects on various risk factors (e.g., pathogenicity, transmissibility).

Handling microorganisms of a particular risk group typically involves working at the accompanying biosafety level. For instance, experiments or sample testing may generate high-concentration aerosols, requiring a higher degree of safety. Biosafety levels (BSLs) prescribe procedures and levels of containment for a specific microorganism. BSLs are ranked from 1 to 4, with BSL-1 being suitable for working with the lowest/least harmful biological agents. Each biosafety level comprises specific requirements for a laboratory with emphasis on facility construction, safety equipment, and laboratory practices. As each level advances, it includes additional biosafety considerations from the preceding level. Table 15.6 shows a summary of the four biosafety levels with their corresponding minimum engineering requirements. Biosafety levels correlate with but do not equate to agent risk groups. The assumption that risk groups and biosafety levels are equated can derail risk assessment, thus missing the prospects to optimize the means of risk mitigation proportionate to the assessed risk. Consequently, a risk assessment will determine the degree of correlation between an agent's risk group classification and biosafety level.

The revisions in the 4th edition of the WHO's Laboratory Biosafety Manual (LBM) provided an insight that is crucial in the need for understanding the evolving pandemic dynamics (e.g., COVID-19) and information and developing new best practices. It focuses on a technology-neutral and cost-effective approach to biosafety, ensuring that laboratory infrastructure, facilities, safety equipment, and work practices are locally relevant, proportionate, and sustainable. Instead of the conventional biosafety levels, core requirements are proposed in the revised LBM which is a combination of essential biosafety elements to be applied and a minimum requirement for safe work during laboratory testing. These include not only the laboratory facility and equipment but also safety practices or procedures and adequately trained and competent laboratory workers which frequently receive less attention (Laboratory Biosafety Manual 4th Edition and Associated Monographs. Geneva: World Health Organization; 2020). Where the local risk assessment requires heightened control measures such as the use of biological safety cabinets, extra personal protective equipment, and infrequently, maximum containment would be used to mitigate the identified risks of exposure to the biological agents under study. The intention is to be more sustainable rather than the discrete classification of BSL. Consequently, it aims to select the applicable set of safety measures to mitigate the identified risks and not rely on predetermined standards that may be an overdesign, depending on the pathogen or activities proposed (Kojima et al., 2018). This paradigm shift will allow more resilience in laboratory design and directing on human capacity building rather than concentrating on the pathogen risk group and biosafety levels as foundation for laboratory biocontainment and risk assessment. An analysis of recent laboratory-acquired infections (LAI) showed that most were caused not by faulty engineering control or equipment failure but rather by human factors such as disregard for insufficient risk assessment, improper use of PPEs, lack of standard operating procedures, and inadequately trained laboratory worker (Kimman et al., 2008). The best-designed and most engineered laboratory is only as good as its least competent worker. Biosafety levels also support the principle of biosecurity. They share a common objective and are complementary, that is, to keep biological materials safe and secured inside the areas where they are used and stored. It is prudent that biosafety and biosecurity in containment levels must be addressed together through a single biorisk management program in compliance with regulatory frameworks and local requirements right from the project conceptualization.

### 15.2.3 Process-Based Risk Assessment

Before the inception stage of building or repurposing a laboratory, it is imperative to identify the processes that will be performed, specifically pathogen manipulation (propagative or non-propagative). The analysis of exposure in the real working condition can be listed and selecting the level of each exposure element (Burzoni et al., 2020). A job safety analysis tool may be used to identify how to perform tasks step by step, any hazards associated with the tasks, and controls to mitigate the hazards. Applying the job safety analysis will provide a process for analyzing work activities that will identify engineering controls, equipment, and personal protective equipment that are necessary to develop work instructions for accomplishing the task. High-risk procedures such as sample mixing, pipetting, centrifugation, and homogenization that produce aerosols can be mitigated by employing engineering controls such as enclosures, access-controlled rooms, biological safety cabinets, and

## TABLE 15.6
## Summary of Biosafety Levels and Minimum Engineering Requirements

| Biosafety Level | Agent | Minimum Engineering Requirement |
|---|---|---|
| Biosafety Level 1 | Well-characterized agents not known to consistently cause disease in immunocompetent adult humans and animals; present minimal potential hazard to laboratory personnel and the environment | Laboratory doors for access control<br>Sink for handwashing<br>Laboratory benchtops impervious to water and resistant to heat and chemicals<br>Eyewash station<br>Windows fitted with screens<br>Lighting adequate for all activities |
| Biosafety Level 2 | Agents associated with human disease and pose moderate danger to laboratory personnel if accidentally inhaled, swallowed, or exposed to the skin | Self-closing doors<br>Sink located near exit<br>Windows sealed or fitted with screens<br>Autoclave available<br>BSCs or other primary containment device used for manipulations of agents that may cause splashes or aerosols<br>There are no specific requirements for ventilation systems |
| Biosafety Level 3 | Indigenous or exotic agents; may cause serious or potentially lethal disease through the inhalation route of exposure | Physical separation from access corridors; restricted access through two consecutive self-closing lockable doors<br>Hands-free sink near exit<br>Windows are sealed<br>Seams, floors, walls, and ceiling surfaces are sealed<br>Ducted air ventilation system with negative airflow into laboratory<br>Facilities are tested annually or after significant modification to ensure operational parameters are met<br>Appropriate communication systems are provided between the laboratory and the outside (e.g., voice, fax, and computer)<br>Emergency communication and emergency access or egress |

(*Continued*)

**TABLE 15.6 (CONTINUED)**
**Summary of Biosafety Levels and Minimum Engineering Requirements**

| Biosafety Level | Agent | Minimum Engineering Requirement |
|---|---|---|
| Biosafety Level 4 | Dangerous and exotic agents that pose high individual risk of aerosol-transmitted laboratory infections and life-threatening disease that are frequently fatal, for which there are no vaccines or treatments;<br>work with a related agent with unknown risk of transmission | Laboratory is in a separate building or a clearly demarcated, isolated, and restricted zone of the building<br>Entry sequence; entry through airlock with airtight door; doors lockable.<br>Walls, floors, ceilings form sealed internal shell<br>Windows are break-resistant and sealed<br>Dedicated, non-recirculating ventilation system required<br>Double-door, pass-through autoclave<br>Full-body, air-supplied, positive pressure suit<br>Automatically activated emergency power source |

Each successive BSL incorporates the minimum requirement of the preceding level(s) and the criteria in the cell.

mechanical ventilation, where air is exchanged and HEPA-filtered through ducting systems and natural ventilation (Li et al., 2019).

Job safety analysis is often related to sample volume and concentration, space requirements, water supply, plumbing, electricity, and waste disposal. Prior to laboratory construction, the design team begins by defining specifications, such as how people will use the laboratory (including what materials or processes will be used), how many people will be working, what their designations are, and how much space is needed. This ensures that everything included in the plan will serve a purpose in supporting the laboratory's users' safety and productivity. As the complexity of laboratory facilities increases, space will be required for engineering controls which serve the facility. In the Philippines during the COVID-19 pandemic, for a diagnostic laboratory to respond promptly, highly specific equipment and devices were essential to process high-volume SARS-COV-2 samples. Various types of high-throughput machines to be utilized in different stages of sample processing for biosafety concerns were carefully evaluated. Appropriate locations for equipment to be installed in the work area and recommend safety measures for end-users were determined. For high-throughput equipment generating large volumes of waste, safe and effective waste disposal methods must be employed. As the pandemic reached its highest, testing laboratories expanded operations, thus, an increase in space was necessary. Locations to process non-activated and inactivated samples and areas to store supplies, materials, and wastes were identified. Pre- and post-inspections for all proposed laboratories were conducted to establish the best locations for BSL-2 room additions, proper placement of biosafety cabinets, and other equipment.

### 15.2.4 Building Sustainable and Resilient Engineered Laboratory

Establishing biological laboratories, especially high-containment facilities, is costly and requires incessant investments of resources and funding to sustain manpower, equipment, infrastructure, certifications/recertifications, and operational needs. Besides funding and technical challenges, there are biosecurity and dual-use risks and local community concerns to deal with in order to sustain operations both in the short and longer term, without compromising biosafety and biosecurity. The pursuit for a sustainable laboratory is an all-inclusive approach to economic, human, and environmental attributes. To sustain laboratory operations, the following may be considered:

1. Sustainability policies anchored on a national strategic/transition plan and regulatory framework create a more holistic perspective (Molero et al., 2020). The laboratory's role in achieving national strategic goals, such as national laboratory network, emergency response, and disease surveillance, must be primarily established.
2. Political support and commitment from relevant government agencies are important for embedding the laboratory in the country's strategies. This encompasses the involvement of the full range of necessary stakeholders. Steadfast commitment must include finance, human resources, materials and supplies, and time frame.
3. At the outset, elucidating what the laboratory's purpose is, and what unfulfilled requirements will address, is discussed early and how it is linked to local, national, or global structures. Laboratory's scope of work is possibly in terms of diagnostics, disease surveillance, research, or biobanking. This sets the pace for the biorisk management program to be implemented, the biosafety and biosecurity measures based on risk-assessment and quality management system to be adapted. This will also forecast the type and volume of work to be done in the laboratory. This is critical in developing requirements for human resources, finance, operations, and infrastructure.
4. The laboratory must be designed in a way that it can adapt to future needs. A design with some flexibility will allow inclusion of features the users might not need at present but could be beneficial later on.
5. Early familiarity with financing issues and cost awareness are key to laboratory financial sustainability starting from conceptualization, design, construction, operation, and maintenance. Financial discussions seek to answer costs to establish and manage the laboratory, one-off costs, ongoing costs, funding sources, and contingency plans.
6. Focus on human resources needs, including where the full range of workers will be sourced and how surge capacity will be properly addressed must be considered. To enhance manpower sustainability, training and learning development, staff retention, and succession plan must be in place.

7. Appropriate infrastructure that is locally driven, in a suitable location, with functional security features and adequate access to utilities is basic for laboratory sustainability. How to best approach imminent failures in a stable power supply, access to water, transport links, harsh environmental conditions, and natural and man-made disasters must be well-thought of early on during the planning process. Identifying these challenges in this aspect will also inform accurate costing and aid in financial planning (Harper et al., 2019).

Resilience is a complementary feature of sustainable laboratories and uses strategies of adaptation and mitigation to improve conventional risk management. Resilient-oriented measures are established based on the resilience engineering principles that are required for recovery after a disaster. Resilience highlights the role of recovery post-disruption to guarantee system survival as well as overall acceptance that it is practically impossible to mitigate all categories of risk simultaneously, and before they occur. The viewpoint is to optimize limited financial and labor resources to prepare the system against a wide variety of threats. Resilience engineering links both system design and risk assessment to prevent the escape of biological hazards (Li et al., 2019).

## 15.3 INVESTMENT IN CAPACITY BUILDING

The COVID-19 pandemic has tested the capabilities of all countries in pathogen containment and rapid diagnostics and response to the crisis. There was a scarcity of laboratories that could perform testing, and some clinical laboratories were converted into molecular labs to meet the testing needs. Low-middle-income countries (LMICs) were forced to invest in building laboratories to support testing and research. However, problems arose when these laboratories faced financial difficulty in maintaining and operating the facilities after restrictions were lifted and laboratory operations were generally decreased.

One of the challenges is that a huge investment was made in these laboratories, but limited support was given in building the competence of the personnel to operate and maintain the facility. One of the most common problems in LMICs is the lack of skilled engineers and laboratory staff who are knowledgeable in laboratory design and maintenance of facilities. Therefore, there is a strong need to invest in the capacity building of local experts in the design, operations, and maintenance of laboratories.

### 15.3.1 Laboratory Design and Certification Training

Designing a laboratory requires collaboration between the architects, engineers, and end-users. More often, end-users are asked to lay out the facilities and let the architects and engineers design and build the facility. This is common at least for the Philippines where consultation among end-users and engineers is not practiced. Another common problem in resource-limited countries is that Western standards may not apply to their local setting. Facilities could not keep up to comply with international standards and invest in engineering controls. In the Philippines, for example, BSL-3 certification is sought from other countries like Singapore to perform BSL-3 certification due to the lack of engineering expertise in the field of biocontainment. As a result, the cost of bringing international experts would be high. Therefore, having a locally trained engineer would reduce the cost of certification and cut the need to fly in experts for facility maintenance.

### 15.3.2 Laboratory Operations and Maintenance Training

Laboratory operation and maintenance planning ensure that the laboratory spaces and equipment are in good working condition. This includes activities such as daily cleaning of the laboratory and maintenance and calibration of critical equipment. Having an established operations and maintenance planning allows the laboratory staff to work safely and efficiently.

However, adding to the cost of building laboratories, especially high-containment laboratories, concerns about sustaining operations and maintenance arise. In laboratories in resource-limited countries, there is a scarcity of mechanical engineers to perform laboratory maintenance. Laboratories often rely on third-party vendors to perform maintenance of biosafety equipment, which can be costly and not sustainable. Initiating an operation and maintenance program will essentially reduce the cost of having to replace a piece of equipment or renovate a facility.

One mechanism to avoid the need to acquire third-party services for equipment calibration and maintenance is to have in-house laboratory engineers who can perform such services. Some hospitals, especially in the Philippines, have established in-house offices that could maintain hospital equipment. However, the ability to perform calibration of safety equipment like BSC is not available. Having an established office that oversees maintenance is a good starting point in building in-house capability. Training for BSC certification can be integrated into their professional development or training program to enhance their capability in laboratory equipment maintenance.

Other than the training for the personnel for BSC calibration, investment in equipment that is needed for BSC calibration and certification is also a consideration. That way, trained staff are fully equipped to service BSCs within the institution. Regular maintenance and calibration should also be sought as part of the quality management system.

### 15.3.3 Standard Operating Procedures Development

More often than not, there is a perception that engineering controls would be sufficient to maintain safety in the laboratory. However, a validated standard operating procedure (SOP) is critical to safely carry out activities in the laboratory. SOPs are a set of documents that describe in detail the laboratory process.

Regular exercises and drills are an essential component of training programs for biorisk management, especially for the maintenance of facilities. Drills and simulation exercises allow the staff to test their readiness in a given scenario such as facility shutdown, etc. It can help identify the gaps and weaknesses in their SOPs and improve the skills of the personnel.

Specialized training is also a valuable investment, especially for LMICs where access to specialized services is not available. For example, biosafety cabinet certification is often performed by a third-party vendor. However, this service can be expensive and relies on the availability of the contractor. An in-house engineer who can perform BSC certification is an advantage.

## 15.4 CONTAINERIZED LABORATORIES

At the start of 2020, the global community was caught off-guard and was vastly unprepared for the emergence of COVID-19 disease. As it became a global pandemic, the volume of samples tested for the SARS-CoV-2 virus grew almost exponentially (Atzrodt et al., 2020). Indeed, the rapid and accurate diagnostics has been vital to its response, surveillance, treatment, and overall management (Peeling et al., 2022). Presently, COVID-19 testing can be found almost ubiquitously; however in the early days of the pandemic, access to laboratory facilities with molecular diagnostics capability was severely limiting (Chau et al., 2020) and in the case of the Philippines was limited to two laboratories (Cena-Navarro et al., 2022).

To augment this gap, hospitals and other governments, universities, and private institutions started to offer COVID-19 molecular diagnostic testing that was compliant with the interim guidelines set by the Department of Health (DOH) (DOH 2020). Like the rest of the world, the Philippine government exerted significant effort to increase its laboratory testing capacity; however, there was a significant lag in the stand-up in COVID-19 testing laboratories in the Philippines. The Philippines imposed a travel ban very early in during the pandemic, and the restrictive testing requirements, low testing capability, and highly restrictive local lockdowns instituted by local governments made the stand-up of laboratories exceedingly difficult. The Philippines is a sprawling archipelago of over 7000 islands. Moving samples between islands was stalled with travel restrictions, and the availability of existing brick and mortar facilities of sufficient design and quality to be converted to a BSL-2 sufficient to safely test the emerging virus was near nonexistent. Where existing facilities could be converted, equipment, reagent, and trained medical technicians were in slim supply. There was neither the time nor the financial resources available to build new brick and mortar laboratories.

Containerized laboratories, also known as modular or mobile laboratories, are a relatively recent innovation that has its roots in the broader history of containerization. The concept of containerization, which involves the use of standardized containers for transporting goods, originated in the coal mining regions of England in the late 18th century (Van et al., 2012). However, it wasn't until after World War II that the system was widely applied, dramatically reducing transport costs and supporting the boom in international trade (Van et al., 2012).

The introduction of container technology was a significant moment in history, and its application to laboratories has opened new possibilities for scientific research and experimentation in diverse locations. The concept of containerized laboratories has been particularly beneficial in resource-limited settings. While developed countries can easily adapt their well-organized routine laboratory services, resource-limited countries often need considerable capacity building as many gaps still exist (Masanza et al., 2010). Efforts have been made by organizations such as the African Field Epidemiology Network (AFENET) to support laboratory capacity development in the Africa region (Masanza et al., 2010). These efforts range from promoting graduate-level training programs to building advanced technical, managerial, and leadership skills, as well as in-service short course training for peripheral laboratory staff (Masanza et al., 2010). Specific projects focus on external quality assurance, basic laboratory information systems, strengthening laboratory management toward accreditation, equipment calibration, harmonization of training materials, networking, and provision of pre-packaged laboratory kits to support outbreak investigation (Masanza et al., 2010).

Modular laboratories, which could be seen as a form of containerized laboratories, have been recognized as cost-effective and sustainable infrastructure for resource-limited settings (Bridges et al., 2014). They provide high-quality laboratory space to support basic science, clinical research projects, or health services, which are often severely lacking in the developing world (Bridges et al., 2014).

However, it's important to note that while these efforts have had a positive effect on laboratory capacity in the region, many opportunities exist for further advancement, especially in areas such as biosafety, biosecurity, and long-term durability of materials in rugged environments.

Containerized laboratories offer several advantages over conventional laboratories that are built in 'brick and mortar buildings', and they have been shown to be cost-effective and sustainable in resource-limited circumstances (Walker et al., 2020). Most especially, they are significantly faster to build and set up, and as a result, laboratory operations can right away commence thereafter. This was particularly the demand in the early days of the pandemic as the laboratory testing capacity for COVID-19 needed to increase right away to accommodate the voluminous number of samples for testing. In the middle of 2020, a preprint was released detailing the specifications of a shipping container laboratory that is optimized for automated COVID-19 diagnostics (Masanza et al., 2010). Undeniably, this concept can potentially address most of the requirements for a rapid setup of a COVID-19 molecular diagnostic laboratory.

In the Philippines, a joint venture between three companies (SciLore LLC., Endec INC., and Global Medical Technologies) was formed called 'SEG'. SEG developed and deployed a turn-key solution to rapidly increase the COVID-19 testing capacity in the Philippines. The containerized laboratories were fabricated from a standard 40 ft. by 20 ft. ISO shipping container in 21 days. In compliance with the DOH interim guidelines (DOH 2020), and the subsequent amendments (DOH 2020), SEG came up with four models. Generally, these are PCR suites designed and configured to adhere to the minimum requirements prescribed by WHO for a biosafety level 2 laboratory. All partitions, fixtures, furniture, and laboratory equipment that are needed to receive samples, extract nucleic acids, run PCR amplification, inactivate biohazards, analyze, and release results are included. The fabricated container laboratory is then shipped and deployed to its intended location in the Philippines in four to five days. Upon arrival, the container laboratory was installed and commissioned for COVID-19 testing in around 19 hours. From its fabrication to its commissioning, it took 26–27 days to complete. Compared to the setup of a conventional laboratory, this is a significant improvement in terms of time and was essential and beneficial during the pandemic.

By the end of 2020, more than 100 laboratories had 'popped-up', been equipped and applied to the Department of Health for a License to Operate as COVID-19 molecular diagnostics (Cena-Navarro RB et al., 2022), 20 of which were containerized laboratories across the country.

## 15.5 LABORATORY DESIGNS FOR LOW-TO-MIDDLE INCOME COUNTRIES

This chapter highlighted the importance of balancing the requirements of engineering designs with cost associated with it. In purview that institutions should follow national standards and international guidance documents to ensure that the facility can be used safely for the purpose it was made for. This chapter also highlighted the importance of following the minimum requirements so as to be at par with what is the requirement for a particular purpose. All of these are anchored in biological risk assessment, wherein institutions should be able to assess each risk associated with the usage of the facility. This will further ensure that for low-to-middle-income countries, cost will not stop the design and functionality of a laboratory but rather be built to specifications. This will make them understand that the cost of having to change PPEs often or instigating administrative controls may even be higher than ensuring that the facility is well-equipped with equipment and facilities to handle pathogens.

## REFERENCES

Atzrodt, C.L., Maknojia, I., McCarthy, R.D.P., Oldfield, T.M., Po, J., Ta, K.T.L., et al. (2020). A guide to COVID-19: A global pandemic caused by the novel coronavirus SARS-CoV-2. *The FEBS Journal*, *287*(17), 3633–3650.

Belgian Biosafety Server. (n.d.). Contained use - International classifications schemes for micro-organisms based on their biological risks | Belgian biosafety server. Belgian Biosafety Server website. Retrieved from https://www.biosafety.be/content/contained-use-international-classifications-schemes-micro-organisms-based-their-biological

Bouchaut, B., & Asveld, L. (2020). Safe-by-design: Stakeholders' perceptions and expectations of how to deal with uncertain risks of emerging biotechnologies in the Netherlands. *Risk Anal.*, *40*(8), 1632–1644.

Bridges, D.J., Colborn, J., Chan, A.S.T., et al. (2014). Modular laboratories—Cost-effective and sustainable infrastructure for resource-limited settings. *Am. J. Trop. Med. Hyg.*, *91*(6), 1074–1078. https://doi.org/10.4269/ajtmh.14-0054

Burzoni, S., Duquenne, P., Mater, G., & Ferrari, L. (2020). Workplace biological risk assessment: Review of existing and description of a comprehensive approach. *Atmosphere*, *11*(7), 1–24.

Government of Canada.(2016).*Canadian Biosafety Handbook (2nd ed.)*.Ottawa, ON, Canada:Government of Canada.Retrieved from https://www.canada.ca/content/dam/phac-aspc/migration/cbsg-nldcb/cbh-gcb/assets/pdf/cbh-gcb-eng.pdf

Cena-Navarro, R., Vitor, R.J., Canoy, R.J., dela Tonga, A., Ulanday, G.E., Silva, M.R.C., & Destura, R.V. (2022 Mar). Biosafety capacity building during the COVID-19 pandemic: Results, insights, and lessons learned from an online approach in the Philippines. *Appl. Biosaf.*, *27*(1), 42–50.

Chau, C.H., Strope, J.D., & Figg, W.D. (2020). COVID-19 clinical diagnostics and testing technology. *Pharmacother. J. Hum. Pharmacol. Drug Ther.*, *40*(8), 857–868.

Department of Health. (2020). AO 2020-0014A: Amendment to the administrative order No. 2020-0014 "guidelines in securing a license to operate a COVID-19 testing laboratory in the Philippines" [Internet]. Department of Health (DOH). [cited 2023 Jan 3]. Available from: https://hfsrb.doh.gov.ph/wp-content/uploads/2021/06/ao2020-0014-A.pdf

Department of Health. (2020). AO 2020-0014B. Amendment to the administrative order No. 2020-0014-A "amendment to the administrative order No. 2020-0014, guidelines in securing a license to operate a COVID-19 Testing Laboratory in the Philippines" [Internet]. Department of Health (DOH). [cited 2023 Jan 3]. Available from: https://hfsrb.doh.gov.ph/wp-content/uploads/2021/10/ao2020-0014-B.pdf

Department of Health. (2020). AO 2020-014: Guidelines in securing a license to operate a COVID-19 testing laboratory in the Philippines [Internet]. Department of Health (DOH); [cited 2023 Jan 3]. Available from: https://doh.gov.ph/sites/default/files/health-update/ao2020-0014.pdf

Dickmann, P., Sheeley, H., & Lightfoot, N. (2015). Biosafety and biosecurity: A relative risk-based framework for safer, more secure, and sustainable laboratory capacity building. *Front. Public Health*, *3*(Oct), 1–6.

Harper, D.R., Ross, E., & Wakefield, B. (2019). The Chatham House sustainable laboratories initiative prior assessment tool.Retrieved from https://www.chathamhouse.org/sites/default/files/publications/research/2019-06-18-Sustainable-Laboratories-Initiative.pdf

Hendrickson, W. (2019). *Construction and major renovations guidelines*. Retrieved from https://www.aphl.org/aboutAPHL/publications/Documents/GH-2019May-Lab-Construction-Reno-Guidelines.pdf

Kimman, T.G., Smit, E., & Klein, M.R. (2008). Evidence-based biosafety: A review of the principles and effectiveness of microbiological containment measures. *Clin. Microbiol. Rev.*, *21*(3), 403–425.

Kojima, K., Booth, C.M., Summermatter, K., Bennett, A., Heisz, M., Blacksell, S.D., & McKinney, M. (2018). Risk-based reboot for global lab biosafety. *Science*, *360*(6386), 260–262.

*Laboratory Biosafety Manual* (Fourth Edition). (2020). World Health Organization. Retrieved from https://apps.who.int/iris/rest/bitstreams/1323419/retrieve

Li, W., Sun, Y., Cao, Q., He, M., & Cui, Y. (2019). A proactive process risk assessment approach based on job hazard analysis and resilient engineering. *J. Loss Prev. Process Ind.*, *59*(November 2018), 54–62.

Maehira, Y., & Spencer, R.C. (2019). Harmonization of biosafety and biosecurity standards for high-containment facilities in low- and middle-income countries: An approach from the perspective of occupational safety and health. *Front. Public Health*, *7*, 249. https://doi.org/10.3389/fpubh.2019.00249

Masanza, M.M., Nqobile, N., Mukanga, D., & Gitta, S.N. (2010). Laboratory capacity building for the International Health Regulations (IHR[2005]) in resource-poor countries: The experience of the African Field Epidemiology Network (AFENET). *BMC Public Health*, *10*(Suppl 1), S8. https://doi.org/10.1186/1471-2458-10-S1-S8

Meechan, P.J., & Potts, J. (2020). *Biosafety in Microbiological and Biomedical Laboratories*.Center for Disease Control and Prevention (CDC); National Institutes of Health (NIH) .Retrieved from https://stacks.cdc.gov/view/cdc/97733

Molero, A., Calabrò, M., Vignes, M., Gouget, B., & Gruson, D. (2020). Sustainability in healthcare: Perspectives and reflections regarding laboratory medicine. *Ann. Lab. Med.*, *41*(2), 139–144.

National Institutes of Health. (2013). Lab module design considerations. https://orf.od.nih.gov/TechnicalResources/Documents/Technical%20Bulletins/13TB/Lab%20Module%20Design%20Considerations%20December%202013%20Bulletin_508.pdf

Patterson, A., Fennington, K., Bayha, R., Wax, D., Hirschberg, R., Boyd, N., & Kurilla, M. (2014). Biocontainment laboratory risk assessment: Perspectives and considerations. *Pathog. Dis.*, *71*(2), 102–108.

Peeling, R.W., Heymann, D.L., Teo, Y.Y., & Garcia, P.J. (2022, Feb 19). Diagnostics for COVID-19: Moving from pandemic response to control. *The Lancet*, *399*(10326), 757–768.

Rutjes, S.A., Vennis, I.M., Wagner, E., Maisaia, V., & Peintner, L. (2023). Biosafety and biosecurity challenges during the COVID-19 pandemic and beyond. *Front. Bioeng. Biotechnol.*, *11*(Mar), 1–8.

van Gelder, P., Klaassen, P., Taebi, B., Walhout, B., van Ommen, R., van de Poel, I., et al. (2021). Safe-by-design in engineering: An overview and comparative analysis of engineering disciplines. *Int. J. Environ. Res. Public Health*, *18*(12), 1–28.

Van Ham, H., & Rijsenbrij, J. (2012). *Development of Containerization* (p. 8). Amsterdam: IOS Press.

Walker, K.T., Donora, M., Thomas, A., Phillips, A.J., Ramgoolam, K., Pilch, K.S., et al. (2020). Contain: An open-source shipping container laboratory optimised for automated COVID-19 diagnostics [Internet]. bioRxiv [cited 2022 Dec 13]. p. 2020.05.20.106625. Available from: https://www.biorxiv.org/content/10.1101/2020.05.20.106625v1

Wheatley, M. (2018). A broad introduction to the design and construction of biosafety laboratories in low-resource settings. Retrieved from https://www.fiiapp.org/wp-content/uploads/2018/07/Una-introducción-al-diseño-y-construcción-de-laboratorios-seguros-en-entornos-con-pocos-recursos-Manual-NRBQ46.pdf

# Index

## B

## C

## D

## E

## G

## J



## K



## L

## M

## N

## O



## P

## Q



## R

## S

## T

## U

## X



## Y



## Z

For Product Safety Concerns and Information please contact our EU representative GPSR@taylorandfrancis.com
Taylor & Francis Verlag GmbH, Kaufingerstraße 24, 80331 München, Germany

www.ingramcontent.com/pod-product-compliance
Lightning Source LLC
LaVergne TN
LVHW020605110826
845149LV00002B/372

* 9 7 8 1 0 3 2 5 4 6 6 9 8 *